D1593033

THE FOURIER INTEGRAL
AND ITS APPLICATIONS

McGraw-Hill Classic Textbook Reissue Series

$\mathcal{I} \mathcal{N} \mathcal{S}.$

THE FOURIER INTEGRAL
AND ITS APPLICATIONS

ATHANASIOS PAPOULIS

Professor of Electrical Engineering
Polytechnic Institute of Brooklyn

McGRAW-HILL
CLASSIC
TEXTBOOK
REISSUE

McGraw-Hill, Inc.
New York St. Louis San Francisco Auckland Bogotá
Caracas Lisbon London Madrid Mexico City Milan
Montreal New Delhi San Juan Singapore
Sydney Tokyo Toronto

THE FOURIER INTEGRAL AND ITS APPLICATIONS

29 30 31 32 33 34 35 36 37 BKMBKM 9 9 8 7 6 5 4
ISBN 0-07-048447-3

ISBN 0-07-048447-3

Library of Congress Card Number 62-10211

Preface

I find it easier in this preface to abandon the impersonal *one* and *we* of the text and address you, the reader, directly.

You might think that this book, covering a topic that has been in use for almost 150 years, is an anachronism. I assure you that I would not have written it if I were not certain that there is, even today, a definite need for such a book.

That this discipline is basic in applied science, you have probably no doubt. The study of the vast area of linear phenomena is considerably simplified with its use. Nevertheless, there is perhaps no applied book that covers exclusively the Fourier integral, although in the last year alone more than ten texts dealing with the Laplace transform were published. In the existing technical literature, the theory of the Fourier integral is treated only incidentally, and the applications appear as separate, unrelated disciplines; the mathematical texts, on the other hand, contain details beyond the interest or background of the applied scientist. In this book I attempt to bridge the gap between these two extremes.

I don't plan to give you, here, an outline of what I propose to cover, because I don't think it will tell you much about the book; in fact, it might mislead you. In an established discipline, a book is not *what* topics it covers, but *how* it covers them, and this you can determine only by reading it.

I would, however, like to say a few words about the method of presentation, because in this area I have tried to develop a personal point of view. Although, with very few exceptions, I am presenting nothing new, I attempted in several instances to give new, simple proofs and to develop the ideas clearly, logically, and without gaps. Nevertheless, I believe, in fact I hope, that you will not find the book too easy to read. Any book, no matter how well-written, if it covers with economy concepts that are new to the reader, must be read with effort.

I made a serious attempt to present the entire material on a uniform level of sophistication; however, the details vary considerably from topic to topic. This relative lack of proportion is intentional. If I judged that an idea might be entirely new to you, I merely outlined it and tried to relate it to your past knowledge; more familiar topics I presented with greater elaboration. Similarly, some applications are only briefly introduced; others are discussed in greater detail. In this difficult and important choice, I tried to anticipate your interest, background, and patience, and was guided mainly by my classroom experience. I also took into consideration the relevant literature.

To give you an example, I devote over two pages to the Paley-Wiener condition because no applied text gives even an outline of its proof, although it is widely quoted. On the other hand, I mention only casually the class of rational transforms, although they have extensive applications in the study of lumped-parameter systems. My reason is that this rather simple area is exhaustively discussed in literally hundreds of books.

Another example of the dilemma with which I was often faced is in the treatment of the singularity functions. To be content with the usual definition of "a function that is zero everywhere except at the origin and its area equals one" would be against my objective of trying to offer at least a sound definition of all new concepts. As you probably know, L. Schwartz has developed a new theory that gives mathematical meaning to the delta function; however, I didn't mention him, even as a reference, because I confess I find his book too difficult to read. Searching the various books on the subject, I decided to develop in the Appendix a very simple, but logically self-contained, and for our purpose adequate version of the theory of distributions and use it to define the delta function. This is, merely, a formal recognition of the fact that the significant characteristics of the delta function are not its specific values but its integral properties.

I have taught this subject for over ten years and it would be difficult to cite all sources that I consulted at various times. I would like, however, to mention that in the preparation of the material on filters I was influenced by K. Küpfmüller's book "Die System Theorie der elektrischen Nachrichten Übertragung" and that my treatment of the integral theorems and Hilbert transforms is based on H. W. Bode's book "Network Analysis and Feedback Amplifier Design."

This material was originally written for a second-semester graduate course in a three-semester sequence on linear systems given by the Electrical Engineering Department of the Polytechnic Institute of Brooklyn. I decided to make it into a book, with John Truxal's encouragement. I thank him for his valuable comments. I would also like to express my appreciation to the National Science Foundation for offering me a fellowship. It was during the tenure of the fellowship that I prepared much of this work.

Athanasios Papoulis

Contents

PART TWO

PART ONE

Chapter 1. Introduction

In this chapter we discuss briefly the various applications of the Fourier integral, and we compare it with the Laplace transform.

1-1. Fourier Analysis

The Fourier integral permits the representation of an arbitrary function $f(t)$, which might consist of a number of completely different analytic pieces in the various parts of the t axis, by a single expression

$$f(t) = \frac{1}{2\pi} \int_{-\infty}^{\infty} F(\omega)e^{j\omega t}\, d\omega \tag{1-1}$$

valid for every t. The quantity $F(\omega)$ is given by

$$F(\omega) = \int_{-\infty}^{\infty} f(t)e^{-j\omega t}\, dt \tag{1-2}$$

and is known as the *Fourier integral* or *Fourier transform* of $f(t)$.

Some authors introduce the factor $1/2\pi$ not in (1-1) but in (1-2), or $1/\sqrt{2\pi}$ in both equations for reasons of symmetry, or the variable $f = \omega/2\pi$ in (1-1) with a possible change of the sign of the exponent; however, all these definitions are essentially equivalent. We have adopted the above because it is commonly used in the engineering literature.

The validity of the expansion (1-1) is often established from a related result in Fourier series. The function $f(t)$ is expanded into a trigonometric series in the $(-T/2, T/2)$ interval, and it is shown that this expansion tends to the integral (1-1) as T tends to infinity. We reproduce briefly this approach. It is known from the theory of Fourier series that an arbitrary function $f(t)$ can be written as a sum

$$f(t) = \sum_{n=-\infty}^{\infty} \alpha_n e^{jn\omega_0 t} \qquad \omega_0 = \frac{2\pi}{T} \tag{1-3}$$

1

where the constants α_n are given by

$$\alpha_n = \frac{1}{T} \int_{-T/2}^{T/2} f(t)e^{-jn\omega_0 t}\, dt \tag{1-4}$$

and the sum (1-3) equals $f(t)$ only for $|t| < T/2$. With $F(\omega)$ defined by the integral (1-2), we observe from (1-4) that

$$T\alpha_n = \int_{-T/2}^{T/2} f(t)e^{-jn\omega_0 t}\, dt \xrightarrow[T\to\infty]{} \int_{-\infty}^{\infty} f(t)e^{-jn\omega_0 t}\, dt = F(n\omega_0) \tag{1-5}$$

Since the constant $\omega_0 = 2\pi/T$ tends to zero with $T \to \infty$, we obtain from (1-3) and (1-5)

$$f(t) = \frac{1}{2\pi} \sum_{n=-\infty}^{\infty} T\alpha_n e^{jn\omega_0 t}\omega_0 \xrightarrow[T\to\infty]{} \frac{1}{2\pi} \int_{-\infty}^{\infty} F(\omega)e^{j\omega t}\, d\omega \tag{1-6}$$

if we interpret the last integral as a limit of a sum. The Fourier integral (1-1) is, thus, established.

The above approach is not satisfactory. Theoretically, it does not give the conditions for the validity of (1-1) and is based on a limited interpretation of the meaning of an integral. Conceptually, it obscures the significance of $F(\omega)$. The point is often made that the student can easily understand the meaning of the series expansion (1-3) and that the passage to the limit is readily accepted. In our experience the representation of $f(t)$ as a sum of exponentials with fundamental tending to zero is certainly not easier to accept than the fact that $f(t)$ can be written directly as an integral of the form (1-1). Furthermore, with the usual restrictions imposed on $f(t)$, the proof of (1-1) is no more difficult than the proof of (1-3).

For the unified derivation of many results it is, in fact, desirable to consider the Fourier series as a special case of the Fourier integral. This is possible if one includes in the analysis the singularity functions. As we show in Chap. 3, the Fourier transform $F(\omega)$ of a periodic function $f(t)$ consists of a sequence of equidistant impulses:

$$F(\omega) = \sum_{n=-\infty}^{\infty} b_n \delta(\omega - n\omega_0) \tag{1-7}$$

Inserting the above sum into (1-1), we obtain

$$f(t) = \frac{1}{2\pi} \sum_{n=-\infty}^{\infty} b_n e^{+jn\omega_0 t}$$

and with $b_n = 2\pi\alpha_n$, the series expansion (1-3) results.

1-2. The Laplace Transform

A quantity related to the Fourier integral $F(\omega)$ is the bilateral Laplace transform

$$F_{II}(p) = \int_{-\infty}^{\infty} e^{-pt}f(t)\, dt \tag{1-8}$$

existing in a vertical strip

$$\gamma_1 < \operatorname{Re} p < \gamma_2 \tag{1-9}$$

where the constants γ_1 and γ_2 depend on $f(t)$. The function $f(t)$ can be expressed in terms of $F_{\mathrm{II}}(p)$ by the inversion formula

$$f(t) = \frac{1}{2\pi j} \int_{Br} e^{pt} F_{\mathrm{II}}(p)\, dp \tag{1-10}$$

similar to (1-1). The path of integration Br is a vertical line $\operatorname{Re} p = \alpha$ in the region (1-9) in which the integral (1-8) converges. In most applications it is assumed that $f(t) = 0$ for $t < 0$, and (1-8) takes the form

$$F_{\mathrm{I}}(p) = \int_0^\infty e^{-pt} f(t)\, dt \tag{1-11}$$

known as unilateral Laplace transform and converging in the half plane

$$\operatorname{Re} p > \gamma \tag{1-12}$$

Clearly, with the exception of the trivial factor j, the functions $F(\omega)$ and $F_{\mathrm{II}}(p)$ seem to be identical; by a Fourier transform, however, one usually means the integral (1-2), where ω is *real*, whereas in (1-8) the variable p can take complex values.

In the last two decades, the Laplace transform has displaced the Fourier integral as the main tool of analysis. This is particularly true in the study of linear systems. One reason seems to be the belief that the Laplace transform can handle a more general class of functions, but this is only apparent. The requirement that (1-8) converges in the strip (1-9) is more restrictive than the assumption that (1-2) exists for every real ω. For analogy we mention the Laurent expansion

$$F(z) = \sum_{n=-\infty}^{\infty} \alpha_n z^n \tag{1-13}$$

converging in a ring

$$a < |z| < b \tag{1-14}$$

and the corresponding Fourier series

$$f(t) = \sum_{n=-\infty}^{\infty} \alpha_n e^{jn\omega_0 t} \tag{1-3}$$

It is possible for a function $f(t)$ to have a Fourier expansion as in (1-3) but no representation in the form

$$f(t) = F(e^{j\omega_0 t})$$

where $F(z)$ is analytic in some ring.

If the imaginary axis is not included in the strip (1-9), i.e., if $\gamma_1 > 0$ or $\gamma_2 < 0$, then $F(\omega)$ does not exist. However, by writing $F_{\mathrm{II}}(p)$ in the form

$$F_{\mathrm{II}}(\alpha + j\omega) = \int_{-\infty}^{\infty} e^{-\alpha t} f(t) e^{-j\omega t}\, dt \tag{1-15}$$

one can interpret it as the Fourier transform of $e^{-\alpha t}f(t)$, where α is considered as a parameter.

The necessity for using Laplace transforms arises whenever one is interested in the analytic properties of $F(\omega)$. We mention below some applications involving such properties.

Evaluation of the Inversion Integral. The evaluation of $f(t)$ by a direct integration of (1-1) along the real axis is, in general, complicated; often the result can be simply found by a suitable modification of the path of integration (Cauchy's theorem). In most cases this is accomplished with the calculus of residues. Sometimes the new path is so chosen that only a portion of it contributes significantly to the value of $f(t)$; this leads to the saddle-point method of integration.

Causal Functions. If a function is zero for negative t, then the real and imaginary parts (or the amplitude and phase) of its Fourier integral $F(\omega)$ are not independent of each other. In fact, if one of these quantities is specified in a certain part of the ω axis and the other in the remaining part of this axis, then $F(\omega)$ can be uniquely found. This is a special form of the problem of determining an analytic function in a certain region of the complex plane, from a partial specification of its real and imaginary parts on the boundary of this region.

Network Theory. In network theory, the various linear systems are characterized by the analytic properties of their system function $H(\omega)$ and the location of its singularities. It is, thus, necessary to define $H(\omega)$ for complex values of ω, i.e., to introduce the Laplace transform of their impulse response.

1-3. Linear Systems

The Fourier transform is an essential tool of analysis of linear time-invariant systems. The principal reason is the fact that, if the input $f(t)$ to such a system is an exponential

$$f(t) = e^{j\omega t}$$

then the response $g(t)$ is proportional to the input

$$g(t) = H(\omega)e^{j\omega t}$$

where the proportionality constant $H(\omega)$ is the system function. It then follows from the linearity of the system and (1-1) that, with $F(\omega)$ the Fourier transform of the input, its response is given by

$$g(t) = \frac{1}{2\pi} \int_{-\infty}^{\infty} F(\omega)H(\omega)e^{j\omega t}\, d\omega \qquad (1\text{-}16)$$

If the system is specified by ordinary or partial differential equations (lumped or distributed parameter systems), then $H(\omega)$ is a rational or transcendental function of ω. In this case, (1-16) is evaluated by a contour integration; i.e., Laplace-transform techniques are applied.

In many applications, the system is characterized only terminally in terms of the amplitude and phase of $H(\omega)$. The system function is not given by a single analytic expression but is specified by a number of pieces, often only graphically. For the determination of $g(t)$ one must, then, perform the integration along the real ω axis. For these applications, the Fourier integral is used and special techniques are developed for the evaluation of (1-16). This approach relates the time and frequency properties of the system and offers an insight into the nature of $g(t)$. It is used even if $H(\omega)$ is given by a single analytic function, when this function has significant values only in a small portion of the ω axis (resonance, modulation, filtering).

1-4. Singularity Functions

The singularity functions are an integral part of Fourier analysis. With their use, many important functions that have no Fourier transform in the ordinary sense are included in the analysis, the derivation of certain results is considerably simplified, and new concepts are introduced. Nevertheless, in most mathematical books the delta function is not mentioned at all. In the technical literature it is generally treated, but not consistently, and mostly with reservations. Thus, although one proves with confidence that the Fourier transform of $\delta(t)$ equals one,

$$\int_{-\infty}^{\infty} \delta(t)e^{-j\omega t}\, dt = 1 \tag{1-17}$$

it is not easy to accept the validity of the inversion formula (1-1)

$$\frac{1}{2\pi}\int_{-\infty}^{\infty} e^{j\omega t}\, d\omega \overset{?}{=} \delta(t) \tag{1-18}$$

because the above integral has no meaning. These difficulties are, of course, due to the inadequacy of the definition of $\delta(t)$. An examination of the applications of the delta function will show that it always appears as an integrand in integrals of the form

$$\int_{-\infty}^{\infty} \delta(t)\phi(t)\, dt \tag{1-19}$$

and that its specific values for a given t are never considered. It is natural, therefore, to define it not as an ordinary function, but by the values of the integrals (1-19). This observation leads to the concept of distribution and clarifies the meaning of $\delta(t)$. In Appendix I† we give a simple, but for our purposes adequate, discussion of the concept of distribution, and we use the delta function throughout the book.

† Throughout the book cross references to equations and sections of Appendixes I and II are identified by the prefixes I- and II- respectively.

1-5. The Fourier Transform in Probability Theory

The Fourier integral is extensively used in probability theory. It appears as the transform of a density (characteristic functions), as the transform of the autocorrelation (power spectrum), and in the spectral decomposition of a stationary random process (written as the Fourier integral of a process with orthogonal increments). The first two cases can be treated deterministically as special topics in Fourier analysis; the last involves stochastic integrals and the necessity of introducing probabilistic concepts, and for this reason it will not be considered in our study.†

Characteristic Functions. A characteristic function $F(\omega)$ is merely the Fourier transform of a positive signal $f(t)$ of area equal to one. Such signals are investigated primarily in probability theory, but they appear also in the analysis of linear systems with monotonic step response; they are the time-domain equivalent of positive real functions used in network theory. In the study of characteristic functions, the special topics of interest are the asymptotic forms of their product (limit theorems) and the conditions for positiveness of $f(t)$.

Power Spectra. Signals with finite power

$$0 < \lim_{T \to \infty} \frac{1}{2T} \int_{-T}^{T} |f(t)|^2 \, dt < \infty \tag{1-20}$$

have, in general, no Fourier transform. They are partially characterized by their autocorrelation

$$R(t) = \lim_{T \to \infty} \frac{1}{2T} \int_{-T}^{T} f(\tau) f(t + \tau) \, d\tau \tag{1-21}$$

and its Fourier transform $S(\omega)$, known as the power spectrum of $f(t)$. These concepts are used mainly in the investigation of stochastic processes, but their properties can be developed independently of any probabilistic considerations.

† A thorough treatment of spectral analysis of random processes is given in A. Blanc-Lapierre and R. Fortet, "Théorie des functions aléatoires," Masson et Cie, Paris, 1953. See also J. L. Doob, "Stochastic Processes," John Wiley & Sons, Inc., New York, 1953.

Chapter 2. Basic Theorems and Examples

In this chapter we shall develop a number of basic theorems and examples that will be extensively used throughout the entire book†; they follow easily from the fundamental theorem and are presented for easy reference.

2-1. The Fourier Integral

Given a function of the real variable t, we form the integral

$$F(\omega) = \int_{-\infty}^{\infty} f(t)e^{-j\omega t}\, dt \tag{2-1}$$

If this integral exists for every real value of the parameter ω, it defines a function $F(\omega)$ known as the *Fourier integral* or *Fourier transform* of $f(t)$. The function $F(\omega)$ is in general complex:

$$F(\omega) = R(\omega) + jX(\omega) = A(\omega)e^{j\phi(\omega)} \tag{2-2}$$

$A(\omega)$ is called the *Fourier spectrum* of $f(t)$, $A^2(\omega)$ its *energy spectrum*, and $\phi(\omega)$ its *phase angle*. In Fig. 2-1 we have shown the various ways of plotting $F(\omega)$; as example we used the function

$$F(\omega) = \frac{1}{\alpha + j\omega} = \frac{\alpha}{\alpha^2 + \omega^2} - j\frac{\omega}{\alpha^2 + \omega^2} = \frac{1}{\sqrt{\alpha^2 + \omega^2}} e^{-j\tan^{-1}(\omega/\alpha)}$$

Fundamental theorem. The following basic equation, known as *inversion formula*, permits the representation of $f(t)$ in terms of its Fourier transform $F(\omega)$:

$$f(t) = \frac{1}{2\pi} \int_{-\infty}^{\infty} F(\omega)e^{j\omega t}\, d\omega \tag{2-3}$$

† E. C. Titchmarsh, "Introduction to the Theory of Fourier Integrals," Oxford University Press, New York, 1937.

A simple formal proof of (2-3) can be readily given if use is made of the identity (I-49)

$$\frac{1}{2\pi} \int_{-\infty}^{\infty} e^{j\omega t}\, d\omega = \delta(t) \tag{2-4}$$

and the equation (I-21)

$$\int_{-\infty}^{\infty} \phi(x)\, \delta(t-x)\, dx = \phi(t) \tag{2-5}$$

defining the *unit impulse* function $\delta(t)$. Indeed, inserting $F(\omega)$, as given by (2-1), into the r.h. side of (2-3), we obtain

$$\frac{1}{2\pi} \int_{-\infty}^{\infty} F(\omega)e^{j\omega t}\, d\omega = \frac{1}{2\pi} \int_{-\infty}^{\infty} e^{j\omega t}\, d\omega \int_{-\infty}^{\infty} f(x)e^{-j\omega x}\, dx$$

Interchanging the order of integration and using (2-4), we have

$$\frac{1}{2\pi} \int_{-\infty}^{\infty} F(\omega)e^{j\omega t}\, d\omega = \frac{1}{2\pi} \int_{-\infty}^{\infty} f(x)\, dx \int_{-\infty}^{\infty} e^{j\omega(t-x)}\, d\omega = \int_{-\infty}^{\infty} f(x)\delta(t-x)\, dx$$

But the last integral equals $f(t)$ [see (2-5)]; hence (2-3) is proved. The above proof is valid only at continuity points of $f(t)$ because only then

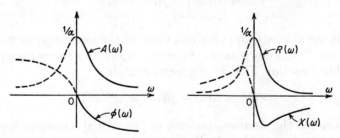

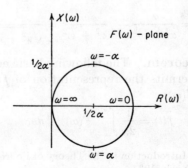

FIGURE 2-1

(2-5) holds. However, the inversion formula (2-3) holds also at discontinuity points if it is assumed that

$$f(t) = \frac{f(t^+) + f(t^-)}{2} \tag{2-6}$$

In Sec. 2-6 we shall reexamine the proof of (2-3); in the remaining part of this section we give without proof sufficient conditions for its validity.†

All functions under consideration will be assumed of *bounded variation*; this, simply stated, means that they can be represented by a curve of finite length in any finite interval of time, a condition satisfied by all functions in our applications.

Condition 1. If $f(t)$ is absolutely integrable in the sense

$$\int_{-\infty}^{\infty} |f(t)| \, dt < \infty \tag{2-7}$$

then its Fourier integral $F(\omega)$ exists and satisfies (2-3).

We emphasize that (2-7) is sufficient but not necessary; there are functions that are not absolutely integrable but have a transform satisfying (2-3). An important example is the function $\sin \omega_0 t/t$, included in the following, more special class.

Condition 2. If $f(t) = g(t) \sin (\omega_0 t + \phi_0)$, where ω_0 and ϕ_0 are arbitrary constants, and if, for $|t| > A > 0$, the function $f(t)/t$ is absolutely integrable and $g(t)$ is monotonically decreasing, then $F(\omega)$ exists and satisfies (2-3).

Another example of this class is the function $1/t^\alpha$, where $1 < \alpha < 0$, as one can easily see.

Comment. If $f(t)$ satisfies Condition 2, then for the validity of the Fourier-integral theorem, the integrals in (2-1) and (2-3) must be given a somewhat limited interpretation known as *Cauchy principal value*, whose meaning is the following:

An integral from minus infinity to plus infinity is actually a limit defined by

$$\int_{-\infty}^{\infty} f(t) \, dt = \lim \int_{-T_1}^{T_2} f(t) \, dt \qquad T_1, T_2 \to \infty \tag{2-8}$$

where the quantities T_1 and T_2 tend to infinity independently of each other. The Cauchy principal value of an integral is the more special limit

$$\int_{-\infty}^{\infty} f(t) \, dt = \lim \int_{-T}^{T} f(t) \, dt \qquad T \to \infty \tag{2-9}$$

While (2-8) implies (2-9), it is possible that (2-9) exists, but not (2-8). For example, if $f(t) = t$, then (2-9) equals zero but (2-8) has no meaning. In fact the integral (2-9) of all odd functions equals zero, whereas

† S. Bochner, "Vorlesungen über Fouriersche Integrale," Chelsea Publishing Company, New York, 1948.

(2-8) might not exist. For functions satisfying Condition 1, the
Fourier integrals can be interpreted in the (2-8) sense; for functions
satisfying Condition 2, the interpretation (2-9) might be necessary.†
We finally remark that, by allowing $f(t)$ and $F(\omega)$ to include singularity functions, we shall in Chap. 3 extend the theory to a more general class of functions that are important in applications but do not
satisfy Condition 1 or 2.

The notation

$$f(t) \leftrightarrow F(\omega)$$

will be used to indicate that the functions $f(t)$ and $F(\omega)$ are related by
the integrals (2-1) and (2-3), and $f(t)$ will be called the *inverse Fourier
transform* of $F(\omega)$; it will always be assumed that $f(t)$ satisfies (2-6).
The variables t and ω will often be referred to as *time* and *frequency*.

2-2. Special Forms of the Fourier Integral

In general the function $f(t)$ is complex; denoting by $f_1(t)$ and $f_2(t)$ its
real and imaginary parts, we have

$$f(t) = f_1(t) + jf_2(t)$$

and

$$F(\omega) = \int_{-\infty}^{\infty} [f_1(t) \cos \omega t + f_2(t) \sin \omega t]\, dt$$

$$- j \int_{-\infty}^{\infty} [f_1(t) \sin \omega t - f_2(t) \cos \omega t]\, dt \quad (2\text{-}10)$$

Therefore

$$R(\omega) = \int_{-\infty}^{\infty} [f_1(t) \cos \omega t + f_2(t) \sin \omega t]\, dt$$

$$X(\omega) = - \int_{-\infty}^{\infty} [f_1(t) \sin \omega t - f_2(t) \cos \omega t]\, dt \quad (2\text{-}11)$$

We can similarly write the inversion formula in a real form; with
$F(\omega) = R(\omega) + jX(\omega)$ and $e^{j\omega t} = \cos \omega t + j \sin \omega t$, we readily obtain from (2-3)

$$f_1(t) = \frac{1}{2\pi} \int_{-\infty}^{\infty} [R(\omega) \cos \omega t - X(\omega) \sin \omega t]\, d\omega \quad (2\text{-}12)$$

$$f_2(t) = \frac{1}{2\pi} \int_{-\infty}^{\infty} [R(\omega) \sin \omega t + X(\omega) \cos \omega t]\, d\omega \quad (2\text{-}13)$$

We shall next examine the form of the above equations for various
special cases.

† In this book all improper integrals will be interpreted as Cauchy principal
values. This will apply also for singularities at finite point; i.e., if $f(t) \to \infty$ for
$t \to t_0$, then the integral of $f(t)$ in an interval (a,b) containing t_0 will be defined by

$$\int_a^b f(t)\, dt = \lim_{\epsilon \to 0} \left[\int_a^{t_0-\epsilon} f(t)\, dt + \int_{t_0+\epsilon}^b f(t)\, dt \right]$$

Real time functions. If $f(t)$ is real, then the real and imaginary parts of $F(\omega)$ are given by

$$R(\omega) = \int_{-\infty}^{\infty} f(t) \cos \omega t \, dt \qquad X(\omega) = -\int_{-\infty}^{\infty} f(t) \sin \omega t \, dt \qquad (2\text{-}14)$$

From (2-14) we conclude that $R(\omega)$ is even and $X(\omega)$ is odd

$$R(-\omega) = R(\omega) \qquad X(-\omega) = -X(\omega) \qquad (2\text{-}15)$$

Therefore
$$F(-\omega) = \overset{*}{F}(\omega) \qquad (2\text{-}16)$$

Conversely, if $F(-\omega) = \overset{*}{F}(\omega)$, then the integrand in (2-13) is an odd function of ω. Hence, $f_2(t) = 0$; i.e., $f(t)$ is real. Thus (2-16) is a necessary and sufficient condition for $f(t)$ to be real.

For real time functions the inversion formula can be written in the form

$$f(t) = \frac{1}{2\pi} \int_{-\infty}^{\infty} [R(\omega) \cos \omega t - X(\omega) \sin \omega t] \, d\omega$$

$$= \frac{1}{\pi} \int_{0}^{\infty} [R(\omega) \cos \omega t - X(\omega) \sin \omega t] \, d\omega \qquad (2\text{-}17)$$

$$= \frac{1}{\pi} \int_{0}^{\infty} A(\omega) \cos [\omega t + \phi(\omega)] \, d\omega$$

$$= \frac{1}{\pi} \operatorname{Re} \int_{0}^{\infty} F(\omega) e^{j\omega t} \, d\omega$$

as we see from (2-12) and (2-2).

Imaginary time functions. If $f(t)$ is purely imaginary,

$$f(t) = jf_2(t),$$

then
$$R(\omega) = \int_{-\infty}^{\infty} f_2(t) \sin \omega t \, dt \qquad X(\omega) = \int_{-\infty}^{\infty} f_2(t) \cos \omega t \, dt \qquad (2\text{-}18)$$

Therefore $R(\omega)$ is odd and $X(\omega)$ is even

$$R(-\omega) = -R(\omega) \qquad X(-\omega) = X(\omega) \qquad F(-\omega) = -\overset{*}{F}(\omega) \qquad (2\text{-}19)$$

The converse is also true: if $F(-\omega) = -\overset{*}{F}(\omega)$, then the inverse transform of $F(\omega)$ is purely imaginary, because the integrand in (2-12) is an odd function of ω.

In the remaining part of this section we shall assume $f(t)$ real; the following special cases are of particular interest.

$f(t)$ **even.** If $f(-t) = f(t)$, then $f(t) \cos \omega t$ is even and $f(t) \sin \omega t$ odd with respect to t; therefore [see (2-14)]

$$R(\omega) = 2 \int_{0}^{\infty} f(t) \cos \omega t \, dt \qquad X(\omega) = 0 \qquad (2\text{-}20)$$

From (2-20) and (2-17) we obtain

$$f(t) = \frac{1}{\pi} \int_0^\infty R(\omega) \cos \omega t \, d\omega \qquad (2\text{-}21)$$

Conversely if the Fourier integral of a real function $f(t)$ is real, then $f(t)$ is even; indeed, assuming $X(\omega) = 0$, we conclude from (2-17) that $f(t)$ is given by (2-21); hence

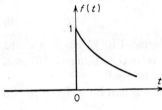

$$f(-t) = f(t).$$

$f(t)$ **odd.** If $f(-t) = -f(t)$, then

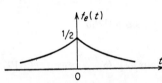

$$R(\omega) = 0$$

$$X(\omega) = -2 \int_0^\infty f(t) \sin \omega t \, dt \qquad (2\text{-}22)$$

as we see from (2-14). In this case the inversion formula (2-17) takes the form

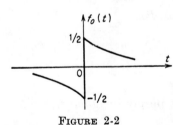

FIGURE 2-2

$$f(t) = -\frac{1}{\pi} \int_0^\infty X(\omega) \sin \omega t \, d\omega \qquad (2\text{-}23)$$

Conversely if the Fourier integral of a real function $f(t)$ is purely imaginary, then $f(t)$ is odd and is given by (2-23).

An arbitrary function $f(t)$ can be decomposed into a sum of an even and an odd function. With $f_e(t)$ and $f_0(t)$ defined by

$$f_e(t) = \frac{f(t) + f(-t)}{2} \qquad f_0(t) = \frac{f(t) - f(-t)}{2} \qquad (2\text{-}24)$$

we have $f_e(-t) = f_e(t)$, $f_0(-t) = -f_0(t)$, and

$$f(t) = f_0(t) + f_e(t) \qquad (2\text{-}25)$$

We further remark that if $R(\omega) + jX(\omega)$ is the Fourier integral of $f(t)$, and $F_{\bar{e}}(\omega)$ and $F_0(\omega)$ the Fourier integrals of $f_e(t)$ and $f_0(t)$ respectively, then $F_e(\omega) = R(\omega)$, $F_0(\omega) = jX(\omega)$; this follows from

$$R(\omega) + jX(\omega) = F_e(\omega) + F_0(\omega)$$

and the fact that $F_e(\omega)$ is real and $F_0(\omega)$ purely imaginary. We thus have the useful equations

$$f_e(t) \leftrightarrow R(\omega) \qquad f_0(t) \leftrightarrow jX(\omega) \qquad (2\text{-}26)$$

$$R(\omega) = 2 \int_0^\infty f_e(t) \cos \omega t \, dt \qquad X(\omega) = -2 \int_0^\infty f_0(t) \sin \omega t \, dt \qquad (2\text{-}27)$$

$$f_e(t) = \frac{1}{\pi} \int_0^\infty R(\omega) \cos \omega t \, d\omega \qquad f_0(t) = -\frac{1}{\pi} \int_0^\infty X(\omega) \sin \omega t \, d\omega \qquad (2\text{-}28)$$

In the following we shall make frequent use of the above decomposition. In Fig. 2-2 we have shown the function $f(t) = U(t)e^{-\alpha t}$ and its even and odd parts; $U(t)$ is the familiar *unit step* function.

Causal time functions. A function will be called *causal* if it equals zero for negative t:

$$f(t) = 0 \qquad t < 0 \qquad (2\text{-}29)$$

This important class of functions is characteristic of the impulse response of causal systems to be developed in Chap. 5. We shall show that $f(t)$ can be expressed in terms of $R(\omega)$ or $X(\omega)$ alone. For $t > 0$ we have $f(-t) = 0$; therefore

$$f(t) = 2f_e(t) = 2f_0(t) \qquad t > 0 \qquad (2\text{-}30)$$

as we readily see from (2-24). From (2-30) and (2-28) we obtain

$$f(t) = \frac{2}{\pi} \int_0^\infty R(\omega) \cos \omega t \, d\omega = -\frac{2}{\pi} \int_0^\infty X(\omega) \sin \omega t \, d\omega \qquad t > 0 \qquad (2\text{-}31)$$

The above is valid only for $t > 0$; for $t = 0$ we have

$$f(0) = \frac{1}{\pi} \int_0^\infty R(\omega) \, d\omega = \frac{f(0^+)}{2} \qquad (2\text{-}32)$$

[see (2-17)]. It is of interest to remark that the functions $R(\omega)$ and $X(\omega)$ are not independent of each other but one of them can be uniquely determined in terms of the other. Indeed, given $R(\omega)$, we find $f(t)$ from (2-31); inserting the resulting expression into the second equation in (2-14) we obtain

$$X(\omega) = -\frac{2}{\pi} \int_0^\infty \int_0^\infty R(y) \cos yt \sin \omega t \, dy \, dt$$

Similarly,

$$R(\omega) = -\frac{2}{\pi} \int_0^\infty \int_0^\infty X(y) \sin yt \cos \omega t \, dy \, dt$$

In Chap. 10 we shall express one of the functions $R(\omega)$ and $X(\omega)$ in terms of the other by explicit equations known as *Hilbert transforms*.

2-3. Simple Theorems†

The following is a list of simple theorems that can be easily derived from Eqs. (2-1) and (2-3); it is assumed that all functions under consideration have Fourier integrals.

A. Linearity. If $F_1(\omega)$ and $F_2(\omega)$ are the Fourier integrals of $f_1(t)$ and $f_2(t)$ respectively, and a_1, a_2 two arbitrary constants, then

$$a_1 f_1(t) + a_2 f_2(t) \leftrightarrow a_1 F_1(\omega) + a_2 F_2(\omega) \tag{2-33}$$

The proof is trivial. The above is also valid for finite sums

$$a_1 f_1(t) + \cdots + a_n f_n(t) \leftrightarrow a_1 F_1(\omega) + \cdots + a_n F_n(\omega)$$

but the extension to infinite sums is not always true because the interchange of integration and summation is possible only if certain conditions are fulfilled.

B. Symmetry. If $F(\omega)$ is the Fourier integral of $f(t)$, then

$$F(t) \leftrightarrow 2\pi f(-\omega) \tag{2-34}$$

The above follows from (2-3) if we write it in the form

$$2\pi f(-t) = \int_{-\infty}^{\infty} F(\omega)e^{-j\omega t}\, d\omega$$

and interchange t and ω. This theorem is a statement of the symmetrical character of Eqs. (2-1) and (2-3).

C. Time scaling. If a is a real constant, then

$$f(at) \leftrightarrow \frac{1}{|a|} F\left(\frac{\omega}{a}\right) \tag{2-35}$$

To prove (2-35), we have for $a > 0$

$$\int_{-\infty}^{\infty} f(at)e^{-j\omega t}\, dt = \frac{1}{a} \int_{-\infty}^{\infty} f(x)e^{-j(\omega/a)x}\, dx = \frac{1}{a} F\left(\frac{\omega}{a}\right)$$

If $a < 0$, the sign in the end result changes because the limits of integration are interchanged.

D. Time shifting. If the function $f(t)$ is shifted by a constant t_0, then its Fourier spectrum remains the same, but a linear term $-t_0\omega$ is added to its phase angle

$$f(t - t_0) \leftrightarrow F(\omega)e^{-jt_0\omega} = A(\omega)e^{j[\phi(\omega) - t_0\omega]} \tag{2-36}$$

† E. A. Guillemin, "The Mathematics of Circuit Analysis," John Wiley & Sons, Inc., New York, 1949.

This follows from

$$\int_{-\infty}^{\infty} f(t - t_0)e^{-j\omega t}\, dt = \int_{-\infty}^{\infty} f(x)e^{-j\omega(t_0 + x)}\, dx = F(\omega)e^{-jt_0\omega}$$

E. Frequency shifting. With ω_0 a real constant, the Fourier integral of $e^{j\omega_0 t}f(t)$ is obtained by shifting $F(\omega)$ by ω_0, as in

$$e^{j\omega_0 t}f(t) \leftrightarrow F(\omega - \omega_0) \tag{2-37}$$

Indeed $\displaystyle\int_{-\infty}^{\infty} f(t)e^{j\omega_0 t}e^{-j\omega t}\, dt = \int_{-\infty}^{\infty} f(t)e^{-j(\omega - \omega_0)t}\, dt = F(\omega - \omega_0)$

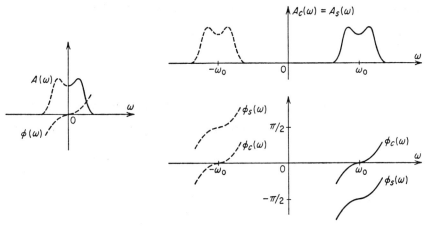

FIGURE 2-3

(2-37) can also be derived from (2-36) and (2-34). From C and E, we conclude that

$$e^{j\omega_0 t}f(at) \leftrightarrow \frac{1}{|a|}\, F\!\left(\frac{\omega - \omega_0}{a}\right)$$

The above theorem can be used to derive the Fourier integral $A_c(\omega)e^{j\phi_c(\omega)}$ of a *modulated signal* $f(t)\cos\omega_0 t$ in terms of the Fourier integral $F(\omega)$ of its *envelope* $f(t)$. Indeed from (2-37) it readily follows that

$$f(t)\cos\omega_0 t \leftrightarrow \frac{F(\omega - \omega_0) + F(\omega + \omega_0)}{2} = A_c(\omega)e^{j\phi_c(\omega)} \tag{2-38}$$

If the envelope $f(t)$ has a low frequency spectrum in the sense

$$A(\omega) = 0 \qquad |\omega| > \omega_0 \tag{2-39}$$

then $A_c(\omega)e^{j\phi_c(\omega)}$ can be simply found by shifting $F(\omega)$ as in Fig. 2-3. We similarly obtain the Fourier integral $A_s(\omega)e^{j\phi_s(\omega)}$ of $f(t)\sin\omega_0 t$:

$$f(t)\sin\omega_0 t \leftrightarrow \frac{F(\omega - \omega_0) - F(\omega + \omega_0)}{2j} = A_s(\omega)e^{j\phi_s(\omega)} \tag{2-40}$$

F. Time differentiation. Taking the nth derivatives of both sides of (2-3), we conclude that

$$\frac{d^n f}{dt^n} \leftrightarrow (j\omega)^n F(\omega) \qquad (2\text{-}41)$$

(2-41) does not guarantee the existence of the transform of $d^n f/dt^n$; it simply says that if that transform exists, it is given by $(j\omega)^n F(\omega)$. Consider the function $\phi(t) = \int_{-\infty}^{t} f(\tau)\, d\tau$. With $\Phi(\omega)$ its transform, we conclude from $d\,\phi(t)/dt = f(t)$ and (2-41) that $j\omega\Phi(\omega) = F(\omega)$. For $\phi(t)$ to satisfy Conditions 1 or 2 of page 9 we must have $\phi(\infty) = 0$, i.e., $\int_{-\infty}^{\infty} f(\tau)\, d\tau = F(0) = 0$. In this case

$$\int_{-\infty}^{t} f(\tau)\, d\tau \leftrightarrow \frac{F(\omega)}{j\omega} \qquad (2\text{-}42)$$

If $F(0) \neq 0$ then $\int_{-\infty}^{t} f(\tau)\, d\tau$ does not belong to the class of functions considered in this chapter. [It is shown in (3-22) that its transform equals $\pi F(0)\,\delta(\omega) + F(\omega)/j\omega$.]

G. Frequency differentiation. Differentiating (2-1), we readily derive

$$(-jt)^n f(t) \leftrightarrow \frac{d^n F(\omega)}{d\omega^n} \qquad (2\text{-}43)$$

H. Conjugate functions. The Fourier integral of the conjugate $\overset{*}{f}(t) = f_1(t) - jf_2(t)$ of a complex function $f(t) = f_1(t) + jf_2(t)$ is given by $\overset{*}{F}(-\omega)$

$$\overset{*}{f}(t) \leftrightarrow \overset{*}{F}(-\omega) \qquad (2\text{-}44)$$

From $\quad F(\omega) = \int_{-\infty}^{\infty} (f_1 + jf_2)e^{-j\omega t}\, dt \qquad \overset{*}{F}(\omega) = \int_{-\infty}^{\infty} (f_1 - jf_2)e^{j\omega t}\, dt$

we have $\qquad \overset{*}{F}(-\omega) = \int_{-\infty}^{\infty} (f_1 - jf_2)e^{-j\omega t}\, dt$

and (2-44) follows.

I. Moment theorem. This theorem relates the derivatives of $F(\omega)$ at the origin to the *moments* of its inverse transform $f(t)$. The nth moment m_n of $f(t)$ is defined by

$$m_n = \int_{-\infty}^{\infty} t^n f(t)\, dt \qquad n = 0, 1, 2, \ldots \qquad (2\text{-}45)$$

and the theorem states that

$$(-j)^n m_n = \frac{d^n F(0)}{d\omega^n} \qquad n = 0, 1, 2, \ldots \qquad (2\text{-}46)$$

The case $n = 0$ follows readily from (2-1); with $\omega = 0$ we obtain

$$m_0 = \int_{-\infty}^{\infty} f(t)\, dt = F(0) = A(0) \tag{2-47}$$

To prove the general case, we expand $e^{-j\omega t}$ and integrate (2-1) termwise; because of (2-45) the equation

$$F(\omega) = \int_{-\infty}^{\infty} f(t) \left[\sum_{n=0}^{\infty} \frac{(-j\omega t)^n}{n!} \right] dt = \sum_{n=0}^{\infty} (-j)^n m_n \frac{\omega^n}{n!} \tag{2-48}$$

results.   Expanding $F(\omega)$ into a series

$$F(\omega) = \sum_{n=0}^{\infty} \frac{d^n F(0)}{d\omega^n} \frac{\omega^n}{n!} \tag{2-49}$$

and equating coefficients of equal powers of ω in (2-48) and (2-49), we deduce (2-46).

Comment.   The above holds only if the termwise integration in (2-48) is valid, and this is possible only if the moments of $f(t)$ are finite; the existence of $d^n F(0)/d\omega^n$ does not guarantee the finiteness of m_n. Thus all derivatives of the rectangular function $p_a(\omega)$ of Example 2-2 are equal to zero at the origin but its inverse $\sin at/\pi t$ has no finite moments for $n \geq 1$.

It is of interest to relate the moments of a real $f(t)$ to the slope of $\phi(\omega)$ and the curvature of $A(\omega)$ at the origin.   To simplify the notations, we shall assume that the area of $f(t)$ equals one:

$$\int_{-\infty}^{\infty} f(t)\, dt = A(0) = 1 \tag{2-50}$$

Expanding the even function $A(\omega)$ and the odd function $\phi(\omega)$ into a series, we obtain [see (2-50)]

$$A(\omega) = 1 + \frac{a_2}{2!} \omega^2 + \cdots \tag{2-51}$$

$$\phi(\omega) = b_1 \omega + \frac{b_3}{3!} \omega^3 + \cdots \tag{2-52}$$

Hence $$e^{j\phi(\omega)} = 1 + j b_1 \omega - \frac{b_1^2}{2!} \omega^2 + \text{higher powers of } \omega$$

Therefore $$F(\omega) = \left(1 + \frac{a_2}{2} \omega^2 + \cdots \right) \left(1 + j b_1 \omega - \frac{b_1^2}{2!} \omega^2 + \cdots \right)$$

$$= 1 + j b_1 \omega + \frac{a_2 - b_1^2}{2!} \omega^2 + \cdots$$

From (2-48) and the above equality we conclude that

$$b_1 = -m_1 \qquad a_2 = m_1^2 - m_2 \tag{2-53}$$

We introduce the quantities

$$\eta = \int_{-\infty}^{\infty} tf(t)\,dt \qquad \sigma^2 = \int_{-\infty}^{\infty} (t - \eta)^2 f(t)\,dt \qquad (2\text{-}54)$$

$\eta = m_1$ is the center of gravity or *mean* of $f(t)$, and σ^2 its central moment of inertia or *dispersion*. Clearly

$$\sigma^2 = \int_{-\infty}^{\infty} t^2 f(t)\,dt - 2\eta \int_{-\infty}^{\infty} tf(t)\,dt + \eta^2 \int_{-\infty}^{\infty} f(t)\,dt = m_2 - m_1{}^2$$

Therefore [see (2-53)]

$$\eta = -b_1 = -\frac{d\phi(0)}{d\omega} \qquad \sigma^2 = -a_2 = -\frac{d^2 A(0)}{d\omega^2} \qquad (2\text{-}55)$$

Thus the negative slope of $\phi(\omega)$ at the origin equals the mean of $f(t)$, and the curvature of $A(\omega)$ equals its dispersion.

2-4. Examples

In this section we shall evaluate the Fourier integrals of a number of simple functions that appear frequently in applications. The results can be obtained directly from (2-1) or (2-3), but the computations are often simplified if use is made of the theorems of the preceding section. The Fourier transforms will be graphically given either by their amplitude and phase or by their real and imaginary parts. We shall make frequent use of the following functions:

1. The function $\sin t/t$ of Fig. 2-4, known also as *Fourier kernel*
2. The tabulated integral

$$Si(t) = \int_0^t \frac{\sin t}{t}\,dt$$

of $\sin t/t$ called *sine integral* and shown in Fig. 2-5
3. The rectangular pulse $p_T(t)$ defined by

$$p_T(t) = U(t + T) - U(t - T) = \begin{cases} 0 & |t| > T \\ 1 & |t| < T \end{cases}$$

and shown in Fig. 2-6
4. The triangular pulse of Fig. 2-7

$$q_T(t) = \begin{cases} 1 - \dfrac{|t|}{T} & |t| < T \\ 0 & |t| > T \end{cases}$$

5. The function

$$\operatorname{sgn} t = \frac{t}{|t|} = \begin{cases} 1 & t > 0 \\ -1 & t < 0 \end{cases}$$

(read sign t) shown in Fig. 2-8

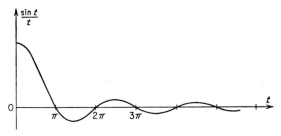

FIGURE 2-4

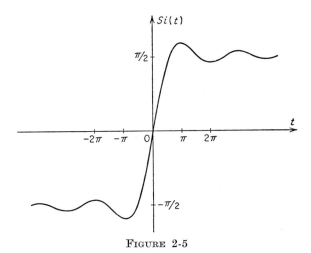

FIGURE 2-5

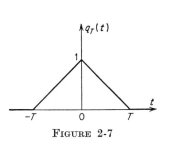

FIGURE 2-6

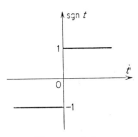

FIGURE 2-7   FIGURE 2-8

Example 2-1. $f(t) = p_T(t)$; from (2-1) we obtain

$$F(\omega) = \int_{-\infty}^{\infty} p_T(t)e^{-j\omega t}\,dt = \int_{-T}^{T} e^{-j\omega t}\,dt = \frac{2\sin\omega T}{\omega}$$

and the pair of Fig. 2-9 results. The solid line is the Fourier spectrum $|2\sin\omega T/\omega|$; the dotted line shows its real part $R(\omega) = 2\sin\omega T/\omega$. From the above and (2-36) we obtain

$$p_T(t - t_0) \leftrightarrow \frac{2\sin\omega T}{\omega} e^{-j\omega t_0} \qquad (2\text{-}56)$$

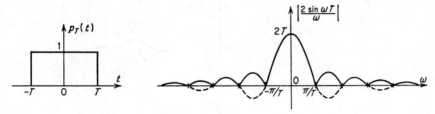

FIGURE 2-9

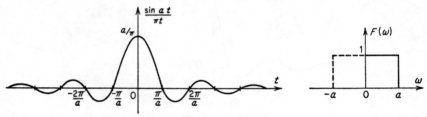

FIGURE 2-10

The phase angle in (2-56) is linear for $\omega \neq n\pi/T$, with jumps equal to π for $\omega = n\pi/T$, because of the change in the sign of $\sin\omega T$ at these points.

Example 2-2. From the above example and the symmetry theorem (2-34), we obtain the pair

$$\frac{\sin at}{\pi t} \leftrightarrow p_a(\omega) \qquad (2\text{-}57)$$

shown in Fig. 2-10.

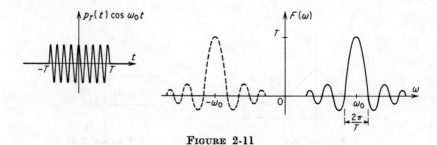

FIGURE 2-11

Example 2-3. The Fourier integral of a pulse-modulated signal

$$f(t) = p_T(t) \cos \omega_0 t$$

is given by

$$F(\omega) = \frac{\sin (\omega - \omega_0)T}{(\omega - \omega_0)} + \frac{\sin (\omega + \omega_0)T}{(\omega + \omega_0)}$$

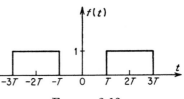

FIGURE 2-12

as we can see from Example 2-1 and Eq. (2-38). The above pair is shown in Fig. 2-11 for $\omega_0 T \gg 1$.

Example 2-4. With $f(t) = p_T(t + 2T) + p_T(t - 2T)$, as in Fig. 2-12, $F(\omega)$ is given by

$$F(\omega) = \frac{2 \sin \omega T}{\omega} (e^{-j2T\omega} + e^{j2T\omega}) = \frac{4 \sin \omega T}{\omega} \cos 2\omega T$$

Example 2-5. Similarly, if $f(t) = p_{T/2}(t + T/2) - p_{T/2}(t - T/2)$, then

$$F(\omega) = \frac{2 \sin (\omega T/2)}{\omega} (e^{j\omega T/2} - e^{-j\omega T/2}) = \frac{4j \sin^2 (\omega T/2)}{\omega}$$

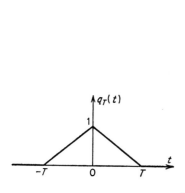
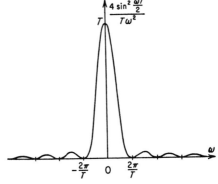

FIGURE 2-13

Example 2-6. The triangular pulse $q_T(t)$ of Fig. 2-13 is obviously the integral of $f(t)/T$, with $f(t)$ as in Example 2-5; therefore [see (2-42)]

$$q_T(t) \leftrightarrow \frac{4 \sin^2 (\omega T/2)}{T\omega^2} \tag{2-58}$$

Example 2-7. From (2-58) and (2-34) we obtain the pair

$$\frac{\sin^2 at}{\pi a t^2} \leftrightarrow q_{2a}(\omega) \tag{2-59}$$

shown in Fig. 2-14. The function $\sin^2 at/\pi at^2$ is called *Fejér kernel* (see also Sec. 2-6). From (2-47) we readily see that

$$\int_{-\infty}^{\infty} \frac{\sin^2 at}{\pi at^2}\, dt = 1$$

since $q_{2a}(0) = 1$.

From (2-59) and (2-38) the pair

$$\frac{2 \sin^2 at}{\pi at^2} \cos \omega_0 t \leftrightarrow q_{2a}(\omega + \omega_0) + q_{2a}(\omega - \omega_0) \qquad (2\text{-}60)$$

of Fig. 2-15 results.

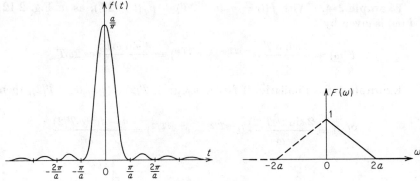

FIGURE 2-14

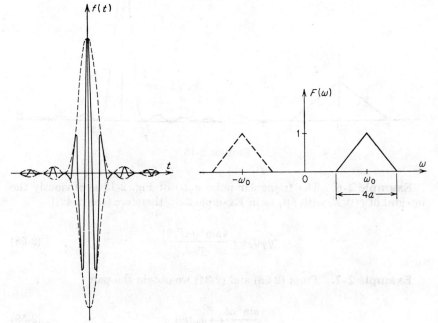

FIGURE 2-15

Example 2-8. $f(t) = e^{-\alpha t}U(t)$, $\alpha > 0$; from (2-1) we obtain

$$F(\omega) = \int_{-\infty}^{\infty} e^{-\alpha t}U(t)e^{-j\omega t}\,dt = \int_{0}^{\infty} e^{-\alpha t}e^{-j\omega t}\,dt = \frac{1}{\alpha + j\omega}$$

The pair

$$e^{-\alpha t}U(t) \leftrightarrow \frac{1}{\alpha + j\omega} = \frac{1}{\sqrt{\alpha^2 + \omega^2}}\,e^{-j\tan^{-1}(\omega/\alpha)} \qquad (2\text{-}61)$$

is shown in Fig. 2-16. From (2-61) a number of useful transforms readily

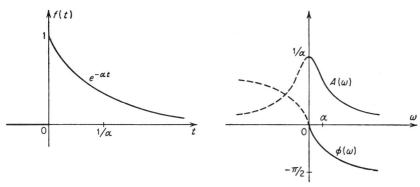

FIGURE 2-16

follow. Indeed, since the even part of $f(t)$ equals $e^{-\alpha|t|}/2$ and the real part of $F(\omega)$ equals $\alpha/(\alpha^2 + \omega^2)$, we conclude from (2-26) that (Fig. 2-17)

$$e^{-\alpha|t|} \leftrightarrow \frac{2\alpha}{\alpha^2 + \omega^2} \qquad (2\text{-}62)$$

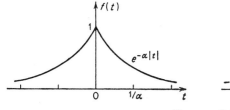

FIGURE 2-17

From (2-61) and theorem (2-43) it follows that

$$\frac{t^{n-1}}{(n-1)!}\,e^{-\alpha t}U(t) \leftrightarrow \frac{1}{(j\omega + \alpha)^n} \qquad (2\text{-}63)$$

Finally, with the help of (2-40), we easily obtain (Fig. 2-18)

$$e^{-\alpha t}\sin\beta t\,U(t) \leftrightarrow \frac{\beta}{(\alpha + j\omega)^2 + \beta^2} \qquad (2\text{-}64)$$

Example 2-9. $f(t) = U(t)(\sin at)/t$. The real part $R(\omega)$ of the Fourier transform of this function can be easily obtained from (2-57) and (2-26); indeed, the even part of $f(t)$ equals $\sin at/2t$, and since $\sin at/2t \leftrightarrow \pi p_a(\omega)/2$, we conclude that

$$R(\omega) = \frac{\pi}{2} p_a(\omega)$$

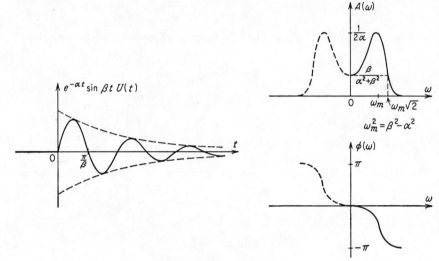

FIGURE 2-18

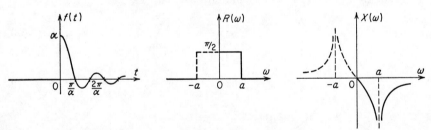

FIGURE 2-19

In Chap. 10 we shall show that $X(\omega)$ is given by (Fig. 2-19)·

$$X(\omega) = \tfrac{1}{2} \ln \left| \frac{\omega - a}{\omega + a} \right|$$

This result can also be obtained from (2-1).

Example 2-10. $f(t) = e^{-\alpha t^2}$. This function is called *Gaussian* and plays an important role in probability theory. To determine its Fourier transform $F(\omega)$, we shall first evaluate its moment m_n and then use (2-48). It is well known that

$$\int_{-\infty}^{\infty} e^{-\alpha t^2} \, dt = \sqrt{\frac{\pi}{\alpha}} \tag{2-65}$$

Differentiating n times with respect to α, we obtain

$$\int_{-\infty}^{\infty} t^{2n} e^{-\alpha t^2} dt = \frac{1.3 \cdots (2n-1)}{2^n} \sqrt{\frac{\pi}{\alpha^{2n+1}}} \tag{2-66}$$

Therefore the even moments of $e^{-\alpha t^2}$ are given by the right-hand side of (2-66). Its odd moments are zero because it is an even function of t; hence [see (2-48)]

$$F(\omega) = \sqrt{\frac{\pi}{\alpha}} \sum_{n=0}^{\infty} \frac{1.3 \cdots (2n-1)}{(2\alpha)^n (2n)!} (-j\omega)^{2n} \tag{2-67}$$

but the above quantity is the series expansion of

$$\sqrt{\frac{\pi}{\alpha}} e^{-\omega^2/4\alpha}$$

Therefore (Fig. 2-20)

$$e^{-\alpha t^2} \leftrightarrow \sqrt{\frac{\pi}{\alpha}} e^{-\omega^2/4\alpha} \tag{2-68}$$

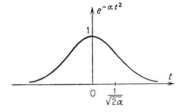 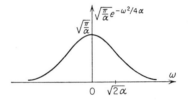

FIGURE 2-20

From (2-68) and (2-14) we obtain the useful identity

$$\int_{-\infty}^{\infty} e^{-\alpha t^2} \cos \omega t \, dt = \sqrt{\frac{\pi}{\alpha}} e^{-\omega^2/4\alpha} \tag{2-69}$$

With $\alpha = \frac{1}{2}$, (2-68) gives the pair

$$e^{-t^2/2} \leftrightarrow \sqrt{2\pi} \, e^{-\omega^2/2}$$

Thus, except for the factor $\sqrt{2\pi}$, the function $e^{-t^2/2}$ is its own Fourier transform. Other functions with this property can be found by solving the differential equation

$$\frac{d^2 f(t)}{dt^2} - t^2 f(t) = -(2n+1) f(t)$$

(see Prob. 7).

2-5. The Convolution Theorem

Next to the inversion formula (2-3), the convolution theorem is the most powerful tool in this analysis. It permits the easy derivation of many important results and will be constantly used throughout the book.

Given two functions $f_1(x)$ and $f_2(x)$, we form the integral

$$f(x) = \int_{-\infty}^{\infty} f_1(y)f_2(x - y)\, dy \qquad (2\text{-}70)$$

This integral defines a function $f(x)$ known as the *convolution* of $f_1(x)$ and $f_2(x)$; the operation (2-70) is often written in the form

$$f(x) = f_1(x) * f_2(x)$$

and it is easily seen that $f_1(x) * f_2(x) = f_2(x) * f_1(x)$.   From (2-70) it readily follows that if the functions $f_1(x)$ and $f_2(x)$ are bounded, then $f(x)$ is continuous.   Similarly if the functions $f_1(x)$ and $f_2(x)$ are positive, then their convolution $f(x)$ is also positive.   Finally if $f_1(x) = 0$ for $x > A_1$, $f_2(x) = 0$ for $x > A_2$, then $f(x) = 0$ for $x > A_1 + A_2$.

An important special case known as *smoothing* is the convolution of an arbitrary function $g(t)$ with the rectangular pulse $p_T(t)$; it is easy to see that

$$g(t) * p_T(t) = \int_{t-T}^{t+T} g(y)\, dy$$

The integral of a function can also be written as a convolution

$$\int_{-\infty}^{t} g(y)\, dy = g(t) * U(t)$$

where $U(t)$ is the unit step function.

Time convolution theorem.   The Fourier transform $F(\omega)$ of the convolution $f(t)$ of two functions $f_1(t)$ and $f_2(t)$ equals the product of the Fourier transforms $F_1(\omega)$ and $F_2(\omega)$ of these two functions.   Thus if

$$f_1(t) \leftrightarrow F_1(\omega) \qquad f_2(t) \leftrightarrow F_2(\omega)$$

then
$$\int_{-\infty}^{\infty} f_1(\tau)f_2(t - \tau)\, d\tau \leftrightarrow F_1(\omega)F_2(\omega) \qquad (2\text{-}71)$$

Proof.   To prove (2-71), we shall form the Fourier integral of $f(t)$ and will show that it equals $F_1(\omega)F_2(\omega)$.   Clearly

$$F(\omega) = \int_{-\infty}^{\infty} e^{-j\omega t}\left[\int_{-\infty}^{\infty} f_1(\tau)f_2(t - \tau)\, d\tau\right] dt \qquad (2\text{-}72)$$

Changing the order of integration, we obtain

$$F(\omega) = \int_{-\infty}^{\infty} f_1(\tau)\left[\int_{-\infty}^{\infty} e^{-j\omega t}f_2(t - \tau)\, dt\right] d\tau$$

From the time-shifting theorem (2-36) we conclude that the bracket above equals $F_2(\omega)e^{-j\omega\tau}$; therefore

$$F(\omega) = \int_{-\infty}^{\infty} f_1(\tau)e^{-j\omega\tau}F_2(\omega)\,d\tau = F_1(\omega)F_2(\omega)$$

and (2-71) is proved.

Comment. In the above proof it was assumed that the order of integration in (2-72) can be changed. This is true if the functions $f_1(t)$ and $f_2(t)$ are *square-integrable* in the sense

$$\int_{-\infty}^{\infty} |f_i(t)|^2\,dt < \infty \qquad i = 1, 2 \tag{2-73}$$

i.e., if $f_1(t)$ and $f_2(t)$ have finite energy.

Frequency convolution theorem. From the above result (2-71) and the symmetry property (2-34) it follows that the Fourier transform $F(\omega)$ of the product $f_1(t)f_2(t)$ of two functions equals the convolution $F_1(\omega) * F_2(\omega)$ of their respective transforms $F_1(\omega)$ and $F_2(\omega)$ divided by 2π:

$$f_1(t)f_2(t) \leftrightarrow \frac{1}{2\pi}\int_{-\infty}^{\infty} F_1(y)F_2(\omega - y)\,dy \tag{2-74}$$

One could also give a direct proof of (2-74) as in the time-convolution theorem.

Parseval's formula. The following basic result, known as *Parseval's formula*, can be easily derived from (2-74); if $F(\omega) = A(\omega)e^{j\phi(\omega)}$ is the Fourier transform of $f(t)$, then

$$\int_{-\infty}^{\infty} |f(t)|^2\,dt = \frac{1}{2\pi}\int_{-\infty}^{\infty} A^2(\omega)\,d\omega \tag{2-75}$$

Indeed, from $f(t) \leftrightarrow F(\omega)$ and theorem (2-44) it follows that $\overset{*}{f}(t) \leftrightarrow \overset{*}{F}(-\omega)$; therefore the Fourier integral of $|f(t)|^2 = f(t)\overset{*}{f}(t)$ is the function $(1/2\pi)F(\omega) * \overset{*}{F}(-\omega)$; i.e.,

$$\frac{1}{2\pi}\int_{-\infty}^{\infty} F(y)\overset{*}{F}[-(\omega - y)]\,dy = \int_{-\infty}^{\infty} |f(t)|^2\,e^{-j\omega t}\,dt \tag{2-76}$$

Putting $\omega = 0$ in (2-76), we obtain (2-75), because

$$F(y)\overset{*}{F}(y) = A^2(y)$$

The following is a more general form of Parseval's formula: if

$$f_1(t) \leftrightarrow F_1(\omega) \qquad f_2(t) \leftrightarrow F_2(\omega)$$

then

$$\int_{-\infty}^{\infty} f_1(t)f_2(t)\,dt = \frac{1}{2\pi}\int_{-\infty}^{\infty} F_1(-\omega)F_2(\omega)\,d\omega \tag{2-77}$$

The proof follows again from (2-74): since

$$f_1(t)f_2(t) \longleftrightarrow \frac{1}{2\pi} F_1(\omega) * F_2(\omega)$$

we conclude from (2-1) that

$$\frac{1}{2\pi} \int_{-\infty}^{\infty} F_1(\omega - x)F_2(x)\,dx = \int_{-\infty}^{\infty} f_1(t)f_2(t)e^{-j\omega t}\,dt \qquad (2\text{-}78)$$

and with $\omega = 0$, (2-77) results.

If the functions $f_1(t)$ and $f_2(t)$ are real, then (2-77) can be written in the form

$$\int_{-\infty}^{\infty} f_1(t)f_2(t)\,dt = \frac{1}{2\pi} \int_{-\infty}^{\infty} \overset{*}{F_1}(\omega)F_2(\omega)\,d\omega \qquad (2\text{-}79)$$

since $F_1(-\omega) = \overset{*}{F_1}(\omega)$ [see (2-16)].

Energy spectra. In Sec. 2-1 we called the quantity $A^2(\omega)$ the energy spectrum of $f(t)$; the justification for this name is based on Parseval's formula (2-75).   Indeed, if we assume $f(t)$ to be the voltage of a source across a 1-ohm resistance, then the quantity

$$\int_{-\infty}^{\infty} f^2(t)\,dt$$

equals the total energy delivered by the source.   According to (2-75) this energy equals the area under the $A^2(\omega)/2\pi$ curve.

Consider the real functions $f_1(t)$ and $f_2(t)$; if we interpret $f_1(t)$ as the voltage of a source and $f_2(t)$ the current that it supplies to some load, then the integral

$$\int_{-\infty}^{\infty} f_1(t)f_2(t)\,dt$$

equals the energy delivered by this source.   With

$$E_{12}(\omega) = \overset{*}{F_1}(\omega)F_2(\omega) \qquad (2\text{-}80)$$

we see from (2-79) that this energy equals the area of $E_{12}(\omega)/2\pi$.   For this reason the quantity $E_{12}(\omega)$ is often called the *cross-energy spectrum* of $f_1(t)$ and $f_2(t)$.

corollaries.   The following two corollaries of the convolution theorem show what happens to a signal if the high-frequency components of its spectrum are eliminated.

1. Suppose that the Fourier transform $F(\omega)$ of a function $f(t)$ is truncated above $|\omega| = a$; the band-limited function

$$F_a(\omega) = \begin{cases} F(\omega) & |\omega| < a \\ 0 & |\omega| > a \end{cases} = F(\omega)p_a(\omega) \qquad (2\text{-}81)$$

results, where $p_a(\omega)$ is a rectangular pulse. Such an operation takes place, for example, if $f(t)$ is the input to an ideal low-pass filter; the Fourier integral of its output is then given by (2-81), as we shall show in Chap. 6. It is of interest to determine the relationship between the inverse transform $f_a(t)$ of $F_a(\omega)$ and the given function $f(t)$. Since the inverse transform of the pulse $p_a(\omega)$ equals sin $at/\pi t$ [see (2-57)], we conclude from (2-81) and (2-71) that

$$f_a(t) = \int_{-\infty}^{\infty} f(\tau) \frac{\sin a(t - \tau)}{\pi(t - \tau)} d\tau$$

Thus truncation of $F(\omega)$ results in convolution of $f(t)$ with the Fourier kernel sin $at/\pi t$

$$f(t) * \frac{\sin at}{\pi t} \leftrightarrow F(\omega)p_a(\omega) \qquad (2\text{-}82)$$

Reasoning as above, we conclude from (2-74) that truncation of $f(t)$ above $|t| = T$ results in convolution of $F(\omega)/2\pi$ with the Fourier transform 2 sin $\omega T/\omega$ of $p_T(t)$

$$f(t)p_T(t) \leftrightarrow \int_{-\infty}^{\infty} F(x) \frac{\sin(\omega - x)T}{\pi(\omega - x)} dx \qquad (2\text{-}83)$$

From (2-83) one concludes that if a bounded function $f(t)$ equals zero outside a finite interval $(-T,T)$, then its Fourier transform $F(\omega)$ must be continuous. Indeed, since $f(t) = 0$ for $|t| > T$, we have $f(t) = f(t)p_T(t)$; therefore

$$F(\omega) = F(\omega) * \frac{\sin \omega T}{\pi\omega}$$

Hence $F(\omega)$ is continuous.

2. Suppose now that we eliminate the components of $F(\omega)$ above $|\omega| = a$ and in addition we favor linearly the low-frequency components; the function $F(\omega)q_a(\omega)$ results, where $q_a(\omega)$ is the triangular pulse of Fig. 2-7. Since the inverse transform of $q_a(\omega)$ equals

$$\frac{2 \sin^2 (at/2)}{\pi at^2}$$

the corresponding operation in the time domain is a convolution of $f(t)$ with the Fejér kernel

$$\int_{-\infty}^{\infty} f(\tau) \frac{2 \sin^2 [a(t - \tau)/2]}{\pi a(t - \tau)^2} d\tau \leftrightarrow F(\omega)q_a(\omega) \qquad (2\text{-}84)$$

2-6. On the Proof of the Fourier-integral Theorem

In Sec. 2-1 we gave a formal proof of (2-3), valid only at points of continuity of $f(t)$. We shall now reconsider this theorem in order to include the discontinuity points and to discuss the Gibbs' phenomenon. The Fourier transform is defined by

$$F(\omega) = \int_{-\infty}^{\infty} f(t)e^{-j\omega t}\, dt \tag{2-1}$$

and our object is to show that

$$f(t) = \frac{1}{2\pi} \int_{-\infty}^{\infty} F(\omega)e^{j\omega t}\, d\omega \tag{2-3}$$

Forming the function

$$f_{\Omega}(t) = \frac{1}{2\pi} \int_{-\Omega}^{\Omega} F(\omega)e^{j\omega t}\, d\omega \tag{2-85}$$

it suffices to show that

$$\lim_{\Omega \to \infty} f_{\Omega}(t) = f(t) \tag{2-86}$$

For this purpose we insert $F(\omega)$ as given in (2-1) into (2-85) and interchange the order of integration:

$$f_{\Omega}(t) = \frac{1}{2\pi} \int_{-\infty}^{\infty} f(\tau)\left[\int_{-\Omega}^{\Omega} e^{j\omega(t-\tau)}\, d\omega \right] d\tau = \int_{-\infty}^{\infty} f(\tau)\frac{\sin \Omega(t-\tau)}{\pi(t-\tau)}\, d\tau \tag{2-87}$$

Thus $f_{\Omega}(t)$ is the weighted average of $f(t)$ with the Fourier kernel $\sin \Omega(t-\tau)/(t-\tau)$ as weight. In (I-47) it is shown that this kernel tends to the impulse function $\delta(t-\tau)$ as Ω tends to infinity:

$$\lim_{\Omega \to \infty} \frac{\sin \Omega(t-\tau)}{\pi(t-\tau)} = \delta(t-\tau) \tag{2-88}$$

Therefore

$$\lim_{\Omega \to \infty} f_{\Omega}(t) = \int_{-\infty}^{\infty} f(\tau)\delta(t-\tau)\, d\tau \tag{2-89}$$

Since the last integral equals $f(t)$ for every continuity point, the desired result (2-86) follows.

Gibbs' phenomenon. We shall next examine the limit of $f_{\Omega}(t)$ at a point t_0 of discontinuity of $f(t)$, and the behavior of $f_{\Omega}(t)$ in the vicinity of this point. Clearly we can write $f(t)$ as a sum of a continuous function $f_c(t)$ and a suitable step function (see Fig. 2-21). To simplify the notation, we assume $t_0 = 0$:

$$f(t) = f_c(t) + [f(0^+) - f(0^-)]U(t) \tag{2-90}$$

From (2-90) and (2-87) we obtain

$$f_{\Omega}(t) = \int_{-\infty}^{\infty} f_c(\tau)\frac{\sin \Omega(t-\tau)}{\pi(t-\tau)}\, d\tau + [f(0^+) - f(0^-)]\int_{-\infty}^{\infty} U(\tau)\frac{\sin \Omega(t-\tau)}{\pi(t-\tau)}\, d\tau \tag{2-91}$$

The first integral in (2-91) tends to $f_c(t)$; it suffices therefore to examine the quantity

$$U_\Omega(t) = \int_{-\infty}^{\infty} U(\tau)\, \frac{\sin \Omega(t-\tau)}{\pi(t-\tau)}\, d\tau = \int_0^\infty \frac{\sin \Omega(t-\tau)}{\pi(t-\tau)}\, d\tau \qquad (2\text{-}92)$$

With $\Omega(t-\tau) = x$, we have

$$U_\Omega(t) = \int_{-\infty}^{\Omega t} \frac{\sin x}{\pi x}\, dx = \int_{-\infty}^0 \frac{\sin x}{\pi x}\, dx + \int_0^{\Omega t} \frac{\sin x}{\pi x}\, dx$$

$$= \frac{1}{2} + \frac{1}{\pi} Si(\Omega t) \qquad (2\text{-}93)$$

The function $U_\Omega(t)$ is shown in Fig. 2-22; since $U_\Omega(0) = \frac{1}{2}$ and $f_c(0) = f(0^-)$ (see Fig. 2-21), we obtain from (2-91)

$$\lim_{\Omega \to \infty} f_\Omega(0) = f_c(0) + \tfrac{1}{2}[f(0^+) - f(0^-)] = \frac{f(0^+) + f(0^-)}{2} \qquad (2\text{-}94)$$

Hence (2-3) is valid if $f(t)$ satisfies (2-6).

Suppose now that Ω is finite. If it is sufficiently large, the first

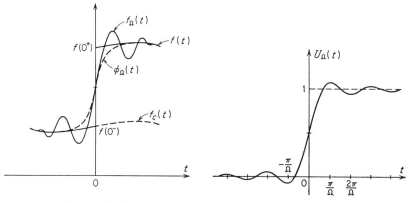

FIGURE 2-21   FIGURE 2-22

integral in (2-91) can be as close as we want to $f_c(t)$; the second integral, however, has always the form of Fig. 2-22. With increasing Ω, only the time scale changes while the ripple remains the same. This behavior of $f_\Omega(t)$ near a discontinuity point of $f(t)$ is called Gibbs' phenomenon.

Fejér kernel. We notice from the preceding discussion that if the Fourier transform $F(\omega)$ of a function $f(t)$ is truncated above $|\omega| = \Omega$, then the corresponding time function $f_\Omega(t) \leftrightarrow F(\omega)p_\Omega(\omega)$ does not give a satisfactory approximation to $f(t)$ in the vicinity of a discontinuity point, no matter how large Ω is. The question arises whether it is possible to find another band-limited function $\phi_\Omega(t)$, that is, a function whose Fourier integral $\Phi(\omega)$ is also zero for $|\omega| > \Omega$, giving a *better* approximation to $f(t)$. If we select for $\Phi(\omega)$ the function

$$\Phi(\omega) = F(\omega)q_\Omega(\omega) \tag{2-95}$$

where $q_\Omega(\omega)$ is a triangular pulse, then [see (2-84)]

$$\phi_\Omega(t) = \int_{-\infty}^{\infty} f(\tau) \frac{2 \sin^2 [\Omega(t - \tau)/2]}{\pi\Omega(t - \tau)^2} \, d\tau \tag{2-96}$$

Thus the function $\phi_\Omega(t)$ equals the weighted average of $f(t)$ with the Fejér kernel as weight. Since this kernel is positive, it can be easily seen that $\phi_\Omega(t)$ rises monotonically near a discontinuity point of $f(t)$ (see Fig. 2-21). The overshoot of the Gibbs' phenomenon is eliminated; however, a slower rise results. We observe that the function $\phi_\Omega(t)$ tends to $f(t)$ as Ω tends to infinity

$$\lim_{\Omega \to \infty} \phi_\Omega(t) = f(t) \tag{2-97}$$

because $\Phi(\omega) \to F(\omega)$. We conclude, therefore, that

$$\lim_{\Omega \to \infty} \frac{2 \sin^2 (\Omega t/2)}{\pi\Omega t^2} = \delta(t)$$

Smoothing. We shall now present another method for approximating, without overshoot, a function $f(t)$ by another band-limited function. We first form the average or *smoothing* function

$$\bar{f}(t) = \frac{1}{2T} \int_{t-T}^{t+T} f(\tau) \, d\tau \tag{2-98}$$

Clearly this function is continuous and equals the convolution of $f(t)$ with the rectangular pulse $p_T(t)/2T$

$$\bar{f}(t) = f(t) * \frac{1}{2T} p_T(t) \tag{2-99}$$

[In Fig. 2-23 we have shown the step function $f(t) = U(t)$ and its average $\bar{U}(t)$.] If the Fourier transform

$$\bar{F}(\omega) = F(\omega) \frac{\sin \omega T}{\omega T}$$

of $\bar{f}(t)$ is truncated above $|\omega| = \Omega$, the function $\bar{F}_\Omega(\omega) = \bar{F}(\omega)p_\Omega(\omega)$ results, and its inverse transform $\bar{f}_\Omega(t)$ is the weighted average of $\bar{f}(t)$ as in (2-87). For Ω sufficiently large, the function $\bar{f}_\Omega(t)$ gives

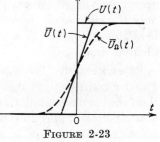

FIGURE 2-23

a satisfactory approximation to $\bar{f}(t)$ without overshoot, since $\bar{f}(t)$ is continuous.

RMS minimization. Consider the class of functions $y(t)$ whose Fourier transform $Y(\omega)$ is band-limited in the sense

$$Y(\omega) = 0 \qquad \text{for } |\omega| > \Omega \qquad (2\text{-}100)$$

If $f(t)$ is approximated by a function of this class, an *RMS error*

$$\epsilon = \int_{-\infty}^{\infty} |f(t) - y(t)|^2 \, dt \qquad (2\text{-}101)$$

results. We shall prove that this error is minimum if the Fourier transform $Y(\omega)$ of $y(t)$ equals the Fourier transform $F(\omega)$ of $f(t)$ in the $(-\Omega,\Omega)$ interval:

$$Y(\omega) = F(\omega) \qquad \text{for } |\omega| < \Omega \qquad (2\text{-}102)$$

Proof. Expanding the square in (2-101), we obtain

$$\epsilon = \int_{-\infty}^{\infty} f^2(t) \, dt + \int_{-\infty}^{\infty} y^2(t) \, dt - 2 \int_{-\infty}^{\infty} f(t)y(t) \, dt \qquad (2\text{-}103)$$

Because of (2-100), (2-79), and (2-75) we easily obtain from (2-103)

$$\begin{aligned}
2\pi\epsilon &= \int_{-\infty}^{\infty} |F(\omega)|^2 \, d\omega + \int_{-\Omega}^{\Omega} |Y(\omega)|^2 \, d\omega \\
&\qquad - \int_{-\Omega}^{\Omega} \overset{*}{F}(\omega) Y(\omega) \, d\omega - \int_{-\Omega}^{\Omega} F(\omega) \overset{*}{Y}(\omega) \, d\omega \\
&= \int_{-\infty}^{-\Omega} |F(\omega)|^2 \, d\omega + \int_{\Omega}^{\infty} |F(\omega)|^2 \, d\omega \\
&\qquad + \int_{-\Omega}^{\Omega} [F(\omega) - Y(\omega)][\overset{*}{F}(\omega) - \overset{*}{Y}(\omega)] \, d\omega
\end{aligned}$$

The first two integrals above are independent of $y(t)$ and the last is nonnegative because the integrand equals

$$|F(\omega) - Y(\omega)|^2$$

Therefore ϵ is minimum if (2-102) is true because only then the last integral equals zero. The minimum value of ϵ is given by

$$\epsilon_{\min} = \frac{1}{2\pi} \int_{-\infty}^{-\Omega} |F(\omega)|^2 \, d\omega + \frac{1}{2\pi} \int_{\Omega}^{\infty} |F(\omega)|^2 \, d\omega \qquad (2\text{-}104)$$

Clearly the optimum function $y(t)$ equals the function $f_\Omega(t)$ of (2-85). Thus, although the function $\phi_\Omega(t)$ [see (2-96)] gives a more satisfactory approximation to $f(t)$ near a discontinuity point, the function $f_\Omega(t)$ is optimum in the sense of minimizing the RMS error ϵ.

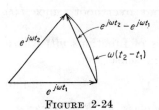

FIGURE 2-24

Bounds on a signal. We shall now show that a signal $f(t)$ and its derivatives are bounded by the absolute moments of its Fourier spectrum $A(\omega)$. With

$$M_n = \frac{1}{2\pi} \int_{-\infty}^{\infty} |\omega|^n A(\omega)\, d\omega$$

we obtain from $f^{(n)}(t) \leftrightarrow (j\omega)^n F(\omega)$

$$|f^{(n)}(t)| = \left| \frac{1}{2\pi} \int_{-\infty}^{\infty} (j\omega)^n F(\omega) e^{j\omega t}\, d\omega \right| \leq M_n \qquad (2\text{-}105)$$

A consequence of (2-105) is the following estimate of the variation of $f(t)$:

$$|f(t_2) - f(t_1)| \leq M_1 |t_2 - t_1| \qquad (2\text{-}106)$$

This can be shown by integrating $f'(t)$ from t_1 to t_2 and using (2-105) or, more directly, from

$$f(t_2) - f(t_1) = \frac{1}{2\pi} \int_{-\infty}^{\infty} F(\omega)[e^{j\omega t_2} - e^{j\omega t_1}]\, d\omega$$

and the obvious inequality (see Fig. 2-24)

$$|e^{j\omega t_2} - e^{j\omega t_1}| = 2 \sin \left| \frac{\omega(t_2 - t_1)}{2} \right| \leq |\omega(t_2 - t_1)|$$

From (2-106) we see that if M_1 is finite then $f(t)$ is continuous. If the spectrum of $f(t)$ equals zero for $|\omega| > \omega_c$ then

$$\int_{-\omega_c}^{\omega_c} |\omega|\, A(\omega)\, d\omega \leq \omega_c \int_{-\omega_c}^{\omega_c} A(\omega)\, d\omega$$

Hence, $M_1 \leq \omega_c M_0$. In this case

$$|f(t_2) - f(t_1)| \leq \omega_c M_0 |t_2 - t_1| \qquad (2\text{-}107)$$

As an application consider a modulated signal $f(t) \cos \omega_0 t$. If the spectrum of its envelope is limited in the band $|\omega| \leq \omega_c$ and $\omega_c \ll \omega_0$, then the variation of $f(t)$ within the period $T = 2\pi/\omega_0$ of the carrier is small. A quantitative estimate of this variation is obtained from (2-107):

$$|f(t + T) - f(t)| \leq \omega_c M_0 T = 2\pi M_0 \frac{\omega_c}{\omega_0}$$

Similar bounds can be established for $F(\omega)$ and its derivatives.

Chapter 3. Singularity Functions and Line Spectra

In Chap. 2 we developed the Fourier transforms of a number of ordinary functions that satisfy the integrability conditions of Sec. 2-1 and are included in the mathematical treatments of the Fourier-integral theory. In applications one uses also functions that are not covered by this analysis, e.g., impulses or periodic signals, and there is a definite need for a more general theory treating the singularity functions. In the technical literature, impulses and their transforms are frequently used; their development, however, is often based on complicated limiting arguments. In an attempt to provide a more satisfactory foundation for the understanding of these functions and their transforms, we give in Appendix I an elementary discussion of the theory of distributions and introduce the impulses as a special case. Our aim, however, is not to develop a Fourier-integral theory of distributions; we plan merely to establish the transform pairs, with the impulse function and its derivatives appearing in the time or frequency domain. We feel that if impulses are accepted at all as a tool of analysis, then one must gain some facility with their use and be able to apply them beyond the trivial applications. We have attempted to show in this and in subsequent chapters that many results involving ordinary functions can be simply derived with the help of impulses. We derive, for example, the Fourier series as a special case of the Fourier integral, and with the same reasoning we give a simple proof of Poisson's sum formula.

In the examples, we have included a number of ordinary functions (Examples 3-4 and 3-8) whose transforms do not contain impulses but have meaning only if they are interpreted as distributions. Although most of the results can be followed with the usual concept of the delta function, certain sums and integrals become meaningless unless they are viewed as generalized limits defined in Appendix I.

3-1. Basic Examples

We shall now develop the transform pairs of a number of singularity functions that are commonly used in applications. The theorems of Chap. 2 are valid, with minor changes (see Example 3-5), also for the functions under consideration if they do not involve products of singularities. We remark that, if an impulse is in the time domain, then its transform can be readily obtained from (2-1); however, (2-3) is no longer true as an ordinary integral, as we see in the first example. Similarly if the impulse is in the frequency domain, the result is more easily derived from (2-3). In obtaining our results, we use the following definition for the impulse function [see (I-21)]: with $\phi(t)$ an arbitrary function, continuous at a given point t_0, $\delta(t)$ is such that

$$\int_{-\infty}^{\infty} \delta(t - t_0)\phi(t)\, dt = \phi(t_0) \tag{3-1}$$

Similarly the nth derivative $d^n\delta/dt^n$ is defined by

$$\int_{-\infty}^{\infty} \frac{d^n\, \delta(t - t_0)}{dt^n}\, \phi(t)\, dt = (-1)^n \frac{d^n\phi(t_0)}{dt^n} \tag{3-2}$$

[see (I-25)].

Example 3-1. $f(t) = \delta(t)$. The Fourier transform of the delta function is readily obtained from (2-1) and (3-1):

$$F(\omega) = \int_{-\infty}^{\infty} e^{-j\omega t}\, \delta(t)\, dt = 1$$

We thus have the pair

$$\delta(t) \leftrightarrow 1 \tag{3-3}$$

shown in Fig. 3-1; the function $k\, \delta(t)$ is represented graphically by a vertical line with the letter k next to it. The transform of the shifted impulse $\delta(t - t_0)$ is given by $e^{-j\omega t_0}$:

$$\delta(t - t_0) \leftrightarrow e^{-j\omega t_0} \tag{3-4}$$

as we see from (2-36) or directly from (2-1) and (3-1). It has a constant amplitude equal to one, and a linear phase. We now examine the validity of the inversion formula (2-3); we must clearly have

$$\delta(t) = \frac{1}{2\pi} \int_{-\infty}^{\infty} e^{j\omega t}\, d\omega = \frac{1}{2\pi} \int_{-\infty}^{\infty} \cos \omega t\, d\omega \tag{3-5}$$

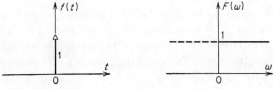

FIGURE 3-1

but these integrals are obviously meaningless unless they are interpreted as distributions in the sense of Appendix I; only then they are equal to $\delta(t)$ and (2-3) is valid [see (I-49)].

Example 3-2. From the above result (3-4) and Eq. (2-34) we obtain the pair

$$e^{j\omega_0 t} \leftrightarrow 2\pi\, \delta(\omega - \omega_0) \qquad 1 \leftrightarrow 2\pi\, \delta(\omega) \qquad (3\text{-}6)$$

shown in Fig. 3-2. Thus the Fourier transform of a constant is an impulse at the origin of area 2π. This result could also be derived directly from (2-3).

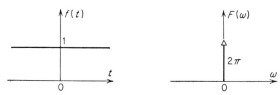

FIGURE 3-2

Example 3-3. From (3-6) and (2-38) the pair

$$\cos \omega_0 t = \tfrac{1}{2}(e^{j\omega_0 t} + e^{-j\omega_0 t}) \leftrightarrow \pi[\delta(\omega - \omega_0) + \delta(\omega + \omega_0)] \qquad (3\text{-}7)$$

of Fig. 3-3 follows. Similarly (Fig. 3-4)

$$\sin \omega_0 t \leftrightarrow j\pi[\delta(\omega + \omega_0) - \delta(\omega - \omega_0)] \qquad (3\text{-}8)$$

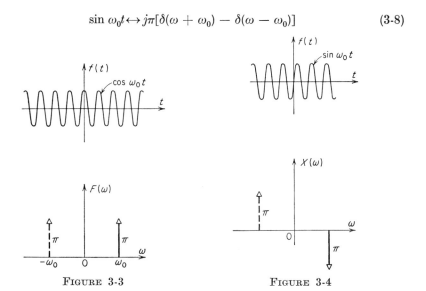

FIGURE 3-3   FIGURE 3-4

Example 3-4. We shall show that the Fourier transform of the function sgn t of Fig. 3-5 equals $2/j\omega$:

$$\operatorname{sgn} t \leftrightarrow \frac{2}{j\omega} \qquad (3\text{-}9)$$

Indeed the inverse transform $f(t)$ of $2/j\omega$ is given by

$$f(t) = \frac{1}{2\pi} \int_{-\infty}^{\infty} \frac{2}{j\omega} e^{j\omega t}\, d\omega = \frac{1}{\pi} \int_{-\infty}^{\infty} \frac{\sin \omega t}{\omega}\, d\omega \qquad (3\text{-}10)$$

but the last integral equals one if $t > 0$ and minus one if $t < 0$ [see (II-57)]; therefore (3-9) is true.   It is of interest to see whether (3-9) can be established directly from (2-1).   Since sgn t is odd, we have [see (2-22)] $R(\omega) = 0$ and

$$X(\omega) = -2 \int_{0}^{\infty} f(t) \sin \omega t\, dt = -2 \int_{0}^{\infty} \sin \omega t\, dt$$

Therefore (3-9) is valid if

$$\int_{0}^{\infty} \sin \omega t\, dt = \frac{1}{\omega} \qquad (3\text{-}11)$$

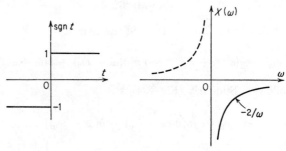

FIGURE 3-5

but this is meaningless; however, in (I-41) we show that, if the quantities in (3-11) are interpreted as distributions, then (3-11) is true.   Thus, although no singularities are involved in the pair (3-9), $2/j\omega$ can be obtained from sgn t only if it is given special meaning.

Example 3-5. $f(t) = U(t)$; the step function $U(t)$ can be written as a sum of a constant equal to $\frac{1}{2}$ and the function sgn $t/2$,

$$U(t) = \tfrac{1}{2} + \tfrac{1}{2} \operatorname{sgn} t \qquad (3\text{-}12)$$

Using, therefore, the results in Examples 3-2 and 3-4, we conclude that the Fourier transform of $U(t)$ equals $\pi\, \delta(\omega) + 1/j\omega$,

$$U(t) \longleftrightarrow \pi\, \delta(\omega) + \frac{1}{j\omega} \qquad (3\text{-}13)$$

with real part $\pi\, \delta(\omega)$ and imaginary part $-1/\omega$ (see Fig. 3-6).   The transform of $e^{j\omega_0 t} U(t)$ is given by

$$e^{j\omega_0 t} U(t) \longleftrightarrow \pi\, \delta(\omega - \omega_0) + \frac{1}{j(\omega - \omega_0)} \qquad (3\text{-}14)$$

[see (2-37)]. From (3-13) and (2-17) we conclude that

$$U(t) = \frac{1}{2\pi} \int_{-\infty}^{\infty} \pi\, \delta(\omega) \cos \omega t\, d\omega + \frac{1}{2\pi} \int_{-\infty}^{\infty} \frac{\sin \omega t}{\omega}\, d\omega$$

$$= \frac{1}{2} + \frac{1}{\pi} \int_{0}^{\infty} \frac{\sin \omega t}{\omega}\, d\omega$$

a result that can be checked directly.

Comment. A superficial application of the differentiation theorem (2-41) to the results in Examples 3-1 and 3-5 leads to wrong results; indeed, since, according to (I-30),

$$\delta(t) = \frac{dU(t)}{dt} \tag{3-15}$$

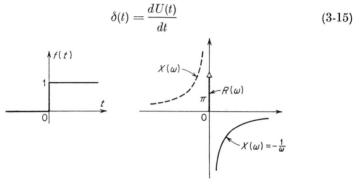

FIGURE 3-6

we conclude from (3-3) and (2-41) that with $F(\omega)$ the Fourier transform of $U(t)$, we must have

$$1 = j\omega F(\omega) \tag{3-16}$$

Therefore

$$F(\omega) = \frac{1}{j\omega} \tag{3-17}$$

a result that is not in agreement with (3-13). The fallacy in this reasoning can be found in the last step. From (3-16) the correct conclusion is not (3-17) but

$$F(\omega) = \frac{1}{j\omega} + k\, \delta(\omega)$$

where k is a constant, because $j\omega\, \delta(\omega) = 0$, as one can see from the property

$$\phi(t)\, \delta(t) = \phi(0)\, \delta(t) \tag{3-18}$$

of the delta function [see (I-23)]. The constant k cannot be determined from (3-16) but can be found from (2-3) with $t = 0$:

$$U(0) = \frac{1}{2} = \frac{1}{2\pi} \int_{-\infty}^{\infty} F(\omega)\, d\omega = \frac{k}{2\pi}$$

Hence $k = \pi$. Thus, in general, if $\omega F_1(\omega) = \omega F_2(\omega)$, it does not follow that $F_1(\omega) = F_2(\omega)$; the correct conclusion is

$$F_1(\omega) = F_2(\omega) + k\,\delta(\omega) \tag{3-19}$$

(The above reminds us of the paradox in algebra of proving, by a zero division, that $2 = 3$.)†

As an application of (3-13) we shall evaluate the transform $G(\omega)$ of the real integral

$$g(t) = \int_{-\infty}^{t} f(\tau)\,d\tau \tag{3-20}$$

in terms of the transform $F(\omega) = R(\omega) + jX(\omega)$ of the integrand $f(t)$. The above integral is, obviously, a convolution of $f(t)$ with the unit step $U(t)$,

$$g(t) = f(t) * U(t) \tag{3-21}$$

Therefore [see (3-13) and (2-71)]

$$G(\omega) = F(\omega)\left[\pi\,\delta(\omega) + \frac{1}{j\omega}\right] = \pi R(0)\,\delta(\omega) + \frac{X(\omega)}{\omega} - j\,\frac{R(\omega)}{\omega} \tag{3-22}$$

because $F(\omega)\,\delta(\omega) = F(0)\,\delta(\omega) = R(0)\,\delta(\omega)$. Using the above result and the inversion formula (2-3), we can write $g(t)$ in terms of $F(\omega)$:

$$g(t) = \frac{R(0)}{2} + \frac{1}{\pi}\int_0^{\infty}\left[\frac{X(\omega)}{\omega}\cos\omega t + \frac{R(\omega)}{\omega}\sin\omega t\right]d\omega \tag{3-23}$$

Example 3-6. The Fourier transform $F(\omega) = R(\omega) + jX(\omega)$ of the ramp function $f(t)$ of Fig. 3-7 can be readily found from (3-13) and the convolution theorem (2-71). Since

$$f(t) = \frac{1}{T}\int_{t-T/2}^{t+T/2} U(\tau)\,d\tau \tag{3-24}$$

we conclude that

$$f(t) = U(t) * \frac{1}{T}\,p_{T/2}(t) \tag{3-25}$$

But the transform of $p_{T/2}(t)$ equals $2\sin(\omega T/2)/\omega$; hence

$$F(\omega) = \left[\pi\,\delta(\omega) + \frac{1}{j\omega}\right]\frac{2\sin(\omega T/2)}{T\omega} = \pi\,\delta(\omega) + \frac{2\sin(\omega T/2)}{jT\omega^2} \tag{3-26}$$

because

$$\frac{2\pi\,\delta(\omega)\sin(\omega T/2)}{T\omega} = \pi\,\delta(\omega)$$

† Similarly, if $\omega^2 F_1(\omega) = \omega^2 F_2(\omega)$, then $F_1(\omega) = F_2(\omega) + K_1\,\delta(\omega) + K_2\,\delta'(\omega)$. This follows easily from (I-26). It can also be deduced from

$$\frac{d^2 f_1(t)}{dt^2} = \frac{d^2 f_2(t)}{dt^2}\,[\text{see }(2\text{-}41)] \qquad f_1(t) = f_2(t) + A + Bt$$

as we can easily see from (3-18). Thus

$$R(\omega) = \pi\,\delta(\omega) \qquad X(\omega) = -\,\frac{2\sin{(\omega T/2)}}{T\omega^2} \tag{3-27}$$

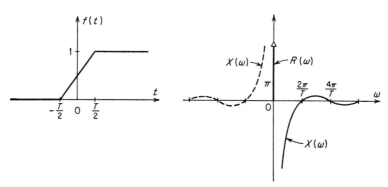

FIGURE 3-7

Example 3-7.   The Fourier transform of a step-modulated input

$$f(t) = U(t)\cos\omega_0 t \tag{3-28}$$

can be obtained from (3-13) and (2-38):

$$
\begin{aligned}
F(\omega) &= \frac{\pi}{2}\left[\delta(\omega-\omega_0)+\delta(\omega+\omega_0)\right] + \frac{1}{2j(\omega-\omega_0)} + \frac{1}{2j(\omega+\omega_0)} \\
&= \frac{\pi}{2}\left[\delta(\omega-\omega_0)+\delta(\omega+\omega_0)\right] + \frac{j\omega}{\omega_0{}^2-\omega^2}
\end{aligned}
\tag{3-29}
$$

and is shown in Fig. 3-8. Similarly if

$$f(t) = U(t)\sin\omega_0 t \tag{3-30}$$

then
$$F(\omega) = \frac{\pi}{2j}\left[\delta(\omega-\omega_0)-\delta(\omega+\omega_0)\right] + \frac{\omega_0}{\omega_0{}^2-\omega^2} \tag{3-31}$$

as in Fig. 3-9.

Example 3-8.   The transform of the nth derivative $d^n\,\delta/dt^n$ of the delta
function can be easily evaluated from (3-2):

$$F(\omega) = \int_{-\infty}^{\infty}\frac{d^n\,\delta}{dt^n}\,e^{-j\omega t}\,dt = (-1)^n\left.\frac{d^n e^{-j\omega t}}{dt^n}\right|_{t=0} = (j\omega)^n$$

We thus have the pair

$$\frac{d^n\,\delta(t)}{dt^n} \leftrightarrow (j\omega)^n \tag{3-32}$$

and from (2-34)

$$t^n \leftrightarrow 2\pi j^n\,\frac{d^n\,\delta(\omega)}{d\omega^n} \tag{3-33}$$

Example 3-9. We shall now show that the Fourier transform of the function $f(t) = |t|$ equals $-2/\omega^2$. From (2-3) and (I-32) follows that

$$f(t) = -\frac{1}{\pi} \int_{-\infty}^{\infty} \frac{\cos \omega t}{\omega^2}\, d\omega = |t| \qquad (3\text{-}34)$$

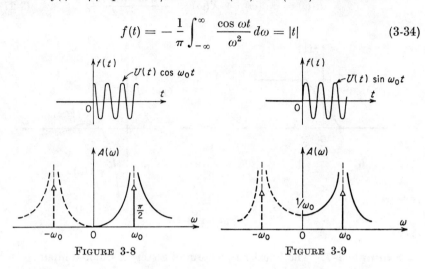

FIGURE 3-8   FIGURE 3-9

Thus, if the above quantities are interpreted as distributions, then the pair

$$|t| \leftrightarrow -\frac{2}{\omega^2} \qquad (3\text{-}35)$$

of Fig. 3-10 results.

The transform of the function $tU(t)$ can be easily obtained from (3-35) and (3-33). We clearly have

$$tU(t) = \frac{t}{2} + \frac{|t|}{2}$$

Hence

$$tU(t) \leftrightarrow j\pi\, \frac{d\,\delta(\omega)}{d\omega} - \frac{1}{\omega^2} \qquad (3\text{-}36)$$

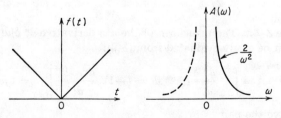

FIGURE 3-10

3-2. Fourier Series

In this section we show that a periodic function $f(t)$, of period T,

$$f(t + T) = f(t) \qquad (3\text{-}37)$$

can be written as a sum of exponentials,

$$f(t) = \sum_{n=-\infty}^{\infty} \alpha_n e^{jn\omega_0 t} \qquad \omega_0 = \frac{2\pi}{T} \tag{3-38}$$

where the constants α_n are given by

$$\alpha_n = \frac{1}{T} \int_{-T/2}^{T/2} f(t) e^{-jn\omega_0 t}\, dt \tag{3-39}$$

It will then follow from (3-38) and (3-6) that the Fourier transform $F(\omega)$ of $f(t)$ is a sequence of equidistant pulses

$$F(\omega) \equiv 2\pi \sum_{n=-\infty}^{\infty} \alpha_n\, \delta(\omega - n\omega_0) \tag{3-40}$$

distance ω_0 apart (see Fig. 3-11). The sum in (3-38) is of course the familiar *Fourier series* expansion of a periodic function and is treated

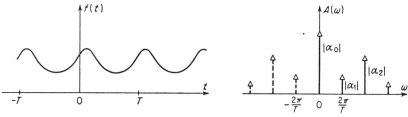

FIGURE 3-11

in the mathematical and technical literature not as part of the Fourier integral but as a separate discipline. We do not intend, here, to develop this theory but to show that the Fourier series can be formally derived as a special case of the Fourier integral. This approach has the added advantage of providing a trivial proof of Poisson's sum formula; furthermore, it facilitates the understanding and proof of the sampling theorems.

To prove that a sequence of equidistant pulses

$$\Phi(\omega) = \sum_{n=-\infty}^{\infty} A_n\, \delta(\omega - n\omega_0)$$

is the Fourier transform of a periodic function $\phi(t)$ is simple; indeed, $\phi(t)$ is given by

$$\phi(t) = \frac{1}{2\pi} \sum_{n=-\infty}^{\infty} A_n e^{jn\omega_0 t}$$

[see (3-6)] and since

$$e^{jn\omega_0[t + (2\pi/\omega_0)]} = e^{jn\omega_0 t}$$

we conclude that $\phi[t + (2\pi/\omega_0)] = \phi(t)$. However, it is not as easy to prove that the Fourier integral $F(\omega)$ of a periodic function $f(t)$ is given by (3-40).   To this end we introduce the pulse train

$$s_T(t) = \sum_{n=-\infty}^{\infty} \delta(t - nT) \tag{3-41}$$

of Fig. 3-12, consisting of a sequence of equidistant pulses $\delta(t - nT)$

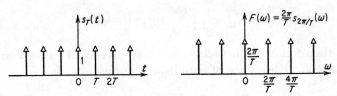

FIGURE 3-12

distance T apart, and we shall show that its Fourier transform is also a similar pulse train

$$\omega_0 s_{\omega_0}(\omega) = \omega_0 \sum_{n=-\infty}^{\infty} \delta(\omega - n\omega_0) \qquad \omega_0 = \frac{2\pi}{T} \tag{3-42}$$

i.e., that

$$s_T(t) \leftrightarrow \omega_0 s_{\omega_0}(\omega) \tag{3-43}$$

It suffices to show that the inverse transform of $(2\pi/T)s_{\omega_0}(\omega)$ equals $s_T(t)$:

$$\frac{1}{T}\sum_{n=-\infty}^{\infty} e^{jn\omega_0 t} = s_T(t) \tag{3-44}$$

We form the function

$$k_N(t) = \frac{1}{T}\sum_{n=-N}^{N} e^{jn\omega_0 t} = \frac{e^{j(N+1)\omega_0 t} - e^{-jN\omega_0 t}}{T(e^{j\omega_0 t} - 1)} = \frac{\sin (N + \frac{1}{2})\omega_0 t}{T \sin (\omega_0 t/2)} \tag{3-45}$$

This function is known as the *Fourier-series kernel* and is obviously

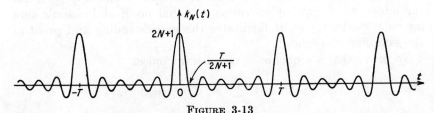

FIGURE 3-13

periodic with period $2\pi/\omega_0 = T$ (Fig. 3-13); therefore, if we prove that in the interval $(-T/2, T/2)$ $k_N(t)$ tends to $\delta(t)$, it will follow that

$$\lim_{N \to \infty} k_N(t) = s_T(t) \tag{3-46}$$

which is equivalent to (3-44).   We have

$$k_N(t) = \frac{\sin (N + \frac{1}{2})\omega_0 t}{T \sin (\omega_0 t/2)} = \frac{\sin (N + \frac{1}{2})\omega_0 t}{Tt} \frac{t}{\sin (\omega_0 t/2)} \tag{3-47}$$

But [see (2-88)]

$$\lim_{N \to \infty} \frac{\sin (N + \frac{1}{2})\omega_0 t}{Tt} = \frac{\pi}{T} \delta(t)$$

and since

$$\frac{t}{\sin (\omega_0 t/2)}$$

is bounded in the $(-T/2, T/2)$ interval, we conclude from (3-18) and (3-47) that for $|t| < T/2$

$$\lim_{N \to \infty} k_N(t) = \frac{\pi}{T} \frac{t}{\sin (\omega_0 t/2)} \delta(t) = \frac{\pi}{T} \frac{t}{\sin (\omega_0 t/2)} \bigg|_{t=0} \delta(t) = \delta(t)$$

We now return to the proof of (3-40). We form the function

$$f_0(t) = \begin{cases} f(t) & |t| < T/2 \\ 0 & |t| > T/2 \end{cases} \tag{3-48}$$

$f(t)$ can be written as a sum

$$f(t) = \sum_{n=-\infty}^{\infty} f_0(t + nT) \tag{3-49}$$

or equivalently as a convolution

$$f(t) = f_0(t) * s_T(t) \tag{3-50}$$

With

$$F_0(\omega) = \int_{-\infty}^{\infty} f_0(t) e^{-j\omega t} \, dt = \int_{-T/2}^{T/2} f(t) e^{-j\omega t} \, dt \tag{3-51}$$

the Fourier integral of $f_0(t)$, we conclude from (3-50), (3-43), (3-18), and (2-71) that

$$F(\omega) = F_0(\omega) \frac{2\pi}{T} \sum_{n=-\infty}^{\infty} \delta(\omega - n\omega_0) = \frac{2\pi}{T} \sum_{n=-\infty}^{\infty} F_0(n\omega_0) \, \delta(\omega - n\omega_0) \tag{3-52}$$

and since [see (3-51) and (3-39)]

$$\alpha_n = \frac{1}{T} \int_{-T/2}^{T/2} f(t) e^{-jn\omega_0 t} \, dt = \frac{F_0(n\omega_0)}{T} \tag{3-53}$$

(3-40) is proved. From (3-53) we see that, except for the factor $1/T$, the coefficients α_n of the Fourier series expansion of $f(t)$ equal the values of the Fourier integral $F_0(\omega)$ of the function $f_0(t)$ at $\omega = n\omega_0 = 2\pi n/T$.

Example 3-10. $f(t)$ equals a sequence of rectangular pulses of width $2a$, as in Fig. 3-14. We have $f_0(t) = p_a(t)$; hence [see (2-56)]

$$F_0(\omega) = \frac{2 \sin a\omega}{\omega}$$

Therefore the Fourier-series coefficients of $f(t)$ are given by

$$\alpha_n = \frac{F_0(n\omega_0)}{T} = \frac{2 \sin na\omega_0}{n\omega_0 T} = \frac{2 \sin (2\pi na/T)}{2\pi n}$$

Partial Sum.   As in Sec. 2-6, it is of interest to relate the partial sum

$$f_N(t) = \sum_{n=-N}^{N} \alpha_n e^{jn\omega_0 t} \tag{3-54}$$

to the function $f(t)$.   Clearly the Fourier integral $F_N(\omega)$ of $f_N(t)$ is given by [see (3-53)]

$$F_N(\omega) = \sum_{n=-N}^{N} 2\pi\alpha_n\,\delta(\omega - n\omega_0) = F_0(\omega) \sum_{n=-N}^{N} 2\pi\,\delta(\omega - n\omega_0)/T$$

and since the last sum is the Fourier transform of the function $k_N(t)$ of (3-45), we conclude from the convolution theorem (2-71) that

$$f_N(t) = \int_{-\infty}^{\infty} f_0(\tau)k_N(t-\tau)\,d\tau = \frac{1}{T}\int_{-T/2}^{T/2} f(\tau)\frac{\sin{(N+\tfrac{1}{2})(t-\tau)\omega_0}}{\sin{[\omega_0(t-\tau)/2]}}\,d\tau$$

Thus $f_N(t)$ equals the weighted average of $f(t)$ with the Fourier kernel $k_N(t)$ as weight.

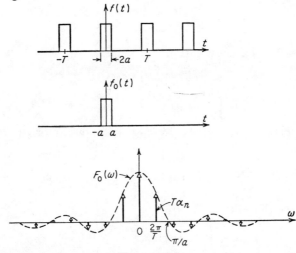

Figure 3-14

Fejér sum.   If $f(t)$ is discontinuous, the trigonometric polynomial $f_N(t)$ does not give a good approximation to $f(t)$ in the vicinity of the discontinuity points, no matter how large N is chosen (Gibbs' phenomenon).   Proceeding as in page 32, we can obtain a better approximation to $f(t)$ by a trigonometric polynomial

$$\phi_N(t) = \sum_{n=-N}^{N} \beta_n e^{jn\omega_0 t}$$

of the same order, if we choose for β_n the values

$$\beta_n = \alpha_n\left(1 - \frac{|n|}{N}\right)$$

where α_n are the Fourier-series coefficients of $f(t)$ [see (2-95)]. It can be easily shown that $\phi_N(t)$ is the weighted average of $f(t)$ with the kernel $Tk_{N/2}^2(t)$ as weight:

$$\phi_N(t) = \frac{1}{T} \int_{-T/2}^{T/2} f(\tau) \frac{\sin^2 [N(t - \tau)\omega_0/2]}{N \sin^2 [(t - \tau)\omega_0/2]} \, d\tau$$

The proof is left as an exercise. Since the above kernel is positive, the Gibbs' phenomenon is eliminated and $\phi_N(t)$ tends smoothly to $f(t)$ as N tends to infinity.

3-3. Poisson's Sum Formula

The Fourier series expansion (3-38) of a periodic function $f(t)$ can be written in the form

$$\sum_{n=-\infty}^{\infty} f_0(t + nT) = \frac{1}{T} \sum_{n=-\infty}^{\infty} e^{jn\omega_0 t} F_0(n\omega_0) \tag{3-55}$$

as we readily see from (3-49) and (3-53). The above suggests an obvious extension. If $\phi(t)$ is an arbitrary function and $\Phi(\omega)$ its Fourier transform, then the following identity, known as *Poisson's formula*, is true:

$$\sum_{n=-\infty}^{\infty} \phi(t + nT) = \frac{1}{T} \sum_{n=-\infty}^{\infty} e^{jn\omega_0 t}\Phi(n\omega_0) \qquad \omega_0 = \frac{2\pi}{T} \tag{3-56}$$

Proof. With $s_T(t)$ as in (3-41), we have

$$\sum_{n=-\infty}^{\infty} \phi(t + nT) = \phi(t) * s_T(t)$$

Therefore the Fourier transform of the above sum is given by [see also (3-52)]

$$\Phi(\omega) \frac{2\pi}{T} \sum_{n=-\infty}^{\infty} \delta(\omega - n\omega_0) = \frac{2\pi}{T} \sum_{n=-\infty}^{\infty} \Phi(n\omega_0)\, \delta(\omega - n\omega_0)$$

and (3-56) follows. The only difference between the ordinary Fourier series (3-55) and the Poisson formula (3-56) is that $f_0(t)$ equals zero for $|t| > T/2$, whereas this is not true for $\phi(t)$. Formula (3-56) is often used in the special form

$$\sum_{n=-\infty}^{\infty} \phi(nT) = \frac{1}{T} \sum_{n=-\infty}^{\infty} \Phi(n\omega_0) \tag{3-57}$$

resulting from (3-56) with $t = 0$.

Comment. If $\phi(t)$ is discontinuous at $t = nT$, then its value in (3-57) should be the average of $\phi(nT^+)$ and $\phi(nT^-)$, because (3-56) resulted from the inversion formula (2-3) valid only if (2-6) is satisfied. If, for example, $\phi(t) = 0$ for $t < 0$, then (3-57) should read

$$\frac{\phi(0^+)}{2} + \sum_{n=1}^{\infty} \phi(nT) = \frac{1}{T} \sum_{n=-\infty}^{\infty} \Phi(n\omega_0) \tag{3-58}$$

or the equivalent

$$\sum_{n=0}^{\infty} \phi(nT) = \frac{\phi(0^+)}{2} + \frac{1}{T} \sum_{n=-\infty}^{\infty} \Phi(n\omega_0) \tag{3-59}$$

Example 3-11. From the pair

$$e^{-\alpha|t|} \leftrightarrow \frac{2\alpha}{\alpha^2 + \omega^2} \qquad \alpha > 0$$

of (2-62) and Eq. (3-57) we obtain with $T = 1$ the interesting formula

$$\sum_{n=-\infty}^{\infty} e^{-\alpha|n|} = \sum_{n=-\infty}^{\infty} \frac{2\alpha}{\alpha^2 + (2\pi n)^2} \tag{3-60}$$

Example 3-12. From

$$e^{-\alpha t^2} \leftrightarrow \sqrt{\frac{\pi}{\alpha}} e^{-\omega^2/4\alpha} \qquad \alpha > 0$$

[see (2-68) and (3-56)], we obtain with $T = 1$ the identity

$$\sum_{n=-\infty}^{\infty} e^{-\alpha(t+n)^2} = \sqrt{\frac{\pi}{\alpha}} \sum_{n=-\infty}^{\infty} e^{j2\pi nt} e^{-\pi^2 n^2/\alpha}$$

which can also be written in the form

$$\sqrt{\frac{\alpha}{\pi}} \sum_{n=-\infty}^{\infty} e^{-\alpha(t+n)^2} = 1 + 2 \sum_{n=1}^{\infty} e^{-\pi^2 n^2/\alpha} \cos 2\pi nt \tag{3-61}$$

(Each side of the above identity equals the *theta function*.)

Application of Poisson's formula to sampled data. Given an arbitrary function $f(t)$, continuous at $t = nT$, we form the function

$$f^*(t) = T \sum_{n=-\infty}^{\infty} f(nT)\delta(t - nT) \tag{3-62}$$

consisting of a sequence of equidistant pulses, as in Fig. 3-15, of area $Tf(nT)$. The Fourier transform $F^*(\omega)$ of $f^*(t)$ is given by [see (3-4)]

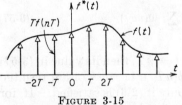

FIGURE 3-15

$$F^*(\omega) = T \sum_{n=-\infty}^{\infty} f(nT)e^{-jnT\omega} \tag{3-63}$$

We shall show that, if $F(\omega)$ is the Fourier transform of $f(t)$, then†

$$F^*(\omega) = \sum_{n=-\infty}^{\infty} F\left(\omega + \frac{2\pi n}{T}\right) \tag{3-64}$$

† J. G. Truxal, "Automatic Feedback Control System Synthesis," McGraw-Hill Book Company, Inc., New York, 1955.

Proof. From the pair $f(t) \leftrightarrow F(\omega)$ and (2-37), the pair

$$f(t)e^{-jxt} \leftrightarrow F(\omega + x)$$

results. With $\varphi(t) = f(t)e^{-jxt}$, we have from (3-57)

$$\sum_{n=-\infty}^{\infty} f(nT)e^{-jxnT} = \frac{1}{T} \sum_{n=-\infty}^{\infty} F\left(x + \frac{2\pi n}{T}\right) \tag{3-65}$$

Substituting ω for x, we obtain from (3-65) and (3-63) the desired result (3-64). Note that if $f(t)$ is causal and $f^*(t)$ is defined by

$$f^*(t) = Tf(0^+)\delta(t) + T \sum_{n=1}^{\infty} f(nT)\delta(t - nT)$$

then its transform satisfies

$$F^*(\omega) = T\frac{f(0^+)}{2} + \sum_{n=-\infty}^{\infty} F\left(\omega + \frac{2\pi n}{T}\right) \tag{3-66}$$

as we easily see from (3-59).

3-4. Periodic-frequency Spectra

If a time function is a sequence of equidistant pulses,

$$f(t) = \sum_{n=-\infty}^{\infty} A_n \delta(t - nT) \tag{3-67}$$

as in Fig. 3-16, then its Fourier transform

$$F(\omega) = \sum_{n=-\infty}^{\infty} A_n e^{-jnT\omega} \tag{3-68}$$

is obviously periodic with period $\Omega = 2\pi/T$.

$$F\left(\omega + \frac{2\pi}{T}\right) = F(\omega)$$

and (3-68) gives its Fourier series expansion, with a reversal in the sign of the exponent; hence

$$A_n = \frac{1}{\Omega} \int_{-\Omega/2}^{\Omega/2} F(\omega)e^{+jnT\omega} \, d\omega \tag{3-69}$$

The following two examples are of particular interest in probability theory.

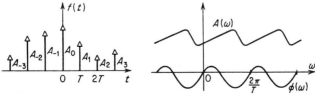

FIGURE 3-16

Example 3-13. *Poisson Distribution.* The function $f(t)$ is given by

$$f(t) = e^{-\lambda} \sum_{n=0}^{\infty} \frac{\lambda^n}{n!} \delta(t - n) \qquad (3\text{-}70)$$

where λ is an arbitrary constant. It is zero for $t < 0$, and its Fourier transform is given by

$$F(\omega) = e^{-\lambda} \sum_{n=0}^{\infty} \frac{\lambda^n}{n!} e^{-jn\omega} = e^{\lambda(e^{-j\omega}-1)} \qquad (3\text{-}71)$$

with amplitude and phase

$$A(\omega) = e^{\lambda(\cos \omega - 1)} \qquad \varphi(\omega) = -\lambda \sin \omega \qquad (3\text{-}72)$$

$f(t)$ is shown in Fig. 3-17 for $\lambda = 2$. The factor $e^{-\lambda}$ is introduced in (3-70) so as to make the total area $F(0)$ of $f(t)$ equal to one.

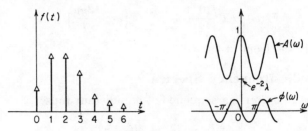

FIGURE 3-17

Example 3-14. *Binomial Distribution.* The meaning of the term binomial will become clear if $f(t)$ is characterized by its Fourier transform

$$F(\omega) = (p + qe^{j\omega})^n \qquad p + q = 1 \qquad (3\text{-}73)$$

Expanding (3-73) and using (3-4), we obtain

$$f(t) = \sum_{k=0}^{n} \binom{n}{k} p^k q^{n-k} \delta(t - k) \qquad (3\text{-}74)$$

Again condition $p + q = 1$ is imposed to make $F(0) = 1$.

3-5. Sampling Theorem†

This is an important theorem in transmission of information. It states that if the Fourier transform of a function $f(t)$ is zero above a certain frequency ω_c (see Fig. 3-18),

$$F(\omega) = 0 \qquad \text{for } |\omega| \geq \omega_c \qquad (3\text{-}75)$$

then $f(t)$ can be uniquely determined from its values

$$f_n = f\left(n \frac{\pi}{\omega_c}\right) \qquad (3\text{-}76)$$

† C. E. Shannon, Communication in the Presence of Noise, *Proc. IRE*, January, 1949.

at a sequence of equidistant points, distance π/ω_c apart.   In fact $f(t)$
is given by

$$f(t) = \sum_{n=-\infty}^{\infty} f_n \frac{\sin (\omega_c t - n\pi)}{\omega_c t - n\pi} \tag{3-77}$$

Proof.   Because of (3-75), the inversion formula (2-?) takes the
form

$$f(t) = \frac{1}{2\pi} \int_{-\omega_c}^{\omega_c} F(\omega)e^{j\omega t}\, d\omega \tag{3-78}$$

Therefore f_n is given by

$$f_n = f\left(n\frac{\pi}{\omega_c}\right) = \frac{1}{2\pi} \int_{-\omega_c}^{\omega_c} F(\omega)e^{jn\pi\omega/\omega_c}\, d\omega \tag{3-79}$$

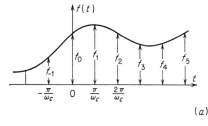
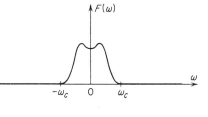

(a)

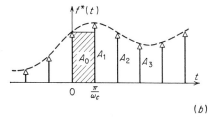
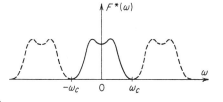

(b)

FIGURE 3-18

We expand the function $F(\omega)$ into a Fourier series in the $(-\omega_c,\omega_c)$
interval:

$$F(\omega) = \sum_{n=-\infty}^{\infty} A_n e^{-jn2\pi\omega/2\omega_c} \qquad -\omega_c < \omega < \omega_c \tag{3-80}$$

where

$$A_n = \frac{1}{2\omega_c} \int_{-\omega_c}^{\omega_c} F(\omega)e^{jn\pi\omega/\omega_c}\, d\omega \tag{3-81}$$

and we conclude from (3-79) that

$$A_n = \frac{\pi}{\omega_c}f_n \tag{3-82}$$

The sum

$$F^*(\omega) = \sum_{n=-\infty}^{\infty} \frac{\pi}{\omega_c}f_n e^{-jn\pi\omega/\omega_c} \tag{3-83}$$

in (3-80) is the periodic repetition of $F(\omega)$ (Fig. 3-18b), and it equals $F(\omega)$ in the $(-\omega_c,\omega_c)$ interval. Clearly [see (3-75)] $F(\omega)$ can be written as a product

$$F(\omega) = p_{\omega_c}(\omega) \sum_{n=-\infty}^{\infty} \frac{\pi}{\omega_c} f_n e^{-jn\pi\omega/\omega_c} \tag{3-84}$$

of $F^*(\omega)$ and the rectangular pulse $p_{\omega_c}(\omega)$. And since [see (2-57)]

$$\frac{\omega_c}{\pi} \frac{\sin(\omega_c t - n\pi)}{\omega_c t - n\pi} \leftrightarrow p_{\omega_c}(\omega) e^{-jn\pi\omega/\omega_c} \tag{3-85}$$

we conclude from (3-84) that

$$f(t) = \sum_{n=-\infty}^{\infty} f_n \frac{\sin(\omega_c t - n\pi)}{\omega_c t - n\pi}$$

Comment. It is of interest to determine the inverse transform $f^*(t)$ of the periodic function $F^*(\omega)$. From (3-83) and (3-4) we obtain

$$f^*(t) = \sum_{n=-\infty}^{\infty} \frac{\pi}{\omega_c} f_n \delta\left(t - \frac{n\pi}{\omega_c}\right)$$

or the equivalent [see (3-18)]

$$f^*(t) = \frac{\pi}{\omega_c} f(t) \sum_{n=-\infty}^{\infty} \delta\left(t - \frac{n\pi}{\omega_c}\right) \tag{3-86}$$

Thus $f^*(t)$ is a sequence of equidistant impulses (Fig. 3-18b) with envelope $(\pi/\omega_c)f(t)$. Notice that the area $(\pi/\omega_c)f_n$ of the nth impulse equals the area of the inscribed rectangle in $f(t)$ (shaded in Fig. 3-18b).

Band-limited Interpolation. Given an arbitrary function $f(t)$ and a constant a, we form the function

$$g(t) = \sum_{n=-\infty}^{\infty} f\left(\frac{n\pi}{a}\right) \frac{\sin(at - n\pi)}{at - n\pi}$$

It is easy to show (see Prob. 19) that $g(n\pi/a) = f(n\pi/a)$ and that the spectrum of $g(t)$ equals zero for $|\omega| > a$. Thus, $g(t)$ offers a band-limited interpolation of $f(t)$.

Frequency sampling. The corresponding sampling theorem in the frequency domain is as follows: if a function $f(t)$ equals zero above $|t| = T$

$$f(t) = 0 \qquad \text{for } |t| > T \tag{3-87}$$

then its Fourier transform $F(\omega)$ can be uniquely determined from its values $F(\pi n/T)$ at a sequence of equidistant points, distance π/T apart. In fact $F(\omega)$ is given by

$$F(\omega) = \sum_{n=-\infty}^{\infty} F\left(\frac{\pi}{T} n\right) \frac{\sin(\omega T - n\pi)}{\omega T - n\pi} \tag{3-88}$$

The proof is similar to the proof of (3-77) and will be omitted.

Chapter 4. Numerical Techniques and Uncertainty Principle

In this chapter we develop numerical techniques for the evaluation of the Fourier transform (2-1) and the inversion formula (2-3). These techniques make use of the special form of these integrals and lead often to simple results. In the chapters on linear systems and filters more detailed methods are presented for determining $f(t)$. In the last two sections we relate the duration of various time signals to the form of their Fourier transform.

4-1. Evaluation of the Fourier Transform

We present two methods for evaluating the integral (2-1). The first method is based on a piecewise polynomial approximation of $f(t)$ and the second on a Fourier-series approach.

Polynomial approximation. If the given function $f(t)$ is sufficiently smooth, then it can be approximated by a small number of polynomial pieces. We shall use only straight segments; the extension to higher-order polynomials can be handled similarly. In Fig. 4-1 we have shown the function $f(t)$ and the approximating polygon $\phi(t)$. Differentiating $\phi(t)$ twice, we obtain a sequence of impulses

$$\phi''(t) = \sum k_i \, \delta(t - t_i) \tag{4-1}$$

o.^{c} area k_i. With $\Phi(\omega)$ the Fourier transform of $\phi(t)$, we conclude from (4-1), (3-4), and (2-41) that

$$(j\omega)^2 \Phi(\omega) = \sum k_i e^{-j\omega t_i} \tag{4-2}$$

Therefore
$$\Phi(\omega) = -\frac{1}{\omega^2} \sum k_i e^{-j\omega t_i} \tag{4-3}$$

We can thus express $F(\omega)$ approximately by a simple sum of exponentials:

$$F(\omega) \simeq -\frac{1}{\omega^2} \sum k_i e^{-j\omega t_i} \tag{4-4}$$

53

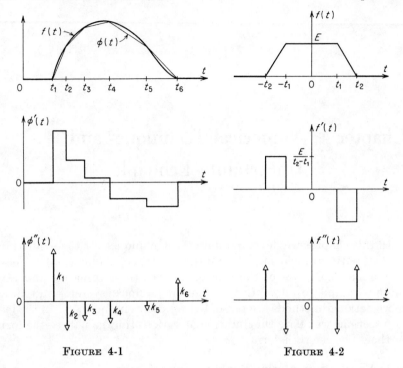

FIGURE 4-1   FIGURE 4-2

As is pointed out in page 40, (4-3) follows from (4-2) only if $\Phi(\omega)$ does not contain the functions $\delta(\omega)$ or $\delta'(\omega)$. The presence of such terms must be determined before differentiation. In most applications $f(t) = 0$ for $|t| > T$, and this guarantees the absence of singularities in $F(\omega)$.

Example 4-1. $f(t)$ is given by the trapezoid of Fig. 4.2. In this case no approximation is necessary, since $f(t)$ is already a polygon. Differentiating twice, we obtain

$$f''(t) = \frac{E}{t_2 - t_1} [\delta(t + t_2) - \delta(t + t_1) - \delta(t - t_1) + \delta(t - t_2)]$$

Hence
$$F(\omega) = \frac{-E}{\omega^2(t_2 - t_1)} (e^{jt_2\omega} - e^{jt_1\omega} - e^{-jt_1\omega} + e^{-jt_2\omega})$$

$$= \frac{4E}{\omega^2(t_2 - t_1)} \sin \frac{\omega(t_2 - t_1)}{2} \sin \frac{\omega(t_2 + t_1)}{2}$$

Fourier series. We shall now determine the Fourier transform $F(\omega)$ of a function $f(t)$ that equals zero outside a finite interval

$$f(t) = 0 \qquad \text{for } |t| > T \tag{4-5}$$

by a Fourier-series method. With

$$\alpha_n = \frac{1}{2T} \int_{-T}^{T} f(t)e^{-jn\omega_0 t}\, dt = \frac{F(n\omega_0)}{2T} \qquad \omega_0 = \frac{\pi}{T} \qquad (4\text{-}6)$$

the coefficients of the Fourier expansion

$$f(t) = \sum_{n=-\infty}^{\infty} \alpha_n e^{jn\omega_0 t} \qquad |t| < T$$

of $f(t)$ in the $(-T,T)$ interval, we conclude from (3-88) that $F(\omega)$ is given by

$$F(\omega) = \sum_{n=-\infty}^{\infty} 2T\alpha_n \frac{\sin(\omega T - n\pi)}{\omega T - n\pi} \qquad (4\text{-}7)$$

Thus the evaluation of $F(\omega)$ is reduced to the simpler problem of determining the sequence of numbers α_n.

This method can be used also to give an estimate of the Fourier transform $F(\omega)$ of an arbitrary function $f(t)$, not satisfying (4-5). We select a sufficiently large constant T and truncate $f(t)$ above $|t| = T$; the function

$$\phi(t) = f(t)p_T(t) \qquad (4\text{-}8)$$

results, whose transform $\Phi(\omega)$ can be found as in (4-7). The functions $F(\omega)$ and $\Phi(\omega)$ are related by

$$\Phi(\omega) = \int_{-\infty}^{\infty} F(x) \frac{\sin T(\omega - x)}{\pi(\omega - x)}\, dx \qquad (4\text{-}9)$$

and for large enough T we have $F(\omega) \simeq \Phi(\omega)$.

Example 4-2. $f(t)$ is given by the cosine pulse

$$f(t) = \tfrac{1}{2}(1 + \cos \omega_0 t) = \tfrac{1}{2} + \tfrac{1}{4}e^{j\omega_0 t} + \tfrac{1}{4}e^{-j\omega_0 t}$$

in the $(-\pi/\omega_0, \pi/\omega_0)$ interval and equals zero elsewhere (Fig. 4-3). It is already written in a Fourier-series form, with $\alpha_0 = \tfrac{1}{2}$, $\alpha_1 = \alpha_{-1} = \tfrac{1}{4}$, and since $T = \pi/\omega_0$, we obtain from (4-7) the transform

$$F(\omega) = \frac{\sin(\omega/\omega_0)\pi}{\omega} + \frac{1}{2}\frac{\sin[(\omega/\omega_0)\pi - \pi]}{\omega - \omega_0} + \frac{1}{2}\frac{\sin[(\omega/\omega_0)\pi + \pi]}{\omega + \omega_0}$$

$$= \frac{1}{\omega_0}\frac{\sin(\omega/\omega_0)\pi}{(\omega/\omega_0)[1 - (\omega/\omega_0)^2]}$$

shown in Fig. 4-3.

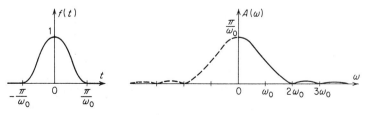

FIGURE 4-3

4-2. Evaluation of the Inversion Integral

The polygonal approximation method of Sec. 4-1 can be used to evaluate $f(t)$ from the real and imaginary parts of $F(\omega)$. The method is particularly suited to functions that equal zero for negative t, as we shall presently show.

Causal time functions. It is shown in Sec. 2-2 that a causal function $f(t)$ can be determined in terms of the real or imaginary part of its transform $F(\omega) = R(\omega) + jX(\omega)$ [see (2-31)]:

$$f(t) = \frac{2}{\pi} \int_0^\infty R(\omega) \cos \omega t \, d\omega = -\frac{2}{\pi} \int_0^\infty X(\omega) \sin \omega t \, d\omega$$

and since [see (2-43)]

$$tf(t) \leftrightarrow j\frac{dF}{d\omega} = -X'(\omega) + jR'(\omega) \qquad t^2 f(t) \leftrightarrow -R''(\omega) - jX''(\omega)$$

we conclude from the above that

$$tf(t) = -\frac{2}{\pi} \int_0^\infty R'(\omega) \sin \omega t \, d\omega \qquad t > 0 \tag{4-10}$$

$$t^2 f(t) = -\frac{2}{\pi} \int_0^\infty R''(\omega) \cos \omega t \, d\omega \qquad t > 0 \tag{4-11}$$

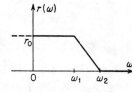
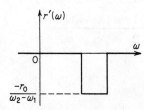
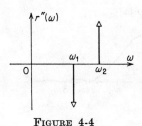

FIGURE 4-4

because the functions $tf(t)$ and $t^2 f(t)$ are also causal. To determine $f(t)$ from $R(\omega)$, we approximate $R(\omega)$ by a sum of trapezoids $r(\omega)$ shown in Fig. 4-4. For $\omega > 0$, the second derivative of $r(\omega)$ consists of two impulses

$$r''(\omega) = \frac{r_0}{\omega_2 - \omega_1} [\delta(\omega - \omega_2) - \delta(\omega - \omega_1)]$$
$$\omega > 0 \tag{4-12}$$

Therefore the time function $g(t)$, whose Fourier transform has as real part $r(\omega)$, is given by

$$g(t) = \frac{2r_0}{\pi(\omega_2 - \omega_1)} \frac{\cos \omega_1 t - \cos \omega_2 t}{t^2} \qquad t > 0 \tag{4-13}$$

as we can readily see from (4-12) and (4-11). This function can be normalized; with

$$\frac{\omega_1}{\omega_2} = k \qquad \omega_2 t = \tau$$

we obtain from (4-13)

$$\frac{\pi}{2r_0\omega_2}\, g\left(\frac{\tau}{\omega_2}\right) = \frac{\cos k\tau - \cos \tau}{(1 - k)\tau^2} \qquad \tau > 0 \qquad (4\text{-}14)$$

The quantity

$$\phi_k(\tau) = \frac{\cos k\tau - \cos \tau}{(1 - k)\tau^2}\, U(\tau) \qquad (4\text{-}15)$$

is a family of curves that depends only on the parameter k and is easily tabulated.   Expressing $g(t)$ in terms of $\phi_k(t)$, we obtain

$$g(t) = \frac{2r_0\omega_2}{\pi}\, \phi_k(\omega_2 t) \qquad (4\text{-}16)$$

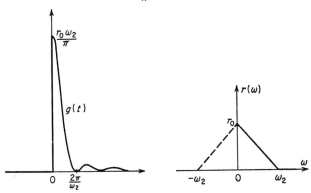

FIGURE 4-5

Special Cases.   If $\omega_1 = 0$, then $k = 0$ and $r(\omega)$ degenerates into the triangle of Fig. 4-5.   Since

$$\phi_0(t) = \frac{1 - \cos t}{t^2}\, U(t) = \frac{2 \sin^2 (t/2)}{t^2}\, U(t)$$

the corresponding time function $g(t)$ is given by

$$g(t) = \frac{4r_0 \sin^2 (\omega_2 t/2)}{\pi\omega_2 t^2}\, U(t) \qquad (4\text{-}17)$$

Notice that $g(t) \geq 0$ (4-18)

If $r(\omega)$ is a rectangle as in Fig. 2-19, then $\omega_1 = \omega_2$, $k = 1$, and since

$$\phi_1(t) = \frac{\sin t}{t}\, U(t)$$

we have (see also Example 2-10)

$$g(t) = \frac{2r_0}{\pi} \frac{\sin \omega_2 t}{t}\, U(t)$$

Positiveness of f(t).   An important problem in the study of linear systems and other areas of application is to give conditions on $F(\omega) = R(\omega) + jX(\omega)$ so that its inverse $f(t)$ will be positive.   The following

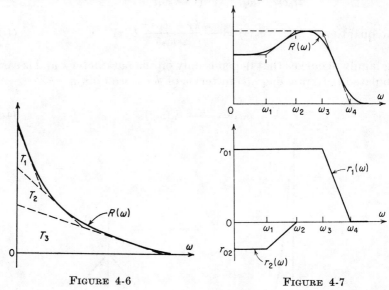

FIGURE 4-6   FIGURE 4-7

useful and not so trivial condition can be very simply obtained from the above results: if $R(\omega)$ is positive and for $\omega > 0$ concave,

$$R(\omega) \geq 0 \qquad \frac{dR(\omega)}{d\omega} \leq 0 \qquad \frac{d^2R}{d\omega^2} \geq 0 \tag{4-19}$$

then $f(t)$ is positive.   Indeed, a function satisfying (4-19) can be approximated arbitrary closely by a sum of positive triangles $R(\omega) \simeq T_1(\omega) + T_2(\omega) + \cdots$ as in Fig. 4-6.   Since for each triangle the corresponding time function is positive [see (4-18)], we conclude that $f(t) \geq 0$.

We return to the problem of evaluating $f(t)$ from $R(\omega)$.   We approximate $R(\omega)$ by a sum of trapezoids†, as in Fig. 4-7,

$$R(\omega) \simeq r_1(\omega) + r_2(\omega) + \cdots \tag{4-20}$$

$f(t)$ is then given by

$$f(t) \simeq g_1(t) + g_2(t) + \cdots \tag{4-21}$$

where $g_i(t)$ is the time function corresponding to $r_i(\omega)$ and can be expressed in terms of the functions $\phi_k(t)$, as in (4-16).   For the curve

† For an estimate of the approximation error see A. Papoulis, An Estimate of the Truncation Error in Fourier Integrals, *IRE-PGCT*, June, 1962.

of Fig. 4-7 we use two terms

$$R(\omega) \simeq r_1(\omega) + r_2(\omega)$$

Hence $\qquad f(t) \simeq \dfrac{2r_{01}\omega_4}{\pi}\, \phi_{k1}(\omega_4 t) + \dfrac{2r_{02}\omega_2}{\pi}\, \phi_{k2}(\omega_2 t)$

where $\qquad\qquad k_1 = \dfrac{\omega_3}{\omega_4} \qquad k_2 = \dfrac{\omega_1}{\omega_2}$

Fourier series. If $F(\omega)$ equals zero above $|\omega| = \omega_c$

$$F(\omega) = 0 \qquad |\omega| > \omega_c \tag{4-22}$$

then its inverse transform $f(t)$ can be found by expanding $F(\omega)$ into a Fourier series in the $(-\omega_c, \omega_c)$ interval:

$$F(\omega) = \sum_{n=-\infty}^{\infty} A_n e^{-jn\pi\omega/\omega_c} \qquad A_n = \frac{1}{2\omega_c}\int_{-\omega_c}^{\omega_c} F(\omega)e^{jn\pi\omega/\omega_c}\,d\omega \tag{4-23}$$

Reasoning as in (4-7), we conclude that $f(t)$ is given by

$$f(t) = \frac{\omega_c}{\pi} \sum_{n=-\infty}^{\infty} A_n \frac{\sin(\omega_c t - n\pi)}{\omega_c t - n\pi} \tag{4-24}$$

Notice that $\qquad\qquad f\!\left(\dfrac{n\pi}{\omega_c}\right) = \dfrac{\omega_c A_n}{\pi} \tag{4-25}$

This method can be used to evaluate approximately $f(t)$, even if $F(\omega)$ is not band-limited. Truncating $F(\omega)$ above a constant ω_c and expanding the resulting function $\Phi(\omega) = F(\omega)p_{\omega c}(\omega)$ as in (4-23), we obtain as its inverse (4-24) a function $\phi(t)$ related to $f(t)$ by

$$\phi(t) = \int_{-\infty}^{\infty} f(\tau)\frac{\sin \omega_c(t - \tau)}{\pi(t - \tau)}\,d\tau \tag{4-26}$$

For sufficiently large ω_c we have $f(t) \simeq \phi(t)$.

4-3. Approximate Evaluation of the Convolution Integral

We shall now discuss a method for evaluating the convolution

$$f(t) = \int_{-\infty}^{\infty} f_1(\tau)f_2(t - \tau)\,d\tau \tag{4-27}$$

between two given functions $f_1(t)$ and $f_2(t)$. A number of numerical techniques, often using the polygonal approximation of Sec. 4-1, have been developed to evaluate the above integral, but they will not be discussed here since they are not related to Fourier analysis. The proposed method does not make direct use of Fourier integrals either; however, the nature of the approximation can best be understood from the properties of the transforms of the given functions. We first

present the formal result and then see under what conditions it leads
to a useful expression.   With $F_1(\omega)$, $F_2(\omega)$, and $F(\omega)$ the transforms
of the functions $f_1(t)$, $f_2(t)$, and $f(t)$ respectively,

$$f_1(t) \leftrightarrow F_1(\omega) \qquad f_2(t) \leftrightarrow F_2(\omega) \qquad f(t) \leftrightarrow F(\omega) \tag{4-28}$$

we have from (2-71)

$$F(\omega) = F_1(\omega)F_2(\omega) \tag{4-29}$$

Denoting by m_k the kth moment of the function $f_1(t)$

$$m_k = \int_{-\infty}^{\infty} t^k f_1(t)\, dt$$

we obtain from (2-48)

$$F_1(\omega) = \sum_{k=0}^{\infty} \frac{m_k}{k!} (-j\omega)^k \tag{4-30}$$

Therefore $F(\omega)$ can be written in the form

$$F(\omega) = F_2(\omega) \sum_{k=0}^{\infty} \frac{m_k}{k!} (-j\omega)^k \tag{4-31}$$

But $(j\omega)^n F_2(\omega)$ is the Fourier integral of $d^n f_2/dt^n$ as we see from (2-41);
hence, if we take the inverse transform of (4-31) termwise, we obtain
the desired result

$$f(t) = m_0 f_2(t) - \frac{m_1}{1!}\frac{df_2}{dt} + \cdots + (-1)^n \frac{m_n}{n!} \frac{d^n f_2}{dt^n} + \cdots \tag{4-32}$$

Thus $f(t)$ is expressed as a series in terms of the derivatives of $f_2(t)$ and
the moments of $f_1(t)$.

In general, the expansion (4-32) is of little use.   For its validity it
was assumed that (4-30) converges for every ω.   Even if this is the
case, the convergence is often
slow, so that a large number of
terms is required for a satisfactory
evaluation of $f(t)$.   However, if
the following assumption is made,
then (4-32) is indeed a useful
result.

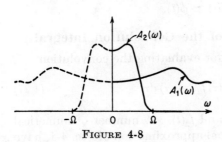

FIGURE 4-8

Assumption. The transform
$F_2(\omega)$ is negligible outside an interval $(-\Omega, \Omega)$, as in Fig. 4-8, and
in this interval $F_1(\omega)$ is sufficiently smooth, so that for $|\omega| < \Omega$ a small
number of terms in (4-30) suffices to approximate adequately $F_1(\omega)$:

$$F_1(\omega) \simeq m_0 + \cdots + \frac{m_n}{n!} (-j\omega)^n \qquad |\omega| < \Omega \tag{4-33}$$

We then obtain from (4-32) and (4-33)

$$f(t) \simeq m_0 f_2(t) - \cdots + (-1)^n \frac{m_n}{n!} \frac{d^n f_2(t)}{dt^n} \tag{4-34}$$

Notice that (4-33) need not be true for every ω, but only for $|\omega| < \Omega$. The above assumption roughly translated in the time domain says that the duration of $f_1(t)$ should be small compared to the duration of $f_2(t)$ (see Sec. 4-4).

Example 4-3. $f_1(t)$ equals a triangular pulse and $f_2(t)$ an exponential as in Fig. 4.9:

$$f_1(t) = E q_T(t) \qquad f_2(t) = e^{-\alpha t} U(t) \qquad \alpha > 0$$

The above assumption is satisfied if $\alpha T \ll 1$. The first three moments of $E q_T(t)$ are given by

$$m_0 = ET \qquad m_1 = 0 \qquad m_2 = \frac{ET^3}{6}$$

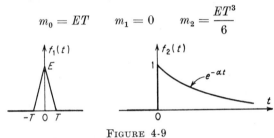

FIGURE 4-9

as we can easily find; hence, retaining only two terms in (4-32), we obtain

$$f(t) \simeq ET \left(1 + \frac{\alpha^2 T^2}{12} \right) e^{-\alpha t} \qquad t \gg T$$

In Chap. 5 we shall show that if the input to a linear system is $f_2(t)$ and its impulse response $f_1(t)$, then the output is given by $f(t)$ as in (4-27). Often the system is characterized not by $f_1(t)$, but by its transform $F_1(\omega) = A_1(\omega) e^{j\varphi_1(\omega)}$. From (2-55) we see that, if $F_1(0) = 1$, then

$$A_1(\omega) = 1 - \frac{\sigma_1^2}{2} \omega^2 + \cdots \qquad \phi(\omega) = -\eta_1 \omega + \cdots \tag{4-35}$$

where η_1 is the center of gravity of $f_1(t)$ and σ_1^2 its central moment of inertia. Assuming that $F_1(\omega)$ is sufficiently smooth in the $(-\Omega, \Omega)$ interval, we obtain from (4-35) and (4-29)

$$F(\omega) \simeq \left(1 - \frac{\sigma_1^2}{2} \omega^2 \right) e^{-j\eta_1 \omega} F_2(\omega) \tag{4-36}$$

Therefore [see (2-41) and (2-36)]

$$f(t) \simeq f_2(t - \eta_1) + \frac{\sigma_1^2}{2} \frac{d^2 f_2(t - \eta_1)}{dt^2} \tag{4-37}$$

Comment.   One might be tempted to conclude from (4-30) and (3-32) that a function $f_1(t)$ can be written as a sum of singularity functions:

$$f_1(t) = \sum_{k=0}^{\infty} (-1)^k \frac{m_k}{k!} \frac{d^k \delta(t)}{dt^k} \tag{4-38}$$

but this result is meaningless, since the above sum for no value of t equals or approximates $f_1(t)$.   However, if $f_1(t)$ is considered not as an ordinary function but as a distribution i.e., if it is to be used as an integrand in integrals of the form (4-27), then (4-38) is valid.

4-4. Duration of a Signal and Uncertainty Principle

In the examples of Sec. 2-4 we noticed that narrow time signals have spectra that are, in some sense, wide, and conversely.   In this section we give a quantitative measure to this observation.   We would like to define the *duration* of a signal and its Fourier transform in a simple and useful way, but no definition is suitable for all signals.   The following possibilities give some insight into the relationship between the spread of a function $f(t)$ and the shape of its transform $F(\omega)$.

RMS duration.   For a measure of the duration of a time signal $f(t)$, one often uses the second moment of $|f(t)|^2$ about a suitably chosen point, e.g., its center of gravity or the position of its maximum.   Since a shift in $f(t)$ results in a linear change in the phase angle of $F(\omega)$, we can take the moment with respect to the origin.   The following identity relates this moment to the *ripple* of $F(\omega) = A(\omega)e^{j\phi(\omega)}$:

$$\int_{-\infty}^{\infty} t^2 |f(t)|^2 \, dt = \frac{1}{2\pi} \int_{-\infty}^{\infty} \left[\left(\frac{dA}{d\omega} \right)^2 + A^2 \left(\frac{d\phi}{d\omega} \right)^2 \right] d\omega \tag{4-39}$$

Proof.   From $f(t) \leftrightarrow F(\omega)$ and (2-43) we obtain the pair

$$(-jt)f(t) \leftrightarrow \frac{dF}{d\omega} = \left(\frac{dA}{d\omega} + jA \frac{d\phi}{d\omega} \right) e^{j\phi} \tag{4-40}$$

From Parseval's formula (2-75) and (4-40) we have

$$\int_{-\infty}^{\infty} t^2 |f(t)|^2 \, dt = \frac{1}{2\pi} \int_{-\infty}^{\infty} \left| \frac{dF}{d\omega} \right|^2 d\omega$$

and since

$$\left| \frac{dF}{d\omega} \right|^2 = \left(\frac{dA}{d\omega} \right)^2 + A^2 \left(\frac{d\phi}{d\omega} \right)^2$$

Eq. (4-39) follows.

From (4-39) we conclude that a high ripple in the frequency spectrum or in the phase angle of $F(\omega)$ results in signals with long duration.   Among all functions with the same amplitude $A(\omega)$, the one that minimizes the left-hand side in (4-39) has zero (linear) phase.

Uncertainty principle. To simplify the notations, we shall assume that the energy of the signals under consideration equals one [see (2-75)]:

$$\int_{-\infty}^{\infty} |f(t)|^2 \, dt = \frac{1}{2\pi} \int_{-\infty}^{\infty} A^2(\omega) \, d\omega = 1 \qquad (4\text{-}41)$$

We define the duration of $f(t)$ by

$$D_t^2 = \int_{-\infty}^{\infty} t^2 \, |f(t)|^2 \, dt \qquad (4\text{-}42)$$

and the duration D_ω of its Fourier transform $F(\omega)$ by

$$D_\omega{}^2 = \int_{-\infty}^{\infty} \omega^2 \, |F(\omega)|^2 \, d\omega \qquad (4\text{-}43)$$

The uncertainty principle states that, if $f(t)$ vanishes at infinity faster than $1/\sqrt{t}$,

$$\lim_{t \to \pm\infty} \sqrt{t} f(t) = 0 \qquad (4\text{-}44)$$

then

$$D_t D_\omega \geq \sqrt{\frac{\pi}{2}} \qquad (4\text{-}45)$$

and that the equality holds only for Gaussian signals

$$f(t) = \sqrt{\frac{\alpha}{\pi}} \, e^{-\alpha t^2} \qquad (4\text{-}46)$$

Proof. The proof of (4-45) follows from Parseval's formula (2-75) and the well-known Schwarz's inequality

$$\left| \int_a^b g_1 g_2 \, dt \right|^2 \leq \int_a^b |g_1|^2 \, dt \int_a^b |g_2|^2 \, dt \qquad (4\text{-}47)$$

where the two sides are equal only if g_1 is proportional to g_2,

$$g_2(t) = k g_1(t) \qquad (4\text{-}48)$$

This inequality can be easily derived from

$$0 \leq \int_a^b [x g_1(t) - g_2(t)]^2 \, dt = x^2 \int_a^b g_1{}^2 \, dt - 2x \int_a^b g_1 g_2 \, dt + \int_a^b g_2{}^2 \, dt$$

Indeed, since the above quadratic is nonnegative for any value of x, its discriminant must be nonpositive and (4-47) is established. If the discriminant equals zero, then for some value $x = k$ of x the quadratic equals zero; this is possible only if $k g_1(t) - g_2(t) \equiv 0$ and (4-48) follows.

We now insert into (4-47) the functions

$$g_1(t) = t f(t) \qquad g_2(t) = \frac{df}{dt}$$

and choose as limits $a = -\infty$, $b = \infty$:

$$\left| \int_{-\infty}^{\infty} tf\frac{df}{dt}\,dt \right|^2 \le \int_{-\infty}^{\infty} |tf|^2\,dt \int_{-\infty}^{\infty} \left| \frac{df}{dt} \right|^2 dt \qquad (4\text{-}49)$$

Integrating the left-hand side by parts we have, because of (4-44) and (4-41),

$$\int_{-\infty}^{\infty} tf\frac{df}{dt}\,dt = t\frac{f^2}{2}\Big|_{-\infty}^{\infty} - \frac{1}{2}\int_{-\infty}^{\infty} |f|^2\,dt = -\frac{1}{2} \qquad (4\text{-}50)$$

From $df/dt \leftrightarrow j\omega\, F(\omega)$ and (2-75) it follows that

$$\int_{-\infty}^{\infty} \left| \frac{df}{dt} \right|^2 dt = \frac{1}{2\pi}\int_{-\infty}^{\infty} |\omega A(\omega)|^2\,d\omega \qquad (4\text{-}51)$$

Inserting (4-51) and (4-50) into (4-49), we obtain

$$\frac{1}{4} \le \int_{-\infty}^{\infty} |tf|^2\,dt\,\frac{1}{2\pi}\int_{-\infty}^{\infty} |\omega A(\omega)|^2\,d\omega$$

and (4-45) is proved.

The equality sign, above, holds if [see (4-48)]

$$df/dt = ktf(t)$$

Solving for $f(t)$, we obtain

$$f(t) = Ce^{kt^2/2}$$

and, with $k = -2\alpha$, (4-46) follows.

Comment. Since $e^{-\alpha t^2}$ is not equal to zero for negative t, even after a shift, the equality sign in (4-45) never holds for causal functions. However, it can be sufficiently closely approached by the function

$$f(t) = \frac{t^n}{n!}\,e^{-t}U(t)$$

if the second moment of $f^2(t)$, in the definition of D_t, is taken with respect to its center of gravity $\eta = n + 1$. It can then be seen that

$$D_t D_\omega = \sqrt{\frac{\pi}{2}}\sqrt{\frac{2n+1}{2n-1}}$$

Linear phase signals.   We now consider real functions $f(t)$ whose Fourier transform $F(\omega)$ has a linear phase angle

$$f(t) \leftrightarrow A(\omega)e^{-j\omega t_0} \qquad (4\text{-}52)$$

These functions are characteristic of the impulse response of linear systems without phase distortion (Chap. 6). It is easy to show that $f(t)$ is symmetrical about $t = t_0$ (Fig. 4-10),

$$f(t) = f(2t_0 - t) \qquad (4\text{-}53)$$

and its value is maximum at $t = t_0$,

$$f_{\max} = f(t_0) = \frac{1}{2\pi} \int_{-\infty}^{\infty} A(\omega)\, d\omega \qquad (4\text{-}54)$$

Indeed, denoting by $f_1(t)$ the inverse transform of $A(\omega)$

$$f_1(t) \leftrightarrow A(\omega) \qquad (4\text{-}55)$$

we conclude from the realness of $A(\omega)$ that $f_1(t)$ is even,

$$f_1(-t) = f_1(t) \qquad (4\text{-}56)$$

and, because $A(\omega) \geq 0$, we have

$$\left| f_1(t) \right| = \frac{1}{2\pi} \int_{-\infty}^{\infty} A(\omega) \left| \cos \omega t \right| d\omega \leq \frac{1}{2\pi} \int_{-\infty}^{\infty} A(\omega)\, d\omega = f_1(0) \qquad (4\text{-}57)$$

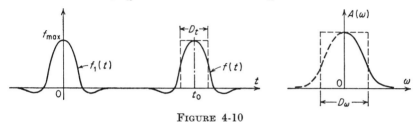

FIGURE 4-10

Hence $f_1(t)$ is maximum for $t = 0$. From (4-52) and (4-55) it follows that

$$f(t) = f_1(t - t_0)$$

(see Fig. 4-10); hence $f(t)$ is symmetrical about t_0, and its maximum $f(t_0)$ is given by (4-54).

A convenient measure of the duration of $f(t)$ is the quantity

$$D_t = \frac{1}{f_{\max}} \int_{-\infty}^{\infty} f(t)\, dt \qquad (4\text{-}58)$$

D_t is the base of a rectangle (Fig. 4-10) with height $f_{\max}$ and area equal to the area under the $f(t)$ curve. In the study of linear systems, D_t is referred to as rise time. If we define D_ω by

$$D_\omega = \frac{1}{A(0)} \int_{-\infty}^{\infty} A(\omega)\, d\omega \qquad (4\text{-}59)$$

i.e., as the base of a rectangle with height

$$A(0) = F(0) = \int_{-\infty}^{\infty} f(t)\, dt \qquad (4\text{-}60)$$

and area equal to the area under the $A(\omega)$ curve, we see from (4-60), (4-59), and (4-58) that

$$D_t D_\omega = 2\pi \qquad (4\text{-}61)$$

[The above definitions of D_t and D_ω are, obviously, different from the corresponding definitions in (4-42) and (4-43).]

Positive time-limited signals. One often associates signals of short duration to spectra that are smooth near the origin. This fact can be stated quantitatively for positive signals of finite duration (see also page 34).

Suppose that (Fig. 4-11)

$$f(t) \geq 0 \quad \text{and} \quad f(t) = 0 \quad \text{for} \begin{cases} t < 0 \\ t > T \end{cases} \tag{4-62}$$

Clearly the Fourier spectrum $A(\omega)$ of $f(t)$ attains its maximum at the origin, as we see from

$$A(\omega) = \left| \int_{-\infty}^{\infty} f(t)e^{-j\omega t}\, dt \right| \leq \int_{-\infty}^{\infty} f(t)\, dt = F(0) = A(0)$$

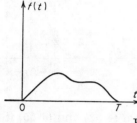 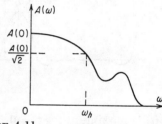

FIGURE 4-11

Denoting by ω_h the first half-power point of $A(\omega)$ (see Fig. 4-11)

$$A(\omega_h) = \frac{A(0)}{\sqrt{2}} \qquad A(\omega) \geq A(\omega_h) \quad \text{for } |\omega| < \omega_h \tag{4-63}$$

we shall show that

$$\omega_h T \geq \frac{\pi}{2} \tag{4-64}$$

The above guarantees that for $|\omega| < \pi/2T$ the spectrum $A(\omega)$ is larger than $A(0)/\sqrt{2}$.

Proof. We shall use the obvious inequalities

$$a^2 + b^2 \geq \frac{(a - b)^2}{2} \qquad |\sin \alpha| + |\cos \alpha| \geq 1 \tag{4-65}$$

From (4-62) we see that the real and imaginary parts of $F(\omega)$ are given by

$$R(\omega) = \int_0^T f(t) \cos \omega t\, dt \qquad X(\omega) = -\int_0^T f(t) \sin \omega t\, dt \tag{4-66}$$

If $0 < \omega \leq \pi/2T$, then for $0 < t < T$ we have

$$\cos \omega t \geq 0 \qquad \sin \omega t \geq 0$$

Hence [see (4-65)] $\cos \omega t + \sin \omega t \geq 1$; therefore

$$R(\omega) - X(\omega) = \int_0^T f(t)(\cos \omega t + \sin \omega t)\, dt \geq \int_0^T f(t)\, dt = A(0) \quad (4\text{-}67)$$

From (4-67) and (4-65) it follows that

$$A^2(\omega) = R^2(\omega) + X^2(\omega) \geq \frac{[R(\omega) - X(\omega)]^2}{2} \geq \frac{A^2(0)}{2}$$

Thus if $0 \leq \omega \leq \pi/2T$, then $A(\omega) \geq A(0)/\sqrt{2} = A(\omega_h)$, and (4-64) is proved.

4-5. Generalization of the Uncertainty Principle†

In many applications one is interested in signals with high concentration in certain parts of the time and frequency domain. In the uncertainty principle as stated in (4-45), we used as a measure of this concentration the second moment of $f^2(t)$ and its energy spectrum $A^2(\omega)$. In this section we shall measure the time and frequency concentration of $f(t)$ by the quantities

$$\alpha^2 = \frac{\displaystyle\int_{-T}^T f^2(t)\, dt}{\displaystyle\int_{-\infty}^\infty f^2(t)\, dt} \qquad \beta^2 = \frac{\displaystyle\int_{-\Omega}^\Omega A^2(\omega)\, d\omega}{\displaystyle\int_{-\infty}^\infty A^2(\omega)\, d\omega} \qquad (4\text{-}68)$$

where T and Ω are given constants. We shall show that, if one of these quantities is specified, the other must remain below a certain maximum depending on ΩT, and we shall find the function for which this maximum is attained. It will be assumed, without loss of generality, that the total energy of $f(t)$ equals one:

$$\int_{-\infty}^\infty f^2(t)\, dt = \frac{1}{2\pi} \int_{-\infty}^\infty A^2(\omega)\, d\omega = 1 \qquad (4\text{-}69)$$

The solution of the above problem will be presented in three parts.

A. Prolate spheroidal wave functions. Consider a signal $f(t)$ whose spectrum $A(\omega)$ is zero for $|\omega| > \Omega$:

$$A(\omega) = 0 \qquad |\omega| > \Omega \qquad (4\text{-}70)$$

We know that $f(t)$ cannot be identically zero outside a finite interval

† The material of this section is based on the article by D. Slepian, H. O. Pollack. and H. T. Landow, Prolate Spheroidal Wave Functions, Fourier Analysis and Uncertainty Principle I and II, *Bell System Tech. J.*, vol. 40, no. 1, pp. 43—84, January, 1961. Figures 4-12, 4-13, and 4-14 are reproduced from this article with permission from the publisher.

$(-T,T)$ (Prob. 39).   Therefore the integral

$$\int_{-T}^{T} f^2(t)\, dt = \alpha^2 \tag{4-71}$$

must be less than one.   What is the maximum value of α, and for what function $f_0(t)$ is it attained?   At the end of the section we shall show that the optimum function $f_0(t)$ satisfies the integral equation

$$\lambda f_0(t) = \int_{-T}^{T} f_0(\tau)\, \frac{\sin \Omega(t - \tau)}{\pi(t - \tau)}\, d\tau \tag{4-72}$$

It is known from the theory of integral equations that (4-72) has nonzero solutions (eigenfunctions) only for certain values of λ (eigenvalues).   Also, because the Fourier kernel $\sin \Omega t/\pi t$ has a positive transform, these eigenvalues are positive.   We denote by λ_0 the maximum eigenvalue of (4-72) and by $\phi(t)$ the corresponding eigenfunction,

$$\lambda_0 \phi(t) = \int_{-T}^{T} \phi(\tau)\, \frac{\sin \Omega(t - \tau)}{\pi(t - \tau)}\, d\tau \tag{4-73}$$

normalized to satisfy the energy requirement (4-69).   We maintain that $\phi(t)$ is band-limited and that its energy in the $(-T,T)$ interval equals λ_0:

$$\int_{-T}^{T} \phi^2(t)\, dt = \lambda_0 \tag{4-74}$$

Indeed, truncating $\phi(t)$ above $|t| = T$, we obtain the function

$$\phi_T(t) = \begin{cases} \phi(t) & |t| < T \\ 0 & |t| > T \end{cases} = \phi(t) p_T(t) \tag{4-75}$$

and (4-73) is equivalent to

$$\lambda_0 \phi(t) = \phi_T(t) * \frac{\sin \Omega t}{\pi t} \tag{4-76}$$

The functions $\phi(t)$ and $\phi_T(t)$ are even; hence their transforms $\Phi(\omega)$ and $\Phi_T(\omega)$ are real and [see (4-76) and (2-71)] they are related by

$$\lambda_0 \Phi(\omega) = \Phi_T(\omega) p_\Omega(\omega) \tag{4-77}$$

because the transform of $(\sin \Omega t)/\pi t$ equals the rectangular pulse $p_\Omega(\omega)$. From (4-77) we conclude that

$$\Phi(\omega) = 0 \qquad \text{for } |\omega| > \Omega \tag{4-78}$$

and [see (2-79)]

$$\int_{-T}^{T} \phi^2(t)\, dt = \int_{-T}^{T} \phi(t) \phi_T(t)\, dt = \frac{1}{2\pi} \int_{-\infty}^{\infty} \Phi(\omega) \Phi_T(\omega)\, d\omega$$

$$= \frac{\lambda_0}{2\pi} \int_{-\Omega}^{\Omega} \Phi^2(\omega)\, d\omega = \lambda_0$$

since the total energy of $\phi(t)$ equals one. From the above it follows that $\phi(t)$ is the optimum function $f_0(t)$

$$f_0(t) = \phi(t)$$

maximizing α^2, and that the maximum value of α^2 equals λ_0

$$\alpha^2 \leq \lambda_0 \qquad (4\text{-}79)$$

With a linear change of variable one can easily show that λ_0 depends only on the product ΩT. The function $\lambda_0(\Omega T)$ is shown in Fig. 4-12.

In the same figure the value of α^2 for the Fourier kernel (sin $\Omega t)/t$ is also shown. It is interesting to note that this value is close to the maximum λ_0.

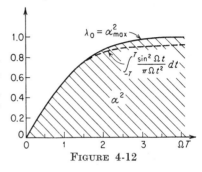

FIGURE 4-12

To every λ_0 there corresponds one optimum function $\phi(t)$. Except for a scale factor, $\phi(t)$ depends only on ΩT. We thus have a one-parameter family of curves known in the literature as *prolate spheroidal wave functions*. They are shown in Fig. 4-13 for $\Omega T = .5, 1, 2,$ and 4.

Comment. The transform $\Phi(\omega)$ of $\phi(t)$ can be expressed in terms of $\phi(t)$. To show this, we take the transforms of both sides of (4-75) and use (2-74):

$$\Phi_T(\omega) = \frac{1}{2\pi} \int_{-\infty}^{\infty} \Phi(y) \frac{2 \sin T(\omega - y)}{(\omega - y)} \, dy \qquad (4\text{-}80)$$

Because of (4-77), the above gives

$$\lambda_0 \Phi_T(\omega) = \int_{-\Omega}^{\Omega} \Phi_T(y) \frac{\sin T(\omega - y)}{\pi(\omega - y)} \, dy \qquad (4\text{-}81)$$

Except for a change in the constants, (4-81) is the same as (4-73). And since this equation has as only solutions the functions $k\phi(t)$, we

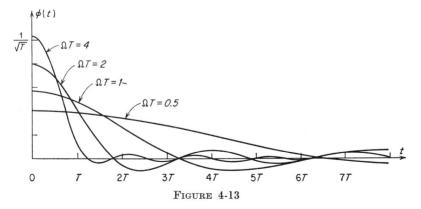

FIGURE 4-13

conclude with a proper change in the independent variable that $\Phi_T(\omega) = k\phi[(T/\Omega)\omega]$. To determine k, we have [see (4-74)]

$$\lambda_0 = \int_{-T}^{T} \phi_T{}^2(t)\, dt = \frac{1}{2\pi} \int_{-\infty}^{\infty} \Phi_T{}^2(\omega)\, d\omega$$

Hence
$$k = \sqrt{2\pi\lambda_0 T/\Omega}$$

and
$$\Phi_T(\omega) = \sqrt{\frac{2\pi\lambda_0 T}{\Omega}}\, \phi\!\left(\frac{T}{\Omega}\,\omega\right)$$

We thus obtain

$$\phi(t) \leftrightarrow \Phi(\omega) = \begin{cases} \sqrt{\dfrac{2\pi T}{\lambda_0 \Omega}}\, \phi\!\left(\dfrac{T}{\Omega}\,\omega\right) & |\omega| < \Omega \\ 0 & |\omega| > \Omega \end{cases}$$

Therefore $\phi(t)$ satisfies the integral equation

$$\sqrt{\frac{2\pi\lambda_0 T}{\Omega}}\, \phi(t) = \int_{-\Omega}^{\Omega} e^{j\Omega t \tau/T} \phi(\tau)\, d\tau$$

B. **Time-limited signals.** Suppose that the function $f(t)$ equals zero outside a finite interval (t_1, t_2). Taking the origin at its center, we have

$$f(t) = 0 \qquad |t| > T = \frac{t_2 - t_1}{2} \tag{4-82}$$

This function cannot be band-limited; hence its energy in a finite band is less than one:

$$\frac{1}{2\pi} \int_{-\Omega}^{\Omega} |F(\omega)|^2\, d\omega = \beta^2 < 1$$

We shall determine the maximum value of β^2 and the optimum function $f_0(t)$ for which this maximum is attained. Reasoning as in section A, we conclude that

$$\beta^2 \leq \lambda_0$$

and that

$$f_0(t) = \frac{\phi_T(t)}{\sqrt{\lambda_0}} \tag{4-83}$$

The factor $1/\sqrt{\lambda_0}$ is introduced to make the energy of the optimum function equal to one. Thus $f_0(t)$ is obtained by truncating $\phi(t)$ above $|t| = T$.

C. **Arbitrary functions.** We now consider the general problem. With

$$\int_{-T}^{T} f^2(t)\, dt = \alpha^2 \qquad \frac{1}{2\pi} \int_{-\Omega}^{\Omega} |F(\omega)|^2\, d\omega = \beta^2 \tag{4-84}$$

we shall assume α^2 (or β^2) given and will obtain the maximum value of β^2 (or α^2) and the maximizing function $f_0(t)$. It can be shown that $f_0(t)$ is a linear combination of the optimum functions of parts A and B

$$f_0(t) = a\phi(t) + b\phi_T(t) \tag{4-85}$$

The proof of (4-85) will not be given. [It is best derived† from the orthogonality properties of the eigenfunction of (4-72)]. The constants a and b are determined from the requirement that the total energy of $f(t)$ equals one and its energy in the $(-T,T)$ interval equals α^2. With the help of (4-74) we easily find

$$a^2 + \lambda_0 b^2 + 2ab\lambda_0 = 1 \qquad \lambda_0(a + b)^2 = \alpha^2$$

Hence
$$a = \sqrt{\frac{1 - \alpha^2}{1 - \lambda_0}} \qquad b = \frac{\alpha}{\sqrt{\lambda_0}} - \sqrt{\frac{1 - \alpha^2}{1 - \lambda_0}} \tag{4-86}$$

To find the value of β corresponding to the optimizing function (4-85), we observe that the energy of $\phi(t)$ and $\phi_T(t)$ in the interval $(-\Omega,\Omega)$ equals one and λ_0^2 respectively [see (4-77)]. Therefore

$$\frac{1}{2\pi} \int_{-\Omega}^{\Omega} |F_0(\omega)|^2 \, d\omega = (a + b\lambda_0)^2 = \beta_{\max}^2 \tag{4-87}$$

Inserting the values of a and b from (4-86) into (4-87), we obtain $\beta_{\max}$ as a function of λ_0 and α. The resulting expression can be simplified if we introduce the arc cosines

$$\cos \theta_0 = \sqrt{\lambda_0} \qquad \cos \theta_1 = \alpha \qquad \cos \theta_2 = \beta_{\max} \tag{4-88}$$

Since $\beta_{\max} = a + b\lambda_0$, we obtain with (4-86)

$$\cos \theta_2 = \frac{\sin \theta_1}{\sin \theta_0} + \cos^2 \theta_0 \left(\frac{\cos \theta_1}{\cos \theta_0} - \frac{\sin \theta_1}{\sin \theta_0} \right) = \cos (\theta_0 - \theta_1)$$

Hence
$$\theta_1 + \theta_2 = \theta_0 \tag{4-89}$$

For an arbitrary $f(t)$ the corresponding β is less than $\beta_{\max}$; therefore the quantities α and β in (4-84) must satisfy

$$\cos^{-1} \alpha + \cos^{-1} \beta \leq \cos^{-1} \sqrt{\lambda_0} \tag{4-90}$$

where the equality sign holds only for the optimum function (4-85).

† F. G. Tricomi, "Integral Equations," Interscience Publishers, Inc., New York, 1957.

If β is specified, the function that makes α maximum is given by $f_0(t) = a_1\phi(t) + b_1\phi_T(t)$. We now have

$$a_1{}^2 + \lambda_0 b_1{}^2 + 2a_1 b_1 \lambda_0 = 1 \qquad (a_1 + \lambda_0 b_1)^2 = \beta^2$$

Therefore

$$b_1 = \sqrt{\frac{1 - \beta^2}{\lambda_0(1 - \lambda_0)}} \qquad a_1 = \beta - \sqrt{\frac{\lambda_0(1 - \beta^2)}{1 - \lambda_0}}$$

The resulting maximum value of α is given by (4-89).

In Fig. 4-14 the curves $\cos^{-1}\alpha + \cos^{-1}\beta = \cos^{-1}\sqrt{\lambda_0}$ are shown for various values of λ_0. These values are taken from Fig. 4-12 for

$$\Omega T = .25, .5, 1, 2$$

By the dashed line in Fig. 4-14, we have shown the relationship between α and β for the Gaussian function $f(t) = e^{-t^2/2}/\sqrt[4]{\pi}$ and for

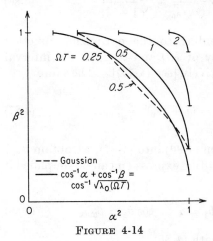

$\Omega T = .5$. For this function we easily find that

$$\alpha^2 = \frac{1}{\sqrt{\pi}} \int_{-T}^{T} e^{-t^2}\, dt = erf\, T$$

$$\beta^2 = \frac{\sqrt{4\pi}}{2\pi} \int_{-\Omega}^{\Omega} e^{-\omega^2}\, d\omega = erf\, \Omega$$

We shall now prove that the optimum function of part A, i.e., the function $f_0(t)$ that maximizes (4-71), satisfies the integral equation (4-72). Denoting by $\alpha_m{}^2$ the energy of $f_0(t)$ in the $(-T, T)$ interval, we conclude that if $f_0(t)$ is truncated above $|t| = T$, the

FIGURE 4-14

resulting function $f_{0T}(t) = f_0(t)p_T(t)$ has total energy $\alpha_m{}^2$. With $F_{0T}(\omega)$ as the transform of $f_{0T}(t)$ we thus have

$$\frac{1}{2\pi} \int_{-\infty}^{\infty} |F_{0T}(\omega)|^2\, d\omega = \alpha_m{}^2 \qquad (4\text{-}91)$$

We maintain that

$$\frac{1}{2\pi} \int_{-\Omega}^{\Omega} |F_{0T}(\omega)|^2\, d\omega \le \alpha_m{}^4 \qquad (4\text{-}92)$$

Indeed, with

$$x(t) = \frac{\Omega}{2\pi T \alpha_m{}^2} F_{0T}\left(\frac{\Omega}{T} t\right)$$

we easily see that $\alpha_m{}^2 x(t) \leftrightarrow f_{0T}(-T\omega/\Omega)$ and

$$\int_{-\infty}^{\infty} |x(t)|^2\, dt = 1 \qquad \int_{-T}^{T} |x(t)|^2\, dt = \frac{1}{2\pi\alpha_m{}^2} \int_{-\Omega}^{\Omega} |F_{0T}(\omega)|^2\, d\omega$$

Thus $x(t)$ is band-limited, its total energy equals one, and if (4-92) were not true, then the energy of $x(t)$ in the $(-T,T)$ interval would exceed $\alpha_m{}^2$, which is impossible; hence, (4-92) is true. Suppose that $f_0(t)$ does not satisfy (4-72); then the function

$$y(t) = \int_{-T}^{T} f_0(\tau) \frac{\sin \Omega(t-\tau)}{\pi(t-\tau)} \, d\tau = f_{0T}(t) * \frac{\sin \Omega t}{\pi t} \tag{4-93}$$

is not proportional to $f_0(t)$. Since $f_0(t)$ is band-limited, we have

$$f_0(t) = \int_{-\infty}^{\infty} f_0(\tau) \frac{\sin \Omega(t-\tau)}{\pi(t-\tau)} \, d\tau$$

Hence

$$\alpha_m{}^2 = \int_{-T}^{T} f_0{}^2(t) \, dt = \int_{-T}^{T} f_0(t) \, dt \int_{-\infty}^{\infty} f_0(\tau) \frac{\sin \Omega(t-\tau)}{\pi(t-\tau)} \, d\tau = \int_{-\infty}^{\infty} f_0(t)y(t) \, dt$$

The last equality was obtained by changing the order of integration and using (4-93). From Schwarz's inequality (4-47) we obtain, since the total energy of $f_0(t)$ equals one,

$$\alpha_m{}^4 = \left[\int_{-\infty}^{\infty} f_0(t)y(t) \, dt \right]^2 < \int_{-\infty}^{\infty} y^2(t) \, dt \tag{4-94}$$

The equality sign is impossible because $y(t)$ is not proportional to $f_0(t)$. With $Y(\omega)$ the transform of $y(t)$, we have from (4-93)

$$Y(\omega) = F_{0T}(\omega)p_\Omega(\omega)$$

Therefore [see (4-92)]

$$\int_{-\infty}^{\infty} y^2(t) \, dt = \frac{1}{2\pi} \int_{-\Omega}^{\Omega} |F_{0T}(\omega)|^2 \, d\omega \le \alpha_m{}^4$$

We thus conclude that $\alpha_m{}^4 < \alpha_m{}^4$, which is absurd; hence $y(t)$ is proportional to $f_0(t)$ and (4-72) is proved.

Applications. An important problem in communication theory is the determination of a time-limited signal $f(t)$ of specified total energy

$$E = \int_{-T}^{T} f^2(t) \, dt \tag{4-95}$$

such that, if it is used as input to a given linear system (Fig. 5-1), the energy of the output $g(t)$ of the system is maximum. With $G(\omega)$ the transform of $g(t)$ and $H(\omega)$ the system function, we have [see (5-22)] $G(\omega) = F(\omega)H(\omega)$, and our problem is to find the function $f(t)$ that maximizes

$$\int_{-\infty}^{\infty} g^2(t) \, dt = \frac{1}{2\pi} \int_{-\infty}^{\infty} |F(\omega)H(\omega)|^2 \, d\omega \tag{4-96}$$

subject to the conditions (4-95).   If the system is a low-pass filter, then $|H(\omega)| = A_0 p_\Omega(\omega)$ [see (6-20)]; hence

$$\int_{-\infty}^{\infty} g^2(t)\, dt = \frac{A_0^2}{2\pi} \int_{-\Omega}^{\Omega} |F(\omega)|^2\, d\omega$$

As we know from part B of the previous discussion, the above integral is maximum for $f(t) = \sqrt{E/\lambda_0}\ \phi_T(t)$, and the problem is solved for this case.

Consider now the general case.†   The energy of the output can be written in the form [see (4-96)]

$$\int_{-\infty}^{\infty} g^2(t)\, dt = \frac{1}{2\pi}\int_{-\infty}^{\infty} \left[\int_{-T}^{T} f(\tau)e^{-j\omega\tau}\, d\tau\right]\left[\int_{-T}^{T} f(x)e^{j\omega x}\, dx\right] |H(\omega)|^2\, d\omega$$

With $L(t)$ the inverse transform of $|H(\omega)|^2$

$$L(t) \leftrightarrow |H(\omega)|^2 \tag{4-97}$$

we obtain from the above, changing the order of integration,

$$\int_{-\infty}^{\infty} g^2(t)\, dt = \int_{-T}^{T}\int_{-T}^{T} f(\tau)f(x)L(x-\tau)\, dx\, d\tau \tag{4-98}$$

Thus our problem is to find a function $f_0(t)$ maximizing the double integral (4-98) subject to (4-95).   It can be shown that this optimum function $f_0(t)$ must satisfy the integral equation

$$\mu f_0(t) = \int_{-T}^{T} f_0(\tau)L(t-\tau)\, d\tau \tag{4-99}$$

It then follows from (4-98) that

$$\int_{-\infty}^{\infty} g^2(t)\, dt = \int_{-T}^{T} f_0(x)\mu f_0(x)\, dx = E\mu$$

Therefore, with μ_0 the maximum eigenvalue of (4-99), $f_0(t)$ is the corresponding eigenfunction and the energy of the output is given by $E\mu_0$.   The proof of (4-99) involves the theory of integral equations and will be omitted.   For a given system, the quantities μ_0 and $f_0(t)$ can be found by solving (4-99) with known techniques‡ (see Prob. 14).

† J. H. H. Chalk, The Optimum Pulse-shape for Pulse Communication, *Proc. Inst. Elec. Engrs. (London)*, vol. 87, pp. 88–92, 1950.

‡ D. C. Youla, The Solution of a Homogeneous Wiener-Hopf Integral Equation Occurring in the Expansion of Second-order Stationary Random Functions, *IRE-PGIT*, pp. 187–193, September, 1957.

Problems

1. Show that if

$$e^{j\phi(t)} \longleftrightarrow F(\omega)$$

and $\phi(t)$ is real, then

$$\cos \phi(t) \longleftrightarrow \frac{F(\omega) + \overset{*}{F}(-\omega)}{2} \qquad \sin \phi(t) \longleftrightarrow \frac{F(\omega) - \overset{*}{F}(-\omega)}{2j}$$

2. Prove that if $f(t)$ is causal, then the real part $R(\omega)$ of its Fourier transform satisfies

$$R(\omega) = \frac{2}{\pi} \int_0^\infty \int_0^\infty R(y) \cos yt \cos \omega t \, dy \, dt \tag{i}$$

3. Prove that

$$\int_{-\infty}^\infty \frac{\sin^3 t}{t^3} \, dt = \frac{3\pi}{4} \qquad \int_{-\infty}^\infty \frac{\sin^4 t}{t^4} \, dt = \frac{2\pi}{3}$$

4. Show that if $f(t)$ is band-limited

$$F(\omega) = 0 \qquad \text{for } |\omega| > \Omega$$

then

$$f(t) * \frac{\sin at}{\pi t} = f(t) \tag{i}$$

for every $a > \Omega$.

Application

$$\frac{1}{\pi} \int_{-\infty}^\infty \frac{\sin \alpha\tau}{\tau} \frac{\sin (t - \tau)}{t - \tau} \, d\tau = \begin{cases} \dfrac{\sin t}{t} & \text{if } \alpha \geq 1 \\[2mm] \dfrac{\sin \alpha t}{t} & \text{if } |\alpha| \leq 1 \end{cases}$$

5. Show that

$$e^{j(at^2 + bt + c)} \longleftrightarrow \sqrt{\frac{\pi j}{a}} \, e^{j[c - (b - \omega)^2 / 4a]} \tag{i}$$

$$\cos (at^2 + bt + c) \longleftrightarrow \sqrt{\frac{\pi}{a}} \cos \left(\frac{\omega^2}{4a} + \frac{b^2}{4a} - c - \frac{\pi}{4} \right) e^{j(b\omega/2a)}$$

$$\sin (at^2 + bt + c) \longleftrightarrow \sqrt{\frac{\pi}{a}} \sin \left(c + \frac{\pi}{4} - \frac{\omega^2}{4a} - \frac{b^2}{4a} \right) e^{j(b\omega/2a)} \tag{ii}$$

$$\sin \omega_0 t \cos \alpha t^2 \longleftrightarrow \tfrac{1}{2} \sqrt{\frac{\pi}{\alpha}} \left[\cos \frac{(\omega + \omega_0)^2}{4\alpha} \, e^{j(\pi/4)} + \cos \frac{(\omega - \omega_0)^2}{4\alpha} \, e^{-j(\pi/4)} \right] \tag{iii}$$

75

6. Show that if $f(t)$ is a solution of the differential equation

$$\frac{d^2x(t)}{dt^2} - t^2x(t) = \lambda x(t)$$

then its Fourier transform $F(\omega)$ is a solution of the same equation.

7. Show that if $x(t)$ is a solution of

$$\frac{d^2x(t)}{dt^2} - t^2x(t) = -(2n + 1)x(t) \qquad \text{(i)}$$

where n is an integer, then except for a constant factor, $x(t)$ is its own transform.

Hint. Try a solution of the form $x(t) = H_n(t)e^{-t^2/2}$, where $H_n(t)$ is a polynomial; show that there is only one solution and use the result of Prob. 6.

8. The Fourier transform of a function $f(t)$ is a unit step $F(\omega) = S(\omega)$; find $f(t)$.

9. The function $f(t)$ is real and its Fourier transform is $F(\omega)$. With

$$\frac{1}{\pi} \int_0^\infty F(\omega)e^{j\omega t}\, d\omega = f_1(t) + jf_2(t)$$

show that

$$f_1(t) = f(t) \qquad f_2(t) = \frac{1}{\pi} \int_{-\infty}^\infty \frac{f(\tau)}{t - \tau}\, d\tau \qquad \text{(i)}$$

10. Find the transform of a finite pulse train

$$f(t) = \sum_{k=0}^{n-1} \delta(t - kT)$$

11. The Fourier spectrum of a sequence of pulses

$$x^*(t) = \sum_{-\infty}^\infty x_n\, \delta(t - nT) \quad \text{equals } A^*(\omega). \quad \text{We form the function}$$

$$y^*(t) = \sum_{-\infty}^\infty y_n\delta(t - nT)$$

where $y_n = \dfrac{1}{N} \displaystyle\sum_{k=n-N+1}^{n} x_k$ is the average of N consecutive values of x_n. Show that the spectrum of $y^*(t)$ is given by

$$\frac{1}{N}\, A^*(\omega)\left|\frac{\sin (NT\omega/2)}{\sin (T\omega/2)}\right| \qquad \text{(i)}$$

Hint. $y^*(t)$ is the convolution of $x^*(t)$ with a train of N equal pulses of area $1/N$.

12. Show that the transform of a train of doublets is a train of impulses:

$$\sum_{n=-\infty}^\infty \delta'(t - nT) \leftrightarrow j\omega_0^2 \sum_{-\infty}^\infty n\, \delta(\omega - n\omega_0) \qquad \omega_0 = \frac{2\pi}{T}$$

Hence

$$\sum_{n=-\infty}^\infty \delta'(t - nT) = -\frac{4\pi}{T^2} \sum_{n=1}^\infty n \sin \frac{2\pi nt}{T}$$

13. Using Poisson's sum formula, prove that

$$\sum_{n=-\infty}^\infty \frac{\sin a(t + nT)}{t + nT} = \omega_c \frac{\sin (2N + 1)\omega_c t}{\sin \omega_c t} \qquad \text{(i)}$$

where $\omega_c = \pi/T$ and N is such that

$$N < \frac{aT}{2\pi} < N + 1 \qquad \text{(ii)}$$

14. The input $f(t)$ to a linear system $H(\omega) = 1/(\alpha + j\omega)$ is a time-limited function [$f(t) = 0$ for $|t| > T$] of total energy E. Determine $f(t)$ so that the energy of the output be maximum.

Hint. Try a solution of (4-99) of the form

$$f(t) = \begin{cases} a \cos \omega t & \text{for } |t| < T \\ \\ 0 & \text{for } |t| > T \end{cases}$$

Solutions

1.
$$\int_{-\infty}^{\infty} \cos \phi(t) e^{-j\omega t}\, dt = \tfrac{1}{2} \int_{-\infty}^{\infty} e^{j[\phi(t)-\omega t]}\, dt + \tfrac{1}{2} \int_{-\infty}^{\infty} e^{-j[\phi(t)+\omega t]}\, dt$$

$$= \frac{F(\omega)}{2} + \frac{\overset{*}{F}(-\omega)}{2}$$

Similarly for the sine.

2. For a causal function we have

$$f(t) = \frac{2}{\pi} \int_0^{\infty} R(y) \cos yt\, dy \qquad R(\omega) = \int_0^{\infty} f(t) \cos \omega t\, dt$$

and (i) follows.

3. From

$$\frac{\sin t}{t} \leftrightarrow \pi p_1(\omega) \qquad \frac{\sin^2 t}{t^2} \leftrightarrow \pi q_2(\omega)$$

and (2-77) we have

$$\int_{-\infty}^{\infty} \frac{\sin^2 t}{t^2} \frac{\sin t}{t}\, dt = \frac{\pi^2}{2\pi} \int_{-\infty}^{\infty} q_2(\omega)p_1(\omega)\, d\omega = \frac{\pi}{2} \int_{-1}^{1} q_2(\omega)\, d\omega = \frac{3\pi}{4}$$

The proof is similar for the second equation.

4. From $F(\omega) = 0$ for $|\omega| > \Omega$ it follows that

$$F(\omega)p_a(\omega) = F(\omega) \qquad \text{for } a > \Omega$$

and since

$$\frac{\sin at}{\pi t} \leftrightarrow p_a(\omega)$$

(i) follows from the convolution theorem (2-71).

5. (i) Since

$$at^2 + bt + c - \omega t = \left[\sqrt{a}\,t + \frac{b-\omega}{2\sqrt{a}}\right]^2 + c - \frac{(b-\omega)^2}{4a}$$

we have

$$\int_{-\infty}^{\infty} e^{j(at^2+bt+c)} e^{-j\omega t}\, dt = e^{j[c-(b-\omega)^2/4a]} \int_{-\infty}^{\infty} e^{j[\sqrt{a}\,t+(b-\omega)/2\sqrt{a}]^2}\, dt$$

but the last integral equals $\sqrt{\pi j/a}$ (see Prob. 30), and (i) is established.
(ii) follows from (i) and Prob. 1.
(iii) follows from (ii) and

$$\sin \omega_0 t \cos \alpha t^2 = \tfrac{1}{2}[\sin (\omega_0 t + \alpha t^2) + \sin (\omega_0 t - \alpha t^2)]$$

6. If $f(t)$ satisfies the given equation, then, taking the Fourier transforms of both sides, we obtain because of (2-41) and (2-43)

$$(j\omega)^2 F(\omega) + \frac{d^2 F(\omega)}{d\omega^2} = \lambda F(\omega)$$

7. Inserting $H_n(t)\, e^{-t^2/2}$ into (i), we obtain $\ddot{H}_n(t) - 2t\dot{H}_n(t) + 2nH_n(t) = 0$. With $H_n(t) = a_0 + \cdots + a_n t^n$ we find the recursion

$$a_{k-2} = \frac{k(k-1)a_k}{2k - 2n - 4} \qquad a_{k-1} = a_{k-3} = \cdots = 0 \qquad k = n, n-2, \ldots$$

Thus (i) has a unique solution and since (see Prob. 6) the transform $X(\omega)$ of $x(t)$ satisfies the same equation, we conclude that $x(t)$ is proportional to $X(t)$. [$H_n(t)$ is a Hermite polynomial.]

8. From (2-34) and (3-13) we obtain

$$f(t) = \tfrac{1}{2}\delta(t) + \frac{j}{2\pi t}$$

9. *Hilbert transforms.* The Fourier transform of $f_1(t) + jf_2(t)$ equals $2F(\omega)S(\omega)$, where $S(\omega)$ is the unit step function. Since (see Prob. 8)

$$f(t) \longleftrightarrow F(\omega) \qquad \delta(t) + \frac{j}{\pi t} \longleftrightarrow 2S(\omega)$$

we have from (2-71)

$$f_1(t) + jf_2(t) = \int_{-\infty}^{\infty} f(\tau)\left[\delta(t - \tau) + \frac{j}{\pi(t - \tau)} \right] d\tau = f(t) + j\frac{1}{\pi}\int_{-\infty}^{\infty} \frac{f(\tau)}{t - \tau}\, d\tau$$

and (i) follows by equating real and imaginary parts.

10.
$$F(\omega) = \sum_{k=0}^{n-1} e^{-jkT\omega} = e^{-j(n-1)T\omega/2}\, \frac{\sin\,(nT\omega/2)}{\sin\,(T\omega/2)}$$

11. (i) follows from Prob. 10 and the convolution theorem.

12. From $s_T(t) \longleftrightarrow \omega_0 s_{\omega 0}(\omega)$ [see (3-43)] and theorem (2-41) it follows that

$$\sum_{n=-\infty}^{\infty} \delta'(t - nT) \longleftrightarrow j\omega\omega_0 s_{\omega 0}(\omega) = j\omega_0^2 \sum_{n=-\infty}^{\infty} n\delta(\omega - n\omega_0)$$

because $\omega\delta(\omega - n\omega_0) = n\omega_0\,\delta(\omega - n\omega_0)$. But the inverse of $j\,\delta(\omega - n\omega_0)$ equals $je^{jn\omega_0 t}/2\pi$; therefore

$$\sum_{n=-\infty}^{\infty} \delta'(t - nT) = \frac{j\omega_0^2}{2\pi} \sum_{n=-\infty}^{\infty} ne^{jn\omega_0 t} = -\frac{\omega_0^2}{2\pi} \sum_{n=-\infty}^{\infty} n \sin n\omega_0 t$$

13. From $(\sin at)/t \longleftrightarrow \pi p_a(\omega)$ and (3-56) we have

$$\sum_{n=-\infty}^{\infty} \frac{\sin a(t + nT)}{t + nT} = \frac{\pi}{T} \sum_{n=-\infty}^{\infty} e^{j2\omega_c nt} p_a(2n\omega_c)$$

But $p_a(2n\omega_c) = 0$ for $|2n\omega_c| > a$; hence, with N as in (ii), the last sum equals

$$\sum_{-N}^{N} e^{j2\omega_c nt} = \frac{\sin\,(2N + 1)\omega_c t}{\sin \omega_c t}$$

and (i) is proved.

14. The inverse transform of $|H(\omega)|^2 = 1/(\alpha^2 + \omega^2)$ is given by [see (2-62)] $L(t) = e^{-\alpha|t|}/2\alpha$; therefore the optimum function $f(t)$ must satisfy the integral equation

$$2\alpha\lambda f(t) = \int_{-T}^{T} f(\tau) e^{-\alpha|t-\tau|} d\tau = \int_{-T}^{t} f(\tau) e^{-\alpha(t-\tau)} d\tau + \int_{t}^{T} f(\tau) e^{\alpha(t-\tau)} d\tau$$

for $|t| < T$. Inserting $a \cos \omega t$ into the above, we easily obtain

$$2\alpha\lambda \cos \omega t = \frac{2\alpha \cos \omega t}{\alpha^2 + \omega^2} + \frac{(\omega \sin \omega T - \alpha \cos \omega T)}{\alpha^2 + \omega^2} e^{-\alpha T}(e^{\alpha t} + e^{-\alpha t})$$

For the above to be true for every $|t| < T$, we must have

$$\omega \sin \omega T - \alpha \cos \omega T = 0 \qquad \lambda = \frac{1}{\alpha^2 + \omega^2}$$

Thus ω is the smallest solution ω_0 of the equation $\tan \omega T = \alpha/\omega$, and the maximum eigenvalue of (4-99) is given by $\lambda_0 = 1/(\alpha^2 + \omega_0{}^2)$. The constant a is determined from the energy requirement (4-95).

PART TWO

Chapter 5. Linear Systems

In the following three chapters we discuss the applications of the Fourier integral to linear systems. In this study we are concerned not with the interior of the system, but only with its terminal properties, i.e., the relationship between a cause $f(t)$ and the resulting effect $g(t)$. The system is specified by a set of assumptions (linearity, time invariance, causality) that correspond to what is often called a "real, linear, passive nondegenerate system with constant coefficients and zero initial conditions." No mention is made of other quantities beyond $f(t)$ and $g(t)$. This chapter deals mostly with general relationships; detailed analysis of specific systems is given in the discussion of filters.

5-1. Definitions

The analysis of most physical systems can be reduced to the investigation of the relationship between certain causes and their effects. Any system can be viewed as a transducer (Fig. 5-1), with the cause $f(t)$ as its *input* and the effect $g(t)$ as its *output* or *response*; $g(t)$ is uniquely determined in terms of $f(t)$. The system is completely characterized terminally if the nature of the dependence of the output on the input is known. For example, $g(t)$ could be the solution of an ordinary differential equation with forcing term $f(t)$, as is the case in lumped parameter systems; it could be the solution of a partial differential equation, as in transmission lines, heat transfer, or radiation; it could also be experimentally established if one has no information about the interior of the system.

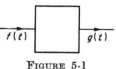

FIGURE 5-1

Thus, for our purposes, a physical system is a set of specifications for determining a function $g(t)$ from a given function $f(t)$:

$$T\{f(t)\} = g(t) \tag{5-1}$$

81

or, to borrow the language of mathematics, a *transformation* of $f(t)$ into $g(t)$. The symbol T in (5-1) designates the law of determining $g(t)$ from $f(t)$ (operator).

In general the determination of a system is indeed complicated. One must know the output to any conceivable input! If the system is linear, however, then its specification is considerably simplified.

Linearity. A system is called linear if: with $g_1(t)$ the output to $f_1(t)$, $g_2(t)$ the output to $f_2(t)$, and a_1 and a_2 two arbitrary constants, the output to $a_1f_1(t) + a_2f_2(t)$ is given by $a_1g_1(t) + a_2g_2(t)$. Using the notation of (5-1) and introducing L for linearity, we can express the above definition by

$$L\{a_1f_1(t) + a_2f_2(t)\} = a_1L\{f_1(t)\} + a_2L\{f_2(t)\} \qquad (5\text{-}2)$$

As simple examples of linear systems we mention the differentiating circuit $L\{f(t)\} = df/dt$ and the delay line $L\{f(t)\} = f(t - t_1)$.

A transformation L of a function into another function satisfying (5-2) is often called *linear operator*. Thus, in regard to the input-output relationship, the terms linear system and linear operator are equivalent.

We denote by $h(t;t_1)$ the response of a linear system to a unit impulse $\delta(t - t_1)$

$$L\{\delta(t - t_1)\} = h(t;t_1) \qquad (5\text{-}3)$$

and will show that, if $h(t;t_1)$ is known, then the system is fully characterized, i.e., that the response $g(t)$ to an arbitrary input $f(t)$ can be found in terms of $h(t;t_1)$. Indeed, expressing $f(t)$ as a sum of impulses $f(t_1)\,\delta(t - t_1)\,dt_1$ [see (I-21)]

$$f(t) = \int_{-\infty}^{\infty} f(t_1)\,\delta(t - t_1)\,dt_1 \qquad (5\text{-}4)$$

we obtain because of (5-2)

$$L\{f(t)\} = \int_{-\infty}^{\infty} f(t_1)\dot{L}\{\delta(t - t_1)\}\,dt_1 = \int_{-\infty}^{\infty} f(t_1)h(t;t_1)\,dt_1 \qquad (5\text{-}5)$$

Hence
$$g(t) = \int_{-\infty}^{\infty} f(t_1)h(t;t_1)\,dt_1 \qquad (5\text{-}6)$$

Comment. In deriving (5-5), we assumed that (5-2) holds for infinitely many terms. Although from definition (5-2) it easily follows that

$$L\left\{\sum_{k=1}^{n} a_kf_k(t)\right\} = \sum_{k=1}^{n} a_k\,L\{f_k(t)\}$$

for n finite, the extension to infinite sums and integrals is an added requirement. In the following we shall interpret linearity in this extended sense.

We note that $h(t;t_1)$ is not one function but a family of functions depending on the parameter t_1. A further major simplification is possible if we assume that the system has constant parameters.

Time invariance. This is a way of saying that the parameters of the system are independent of time; e.g., if the defining law is a differential equation, then its coefficients are constant. Terminally a system is said to be with *constant parameters*, if, with $g(t)$ its response to $f(t)$, the response to $f(t - t_1)$ equals $g(t - t_1)$:

$$L\{f(t - t_1)\} = g(t - t_1) \tag{5-7}$$

where t_1 is an arbitrary constant.

Denoting by $h(t)$ the response of such a system to an impulse $\delta(t)$,

$$L\{\delta(t)\} = h(t) \tag{5-8}$$

we see from (5-7) that its response to $\delta(t - t_1)$ is given by

$$h(t;t_1) = h(t - t_1) \tag{5-9}$$

Therefore the output to an arbitrary input $f(t)$ can be expressed in terms of $h(t)$ alone [see (5-6)]:

$$g(t) = \int_{-\infty}^{\infty} f(\tau)h(t - \tau)\, d\tau = \int_{-\infty}^{\infty} f(t - \tau)h(\tau)\, d\tau \tag{5-10}$$

We thus reach the amazing conclusion that a linear system with constant parameters, i.e., a system whose response satisfies (5-2) and (5-7), is uniquely determined from the knowledge of a single function $h(t)$. In the following we shall consider only linear systems with constant parameters, and for brevity we shall call them linear.

Eigenfunctions. The importance of the Fourier integral in the analysis of linear systems is due to the fact that, if the input is an exponential $e^{j\omega t}$, then the output is also an exponential proportional to the input

$$L\{e^{j\omega t}\} = k e^{j\omega t} \tag{5-11}$$

To prove (5-11), we denote by $e(t)$ the response to $e^{j\omega t}$

$$L\{e^{j\omega t}\} = e(t) \tag{5-12}$$

and we have from (5-7), with t_1 a constant,

$$L\{e^{j\omega(t+t_1)}\} = e(t + t_1)$$

But [see (5-2)]

$$L\{e^{j\omega(t+t_1)}\} = L\{e^{j\omega t_1}e^{j\omega t}\} = e^{j\omega t_1}e(t)$$

Hence

$$e(t + t_1) = e^{j\omega t_1}e(t)$$

and with $t = 0$

$$e(t_1) = e(0)e^{j\omega t_1}$$

But t_1 is arbitrary; therefore

$$e(t) = e(0)e^{j\omega t}$$

Thus the output is proportional to the input, with $e(0)$ the proportionality constant. The quantity $e(0) = k$ is in general complex and depends on the parameter ω; it will be denoted by $H(\omega)$:

$$e(0) = H(\omega) = A(\omega)e^{j\phi(\omega)} = R(\omega) + jX(\omega)$$

Inserting $e(t)$ into (5-12), we obtain (Fig. 5-2)

$$L\{e^{j\omega t}\} = H(\omega)e^{j\omega t} \tag{5-13}$$

In the next section we shall show that $H(\omega)$, known as system function, is the Fourier transform of the impulse response $h(t)$ and will rederive (5-13).

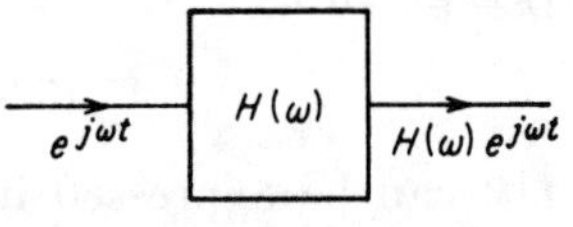

In the language of operators, a function $f(t)$ satisfying the equation

$$L\{f(t)\} = kf(t)$$

FIGURE 5-2

is called *eigenfunction* and the corresponding value of k, *eigenvalue*. From the above discussion it follows that exponentials as in (5-13) are eigenfunctions of linear time-invariant operators.

In general we shall allow the functions $f(t)$ and $g(t)$ to be complex; however, since we are interested in real systems, we shall make the obvious assumption that, if the input $f(t)$ is real, then the output

$$g(t) = L\{f(t)\}$$

is also real. From the *realness* of L it easily follows that if

$$L\{f_1(t) + jf_2(t)\} = g_1(t) + jg_2(t)$$

then $$L\{f_1(t)\} = g_1(t) \qquad L\{f_2(t)\} = g_2(t)$$

since $f_1(t)$ and $f_2(t)$ are real. From the above and (5-13) we conclude that if the input is a sine wave

$$f(t) = \cos \omega t = \operatorname{Re} e^{j\omega t}$$

then the output is given by

$$g(t) = \operatorname{Re} H(\omega)e^{j\omega t} = R(\omega) \cos \omega t - X(\omega) \sin \omega t$$

Thus

$$L\{\cos \omega t\} = R(\omega) \cos \omega t - X(\omega) \sin \omega t = A(\omega) \cos [\omega t + \phi(\omega)] \tag{5-14}$$

Stability. A linear system is called stable if its response to any bounded input is bounded: If $|f(t)| < M$, then $|g(t)| < MI$, where I is a constant independent of the input. The above definition is equivalent to the absolute integrability of the impulse response $h(t)$ of the given system

$$I = \int_{-\infty}^{\infty} |h(t)| \, dt < \infty \tag{5-15}$$

Indeed, if (5-15) is true, then [see (5-10)]

$$|g(t)| = \left| \int_{-\infty}^{\infty} f(t - \tau) h(\tau) \, d\tau \right| \leq M \int_{-\infty}^{\infty} |h(\tau)| \, d\tau = MI$$

To prove the converse, it suffices to show that, if $I = \infty$, then a bounded function can be found causing an unbounded response. Such a function is shown in Fig. 5-3 and is given by

$$f(-t) = \frac{h(t)}{|h(t)|}$$

Clearly $|f(t)| = 1$, but the corresponding response is infinite at $t = 0$, as we see from (5-10),

$$g(0) = \int_{-\infty}^{\infty} f(-\tau) h(\tau) \, d\tau$$

$$= \int_{-\infty}^{\infty} \frac{h^2(t)}{|h(t)|} \, dt$$

$$= \int_{-\infty}^{\infty} |h(t)| \, dt$$

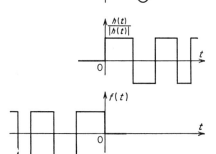

FIGURE 5-3

Causality. A physical passive system has the property that if the input is zero for $t < t_1$, then the output is also zero for $t < t_1$. If $f(t) = 0$ for $t < t_1$, then

$$g(t) = L\{f(t)\} = 0 \qquad \text{for } t < t_1 \tag{5-16}$$

This condition we call *causality*. From (5-16) it follows that

$$h(t) = 0 \qquad \text{for } t < 0 \tag{5-17}$$

i.e., the impulse response is causal in the sense of Sec. 2-2. For causal systems, the response $g(t)$ is given by

$$g(t) = \int_{-\infty}^{t} f(\tau) h(t - \tau) \, d\tau = \int_{0}^{\infty} f(t - \tau) h(\tau) \, d\tau \tag{5-18}$$

as we readily see from (5-10).

Comment. The response of a general system at a certain time t_1 is specified only if the input $f(t)$ is given for every t; however, if the system is causal, then $g(t)$ can be uniquely found for $t = t_1$ if $f(t)$ is known only in the $(-\infty, t_1)$ interval. This follows from (5-18).

5-2. The System Function

The Fourier transform $H(\omega)$ of the impulse response $h(t)$ of a linear system is called system function

$$H(\omega) = \int_{-\infty}^{\infty} h(t)e^{-j\omega t}\, dt = R(\omega) + jX(\omega) = A(\omega)e^{j\phi(\omega)} \qquad (5\text{-}19)$$

and is often specified by its *attenuation* $\alpha(\omega)$ and *phase shift* or *phase lag* $\theta(\omega)$, defined by

$$\alpha(\omega) = -\ln A(\omega) \qquad \theta(\omega) = -\phi(\omega) \qquad H(\omega) = e^{-\alpha(\omega)}e^{-j\theta(\omega)} \qquad (5\text{-}20)$$

From the inversion formula (2-17) we have

$$h(t) = \frac{1}{\pi} \int_0^{\infty} A(\omega) \cos\left[\omega t - \theta(\omega)\right] d\omega \qquad (5\text{-}21)$$

With $F(\omega)$ and $G(\omega)$ the Fourier transforms of the input $f(t)$ and the output $g(t)$

$$f(t) \leftrightarrow F(\omega) \qquad g(t) \leftrightarrow G(\omega)$$

we obtain from (5-10) the important formula

$$G(\omega) = F(\omega)H(\omega) \qquad (5\text{-}22)$$

Hence $g(t)$ can be written in the form [see (2-3)]

$$g(t) = \frac{1}{2\pi} \int_{-\infty}^{\infty} F(\omega)H(\omega)e^{j\omega t}\, d\omega \qquad (5\text{-}23)$$

If $f(t) = e^{j\omega_0 t}$, then [see (3-6)] $F(\omega) = 2\pi\,\delta(\omega - \omega_0)$; therefore [see (5-22) and (3-18)]

$$G(\omega) = 2\pi\,\delta(\omega - \omega_0)H(\omega) = 2\pi H(\omega_0)\,\delta(\omega - \omega_0)$$

Thus the response to $e^{j\omega_0 t}$ is given by

$$g(t) = H(\omega_0)e^{j\omega_0 t}$$

as it was also proved in Sec. 5-1, Eq. (5-13). A special case of the above is the response to a constant; from (5-13) and the linearity of the system we conclude that

$$L\{C\} = CH(0) = CA(0) \qquad (5\text{-}24)$$

Periodic Inputs. If the input $f(t)$ to a linear system is periodic, then the output $g(t)$ is also periodic. Indeed, expanding $f(t)$ into a Fourier series

$$f(t) = \sum_{-\infty}^{\infty} a_n e^{jn\omega_0 t} \qquad (5\text{-}25)$$

we obtain from (5-13) and (5-2)

$$g(t) = \sum_{-\infty}^{\infty} a_n H(n\omega_0)e^{jn\omega_0 t} \qquad (5\text{-}26)$$

Comment. Often a periodic input is applied to a system at $t = 0$; i.e., $f(t) = 0$ for $t < 0$, and for $t > 0$ it is given by (5-25). In this case the expression (5-26) gives the asymptotic form of $g(t)$, i.e., the so-called *steady-state* response (see Sec. 9-4).

In the remaining part of this chapter we shall consider only causal systems. As we have shown in (2-31), $h(t)$ can then be expressed in terms of $R(\omega)$ or $X(\omega)$ alone:

$$h(t) = \frac{2}{\pi} \int_0^\infty R(\omega) \cos \omega t \, d\omega = -\frac{2}{\pi} \int_0^\infty X(\omega) \sin \omega t \, d\omega \qquad t > 0 \quad (5\text{-}27)$$

Step response. The response of a system to a unit step is denoted by $a(t)$

$$L\{U(t)\} = a(t) \tag{5-28}$$

From (5-18) we obtain

$$a(t) = \int_0^t h(\tau) \, d\tau \tag{5-29}$$

and with $t = \infty$

$$a(\infty) = \int_0^\infty h(\tau) \, d\tau = H(0) = R(0) \tag{5-30}$$

Since [see (3-13)]

$$U(t) \leftrightarrow \pi \, \delta(\omega) + \frac{1}{j\omega}$$

we conclude from (5-22) that the Fourier transform of $a(t)$ is given by [see (3-18)]

$$\left[\pi \, \delta(\omega) + \frac{1}{j\omega}\right][R(\omega) + jX(\omega)] = \pi R(0) \, \delta(\omega) + \frac{X(\omega)}{\omega} - j\frac{R(\omega)}{\omega}$$

But $a(t)$ is obviously causal; hence [see (5-27)], for $t > 0$,

$$a(t) = \frac{2}{\pi} \int_0^\infty \frac{R(\omega)}{\omega} \sin \omega t \, d\omega \tag{5-31}$$

and

$$a(t) = \frac{1}{\pi} \int_{-\infty}^\infty \left[\pi R(0) \, \delta(\omega) + \frac{X(\omega)}{\omega}\right] \cos \omega t \, d\omega = R(0) + \frac{2}{\pi} \int_0^\infty \frac{X(\omega)}{\omega} \cos \omega t \, d\omega \tag{5-32}$$

We shall now express the response $g(t)$ to an arbitrary input $f(t)$, in terms of the step response $a(t)$, by an integral similar to (5-18). Clearly the function $f(t)$ can be written in the form

$$f(t) = f(-\infty) + \int_{-\infty}^t f'(\tau) \, d\tau = f(-\infty) + \int_{-\infty}^\infty f'(\tau) U(t - \tau) \, d\tau$$

and from (5-24) and (5-28) we obtain

$$L\{f(t)\} = L\{f(-\infty)\} + \int_{-\infty}^{\infty} f'(\tau)L\{U(t-\tau)\}\, d\tau$$

$$= A(0)f(-\infty) + \int_{-\infty}^{\infty} f'(\tau)a(t-\tau)\, d\tau$$

Thus

$$g(t) = A(0)f(-\infty) + \int_{-\infty}^{\infty} f'(\tau)a(t-\tau)\, d\tau \qquad (5\text{-}33)$$

This result is general, since no use of the causality condition was made. For causal systems, $a(t-\tau) = 0$ for $\tau > t$; hence $g(t)$ is given by

$$g(t) = A(0)f(-\infty) + \int_{-\infty}^{t} f'(\tau)a(t-\tau)\, d\tau \qquad (5\text{-}34)$$

Comment. The above is valid even if $f(t)$ is discontinuous, provided $f'(t)$ contains the appropriate impulses. Thus, if $f(t) = 0$ for $t < 0$ and its derivative for $t > 0$ is given by $f_i'(t)$, then

$$f'(t) = f(0^+)\, \delta(t) + f_i'(t)$$

Hence

$$g(t) = \int_{-\infty}^{\infty} [f(0^+)\, \delta(\tau) + f_i'(\tau)]a(t-\tau)\, d\tau$$

$$= f(0^+)a(t) + \int_{0}^{t} f_i'(\tau)a(t-\tau)\, d\tau$$

Step-modulated input. An input of particular interest is the function

$$f(t) = U(t)\cos\omega_0 t$$

whose Fourier transform $F(\omega)$ is given by [see (3-29)]

$$F(\omega) = \frac{\pi}{2}[\delta(\omega - \omega_0) + \delta(\omega + \omega_0)] + \frac{j\omega}{\omega_0^2 - \omega^2} \qquad (5\text{-}35)$$

From (5-35) and (5-22) we obtain

$$G(\omega) = \frac{\pi}{2}[R(\omega_0)\, \delta(\omega - \omega_0) + R(-\omega_0)\, \delta(\omega + \omega_0)] + \frac{X(\omega)\omega}{\omega^2 - \omega_0^2}$$

$$+ j\left\{\frac{\pi}{2}[X(\omega_0)\, \delta(\omega - \omega_0) + X(-\omega_0)\, \delta(\omega + \omega_0)] + \frac{R(\omega)\omega}{\omega_0^2 - \omega^2}\right\} \qquad (5\text{-}36)$$

Clearly $g(t)$ is causal, because $f(t) = 0$ for $t < 0$; therefore [see (5-36) and (5-27)]

$$g(t) = R(\omega_0)\cos\omega_0 t + \frac{2}{\pi}\int_{0}^{\infty} \frac{X(\omega)\omega}{\omega^2 - \omega_0^2}\cos\omega t\, d\omega$$

$$= -X(\omega_0)\sin\omega_0 t + \frac{2}{\pi}\int_{0}^{\infty} \frac{R(\omega)\omega}{\omega^2 - \omega_0^2}\sin\omega t\, d\omega \qquad (5\text{-}37)$$

We thus expressed the response to a step-modulated input in terms of the real and imaginary parts of the system function.

5-3. Evaluation of the Step Response

We shall now determine the step response of a linear system in terms of its system function

$$H(\omega) = R(\omega) + jX(\omega) = A(\omega)e^{j\phi(\omega)}$$

Because of the assumed causality, $H(\omega)$ need be only partially known (see also Chap. 10).

Given $R(\omega)$. It was shown in Sec. 2-2 that, if $R(\omega)$ is given, then $H(\omega)$ is uniquely determined. However, the function $h(t)$ can be found directly from $R(\omega)$ [see (5-27)] without the necessity of determining $H(\omega)$. Thus

$$h(t) = \frac{2}{\pi} \int_0^\infty R(\omega) \cos \omega t \, d\omega \qquad t > 0$$

For the numerical evaluation of the above integral, we approximate $R(\omega)$ by a simple set of *standard parts*

$$R(\omega) = r_1(\omega) + r_2(\omega) + \cdots \tag{5-38}$$

whose corresponding impulse responses $h_1(t)$, $h_2(t)$, ... are known. From (5-38) and (5-27) it follows that $h(t)$ is obtained by a simple addition

$$h(t) = h_1(t) + h_2(t) + \cdots \tag{5-39}$$

As it was shown in Sec. 4-2, a convenient choice† of $r(\omega)$ is the trapezoid of Fig. 5-4, whose impulse response $h(t)$ is given by [see (4-16)]

$$h(t) = \frac{2r_0\omega_2}{\pi} \phi_k(\omega_2 t) \tag{5-40}$$

where [see (4-15)]

$$\phi_k(t) = \frac{\cos kt - \cos t}{(1-k)t^2} U(t) \qquad k = \frac{\omega_1}{\omega_2}$$

The corresponding step response $a(t)$ is found directly from (5-40) and (5-29). It can easily be seen that

$$\frac{\pi}{2r_0} a\left(\frac{t}{\omega_2}\right) = Si(kt) + \frac{1}{1-k}\left[Si(t) - Si(kt) + \frac{\cos t - \cos kt}{t}\right] \tag{5-41}$$

where $Si(t)$ is the sine integral (Sec. 2-4). The right-hand side of (5-41) depends only on the parameter k. Denoting it by $n_k(t)$, we conclude

---

† V. V. Solodovnikov, "Introduction to the Statistical Dynamics of Automatic Control Systems" (translated from Russian), Dover Publications, New York, 1960.

from (5-41) that

$$a(t) = \frac{2r_0}{\pi}\, n_k(\omega_2 t) \tag{5-42}$$

If $R(\omega)$ is approximated by a sum of trapezoids (see Fig. 4-7), then the resulting step response is a sum of functions of the form (5-42).

Overshoot of step response. In many applications one is interested in determining the overshoot

$$\frac{a_{\max} - a(\infty)}{a(\infty)}$$

of the step response $a(t)$ above its final value $a(\infty)$ (Fig. 5-5). For its determination, the function $a(t)$ should in general be evaluated; however, if certain assumptions are made about the real part $R(\omega)$ of the

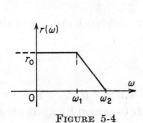

FIGURE 5-4   FIGURE 5-5

system functions $H(\omega)$, then simple useful upper limits to the maximum $a_{\max}$ of $a(t)$ can be given. We shall discuss three cases.

Case 1. $R(\omega)$ is positive and, for $\omega > 0$, concave (Fig. 5-6):

$$R(\omega) \geq 0 \qquad \frac{dR(\omega)}{d\omega} \leq 0 \qquad \frac{d^2R(\omega)}{d\omega^2} \geq 0 \tag{5-43}$$

It was shown in Sec. 4-2 that, if $R(\omega)$ satisfies (5-43) [see (4-19)], then the corresponding impulse response is positive

$$h(t) \geq 0 \tag{5-44}$$

and since $a(t)$ is the integral of $h(t)$, it increases monotonically to its final value $a(\infty)$; i.e., the overshoot equals zero.

Case 2. $R(\omega)$ is positive decreasing (Fig. 5-7):

$$R(\omega) \geq 0 \qquad \frac{dR(\omega)}{d\omega} \leq 0 \tag{5-45}$$

We shall show that, in this case, $a(t)$ is positive:

$$a(t) \geq 0 \tag{5-46}$$

and the overshoot is less than 0.18:

$$a_{\max} \leq 1.18a(\infty) = 1.18R(0) \tag{5-47}$$

Indeed, from (5-31) we have

$$\frac{\pi}{2} a(t) = \int_0^\infty \frac{R(\omega)}{\omega} \sin \omega t \, d\omega$$

$$= \int_0^{\pi/t} \frac{R(\omega)}{\omega} \sin \omega t \, d\omega + \int_{\pi/t}^{2\pi/t} \frac{R(\omega)}{\omega} \sin \omega t \, d\omega + \cdots \quad (5\text{-}48)$$

With

$$I_n = \int_{(n-1)\pi/t}^{n\pi/t} \frac{R(\omega)}{\omega} \sin \omega t \, d\omega \qquad n = 1, 2, \ldots$$

we easily conclude from (5-45) that

$$|I_1| \geq |I_2| \geq |I_3| \geq \cdots$$

and that the integrals I_1, I_3, I_5, ... are positive and the integrals I_2, I_4, I_6, ... negative. Therefore

$$I_1 + I_2 + I_3 + \cdots \leq I_1 > 0 \qquad (5\text{-}49)$$

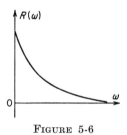

FIGURE 5-6

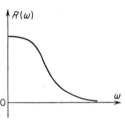

FIGURE 5-7

The sum $I_1 + I_2 + I_3 + \cdots$ equals $\pi a(t)/2$ [see (5-48)]; hence

$$0 \leq \frac{\pi}{2} a(t) \leq \int_0^{\pi/t} \frac{R(\omega)}{\omega} \sin \omega t \, d\omega \leq R(0) \int_0^{\pi/t} \frac{\sin \omega t}{\omega} \, d\omega \qquad (5\text{-}50)$$

because [see (5-45)] $R(\omega) \leq R(0)$. But

$$\int_0^{\pi/t} \frac{\sin \omega t}{\omega} \, d\omega = \int_0^\pi \frac{\sin x}{x} \, dx = Si(\pi) = 1.18\pi/2$$

as one can see from the tables of the sine integral; therefore [see (5-30)]

$$0 \leq a(t) \leq 1.18R(0) = 1.18a(\infty)$$

Case 3. $R(\omega)$ is positive, with a single maximum R_{max} as in Fig. 5-8. We shall show that in this case the overshoot is less than $0.18R_{max}/R(0)$:

$$a_{max} \leq 1.18a(\infty) \frac{R_{max}}{R(0)} \qquad (5\text{-}51)$$

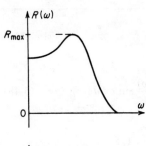

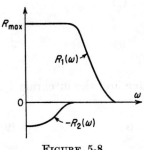

FIGURE 5-8

We can write $R(\omega)$ as a difference of two positive decreasing functions $R_1(\omega)$ and $R_2(\omega)$ (see Fig. 5-8):

$$R(\omega) = R_1(\omega) - R_2(\omega)$$

With $a_1(t)$ and $a_2(t)$ the corresponding step responses, we have

$$a(t) = a_1(t) - a_2(t)$$

From the result of Case 2 we conclude that

$$a_2(t) \geq 0$$

and $a_{1\,max} \leq 1.18 R_1(0) = 1.18 R_{max}$

because $R_1(0) = R_{max}$.   Therefore

$$a(t) \leq a_1(t) \leq a_{1\,max} \leq 1.18 R_{max}$$
$$= 1.18 R(0) \frac{R_{max}}{R(0)}$$

and since $R(0) = a(\infty)$, the desired result (5-51) follows.   The above estimate (5-51) can be written in terms of the maximum A_{max} of $A(\omega)$.   Indeed, since $R(\omega) = A(\omega) \cos \phi(\omega)$, we conclude that

$$R_{max} \leq A_{max}   \qquad R(0) = A(0)$$

Therefore $$a_{max} \leq 1.18 a(\infty) \frac{A_{max}}{A(0)} \tag{5-52}$$

Given $A(\omega)$.   In many applications, particularly in the study of filters, one is interested in relating the time properties of a system to the amplitude $A(\omega)$ or phase $\phi(\omega)$ of its system function $H(\omega)$.   Since a physical system is causal, $A(\omega)$ and $\phi(\omega)$ are not independent of each other.   In fact for minimum-phase-shift functions, $H(\omega)$ is uniquely determined from $A(\omega)$, as is shown in Chap. 10.   Therefore $h(t)$ can, in principle, be found from $A(\omega)$; this is not an easy problem, however, since there is no direct relationship between $h(t)$ and $A(\omega)$ similar to (5-27).   Furthermore, an expansion of $A(\omega)$ into a sum of simple functions cannot be used, because the resulting time responses do not in general add.   One could express the attenuation $\alpha(\omega) = \ln A(\omega)$ as a sum of standard parts

$$\alpha(\omega) = \alpha_1(\omega) + \alpha_2(\omega) + \cdots \tag{5-53}$$

With $H_1(\omega)$, $H_2(\omega)$, ... and $h_1(t)$, $h_2(t)$, ... the corresponding system functions and their inverse transforms, it would then follow that

$$H(\omega) = H_1(\omega) H_2(\omega) \cdots \tag{5-54}$$

and $$h(t) = h_1(t) * h_2(t) * \cdots \tag{5-55}$$

Thus, unlike the simple addition (5-39), the evaluation of $h(t)$ from (5-55) would involve repeated convolutions. Another complication of this approach is the difficulty of finding standard parts that are sufficiently simple to permit an easy expansion of $\alpha(\omega)$ in the form (5-53) and that also have simple time properties. A number of methods based on (5-53) have been suggested, but they are not particularly attractive and will not be discussed here.

An alternative method for determining $h(t)$ from $A(\omega)$ is to abandon the causality requirement and associate to $A(\omega)$ a conveniently chosen $\phi(\omega)$ so that $h(t)$ can be easily evaluated from (5-21). One can thus gain a useful insight on the relationship between various frequency characteristics and the corresponding time responses. It is true, of course, that the resulting system function is not realizable by a passive physical system because $h(t)$ is not zero for negative t. However, it is shown in the next chapter that, if sufficient time delay is allowed, then with minor modifications on $A(\omega)$, one can obtain causal responses. This approach is adopted in the following two chapters dealing with the analysis of low-pass and bandpass filters.

Chapter 6. Low-pass Filters

The term *filter* is used to describe linear systems whose amplitude characteristic $A(\omega)$ is negligible in certain parts of the frequency axis. If $A(\omega)$ is small in some sense for $\omega > \omega_c$, then the filter is called low-pass, and ω_c its cutoff frequency. In this chapter we deal at some length with the properties of low-pass filters,† not only because of their importance in applications, but also because the results can be used to analyze more general filters and, in fact, to obtain the inverse Fourier transform of functions with band-limited spectra. The discussion could be considered an elaboration of the techniques of Chap. 4 to more special situations.

6-1. Definitions

We shall write the system function of a filter in the form

$$H(\omega) = A(\omega)e^{-j\theta(\omega)} \tag{6-1}$$

Its impulse response can be evaluated from

$$h(t) = \frac{1}{\pi} \int_0^\infty A(\omega)\cos\left[\omega t - \theta(\omega)\right] d\omega \tag{6-2}$$

The Fourier transform of the step response $a(t)$ is given by [see (5-22)]

$$\left[\pi\,\delta(\omega) + \frac{1}{j\omega}\right]A(\omega)e^{-j\theta(\omega)} = \pi A(0)\,\delta(\omega) + \frac{A(\omega)}{\omega}\,e^{-j[\theta(\omega)+\pi/2]}$$

Therefore
$$a(t) = \frac{A(0)}{2} + \frac{1}{\pi}\int_0^\infty \frac{A(\omega)}{\omega}\sin\left[\omega t - \theta(\omega)\right]d\omega \tag{6-3}$$

We remark that, since $A(\omega)$ and $\theta(\omega)$ are assumed independent of each other, the systems under consideration are not causal; thus $h(t)$ does not equal zero for negative t, and $a(t)$ is related to it by

$$a(t) = \int_{-\infty}^t h(\tau)\,d\tau \tag{6-4}$$

† K. Küpfmüller, "Die Systemtheorie der elektrischen Nachrichten Übertragung," S. Hirzel Verlag, Leipzig, 1952.

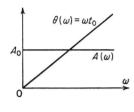
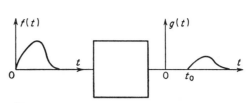

Figure 6-1

and not by (5-29). A filter is called *distortionless* if its response $g(t)$ to an arbitrary input $f(t)$ has the same form as the input

$$g(t) = A_0 f(t - t_0) \tag{6-5}$$

With $F(\omega)$ and $G(\omega)$ the Fourier transforms of $f(t)$ and $g(t)$, we obtain from (6-5) and (2-36)

$$G(\omega) = A_0 e^{-j\omega t_0} F(\omega) \tag{6-6}$$

Therefore [see (5-22)] the system function of a distortionless filter is given by

$$H(\omega) = A_0 e^{-j\omega t_0} \tag{6-7}$$

It has a constant amplitude and a linear phase shift

$$A(\omega) = A_0 \qquad \theta(\omega) = \omega t_0 \tag{6-8}$$

as in Fig. 6-1.

We say that a filter is *amplitude-distorted* if $A(\omega)$ is not a constant, and *phase-distorted* if $\theta(\omega)$ is not linear. We shall examine separately the effect of amplitude and phase distortion on the time properties of the filter.

Linear phase filters. The amplitude and phase shift of a filter without phase distortion is shown in Fig. 6-2 and its impulse response $h(t)$ is given by

$$h(t) = \frac{1}{\pi} \int_0^\infty A(\omega) \cos \omega(t - t_0) \, d\omega \tag{6-9}$$

It is symmetrical about t_0,

$$h(t_0 + t) = h(t_0 - t) \tag{6-10}$$

and its maximum is reached at $t = t_0$,

$$h_{\max} = h(t_0) = \frac{1}{\pi} \int_0^\infty A(\omega) \, d\omega \tag{6-11}$$

[see (4-53) and (4-54)]. The step response $a(t)$ is obtained from (6-3)

$$a(t) = \frac{A(0)}{2} + \frac{1}{\pi} \int_0^\infty \frac{A(\omega) \sin \omega(t - t_0)}{\omega} \, d\omega \tag{6-12}$$

and it satisfies

$$a(\infty) = \int_{-\infty}^{\infty} h(\tau)\, d\tau = A(0) \qquad a(t_0) = \frac{A(0)}{2} \tag{6-13}$$

From the symmetry of $h(t)$ it follows that $a(t)$ is symmetrical about its mid-point $a(t_0)$, its slope is maximum at $t = t_0$ and is also given by

$$a(t) = \frac{A(0)}{2} + \int_0^{t-t_0} h(t_0 + \tau)\, d\tau \tag{6-14}$$

The location of the center of symmetry

$$t_0 = \theta'(0)$$

of $h(t)$ we call *delay time*, and we define the *rise time* t_r of $a(t)$ as the time it needs to reach the final value $a(\infty)$ starting from zero and rising linearly with constant slope $h_{\max}$ (see Fig. 6-2)

$$t_r h_{\max} = a(\infty) = A(0) \tag{6-15}$$

t_r is the duration of $h(t)$ as defined in (4-58). With

$$D_\omega = \frac{1}{A_0} \int_{-\infty}^{\infty} A(\omega)\, d\omega \tag{6-16}$$

[see (4-59)], t_r is given by

$$t_r = \frac{A(0)\pi}{\displaystyle\int_0^{\infty} A(\omega)\, d\omega} = \frac{2\pi}{D_\omega} \tag{6-17}$$

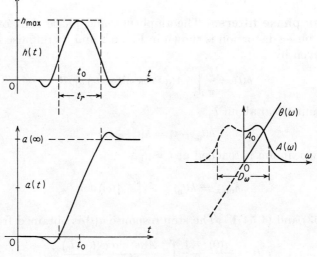

Figure 6-2

All-pass filters.   A filter with constant amplitude (Fig. 6-3)

$$|H(\omega)| = A_0 \qquad (6\text{-}18)$$

is often called all-pass, although the term is mostly used for causal systems. From Parseval's formula and (6-18) it readily follows that the energy of the output to such a filter is proportional to the energy of its input

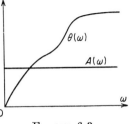

Figure 6-3

$$\int_{-\infty}^{\infty} g^2(t)\, dt = A_0^{\,2} \int_{-\infty}^{\infty} f^2(t)\, dt \qquad (6\text{-}19)$$

because

$$\int_{-\infty}^{\infty} |G(\omega)|^2\, d\omega = A_0^{\,2} \int_{-\infty}^{\infty} |F(\omega)|^2\, d\omega$$

6-2. Amplitude Distortion

We shall now develop the time properties of a number of filters with linear phase shift $\theta(\omega) = \omega t_0$.

A. Ideal low-pass filter.   A filter whose amplitude is constant for $|\omega| < \omega_c$ and zero for $|\omega| > \omega_c$ is called ideal low-pass (Fig. 6-4).

$$A(\omega) = \begin{cases} A_0 & \text{for } |\omega| < \omega_c \\ 0 & \text{for } |\omega| > \omega_c \end{cases} = A_0 p_{\omega_c}(\omega)$$

Its system function is given by

$$H(\omega) = A_0 p_{\omega_c}(\omega) e^{-j\omega t_0} \qquad (6\text{-}20)$$

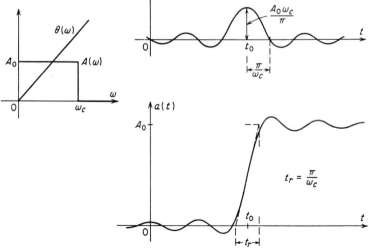

Figure 6-4

where $p_{\omega_c}(\omega)$ is a rectangular pulse, and its impulse response by

$$h(t) = \frac{A_0}{\pi} \int_0^{\omega_c} \cos \omega(t - t_0)\, d\omega = \frac{A_0 \sin \omega_c(t - t_0)}{\pi(t - t_0)} \qquad (6\text{-}21)$$

with $h_{\max} = A_0 \omega_c / \pi$ and rise time $t_r = \pi / \omega_c$. To obtain the step response $a(t)$, we use (6-21) and (6-14):

$$a(t) = \frac{A_0}{2} + \frac{A_0}{\pi} \int_0^{t-t_0} \frac{\sin \omega_c \tau}{\tau}\, d\tau = \frac{A_0}{2}\left\{ 1 + \frac{2}{\pi} Si[\omega_c(t - t_0)]\right\} \qquad (6\text{-}22)$$

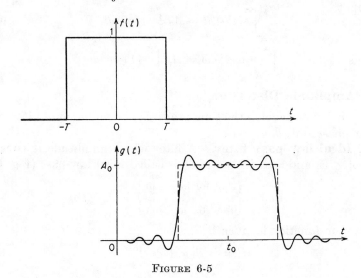

FIGURE 6-5

where $Si(x)$ is the sine integral.   The functions $h(t)$ and $a(t)$ are shown in Fig. 6-4.   If the input is a rectangular pulse (Fig. 6-5)

$$f(t) = p_T(t) = U(t + T) - U(t - T)$$

then the output is given by

$$g(t) = a(t + T) - a(t - T)$$

as shown in Fig. 6-5.   The response to an arbitrary input can be found from (5-10) or (5-23).

We shall denote by $H_i(\omega)$ the system function of an ideal low-pass filter whose amplitude equals one in the bandpass

$$H_i(\omega) = p_{\omega_c}(\omega) e^{-j\omega t_0} \qquad (6\text{-}23)$$

and by $g_i(t)$ its response to an arbitrary input $f(t)$:

$$g_i(t) = \frac{1}{\pi} \int_{-\infty}^{\infty} f(\tau) \frac{\sin \omega_c(t - t_0 - \tau)}{t - t_0 - \tau}\, d\tau \qquad (6\text{-}24)$$

Comment. If the input has a low-frequency spectrum in the sense

$$F(\omega) = 0 \qquad \text{for } |\omega| > \omega_c \qquad (6\text{-}25)$$

then the transform $G_i(\omega)$ of the output is given by

$$G_i(\omega) = F(\omega)H(\omega) = F(\omega)e^{-j\omega t_0}$$

Hence

$$g_i(t) = f(t - t_0) \qquad (6\text{-}26)$$

i.e., the filter acts as an ideal delay line to inputs satisfying (6-25).

We shall next determine the response $g(t)$ to a periodic input

$$f(t) = \sum_{n=-\infty}^{\infty} a_n e^{jn\omega_0 t}$$

We observe that the response to $e^{jn\omega_0 t}$ equals $A_0 e^{jn\omega_0(t-t_0)}$ if $n\omega_0 < \omega_c$,

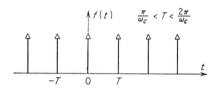

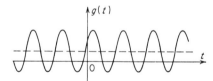

FIGURE 6-6

and is zero if $n\omega_0 > \omega_c$; hence

$$g(t) = A_0 \sum_{n=-N}^{N} a_n e^{jn\omega_0(t-t_0)} \qquad (6\text{-}27)$$

where N is the largest harmonic contained in the passband $|\omega| < \omega_c$

$$N\omega_0 < \omega_c < (N+1)\omega_0$$

With $f_N(t)$ the partial sum of the Fourier series expansion of $f(t)$, we conclude from (6-27) and (3-54) that

$$g(t) = A_0 f_N(t - t_0)$$

If the input $f(t)$ is periodic for $t > 0$ and equals zero for $t < 0$, then the response $g(t)$ is not given by (6-27); however, it tends to the sum in (6-27) as t tends to infinity (steady state).

As an example of (6-27) we shall determine the response $g(t)$ to a train of pulses (Fig. 6-6)

$$f(t) = s_T(t) = \frac{1}{T} \sum_{n=-\infty}^{\infty} e^{jn\omega_0 t} \qquad \omega_0 = \frac{2\pi}{T}$$

[see (3-41) and (3-44)].   Assuming $\omega_0 < \omega_c < 2\omega_0$, we have $N = 1$; hence

$$g(t) = \frac{A_0}{T}[e^{-j\omega_0(t-t_0)} + 1 + e^{j\omega_0(t-t_0)}] = A_0 \frac{1 + 2\cos\omega_0(t - t_0)}{T}$$

as in Fig. 6-6.

Ideal High-pass Filter.   The frequency characteristics of an ideal high-pass filter are given by

$$A(\omega) = \begin{cases} 0 & \text{for } |\omega| < \omega_c \\ A_0 & \text{for } |\omega| > \omega_c, \theta(\omega) = \omega t_0 \end{cases} \tag{6-28}$$

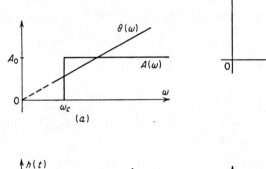

(a)

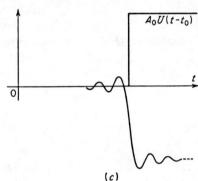

(c)

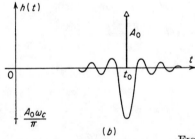

(b)

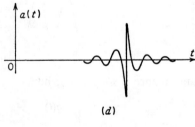

(d)

FIGURE 6-7

(see Fig. 6-7a).   Its system function $H(\omega)$ can be written in the form

$$H(\omega) = [A_0 - A_0 p_{\omega_c}(\omega)]e^{-j\omega t_0} \tag{6-29}$$

as we can easily see from (6-28).   From (6-29) and (6-21) we conclude that

$$h(t) = A_0\,\delta(t - t_0) - A_0\frac{\sin\omega_c(t - t_0)}{\pi(t - t_0)} \tag{6-30}$$

as in Fig. 6-7b.   Integrating (6-30), we obtain the step response $a(t)$ [see also (6-22)],

$$a(t) = A_0 U(t - t_0) - \frac{A_0}{2}\left\{1 + \frac{2}{\pi}Si[\omega_c(t - t_0)]\right\} \tag{6-31}$$

shown in Fig. 6-7d.   It equals the difference of $A_0 U(t - t_0)$ and the step response (6-22) of the ideal low-pass filter (Fig. 6-7c).

B. Cosine filter. The amplitude of the filter of Fig. 6-8 varies sinusoidally in the $(-\omega_c,\omega_c)$ interval and equals zero outside this interval

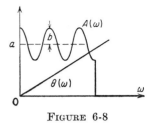

$$A(\omega) = \begin{cases} a + b\cos\dfrac{n\pi}{\omega_c}\,\omega & \text{for } |\omega| < \omega_c \\[2mm] 0 & \text{for } |\omega| > \omega_c \end{cases}$$

FIGURE 6-8

Therefore its system function is given by

$$H(\omega) = \left(a + \frac{b}{2}e^{jn\pi\omega/\omega_c} + \frac{b}{2}e^{-jn\pi\omega/\omega_c}\right)e^{-j\omega t_0}p_{\omega_c}(\omega) \qquad (6\text{-}32)$$

From (6-32) and (6-20) we see that the above filter is equivalent to three ideal low-pass filters, with the same cutoff frequency ω_c and delay

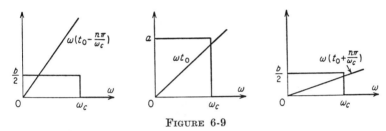

FIGURE 6-9

times t_0, $t_0 - n\pi/\omega_c$, and $t_0 + n\pi/\omega_c$ respectively (Fig. 6-9), connected in parallel. Therefore its response $g(t)$ to an arbitrary input $f(t)$ is a sum of three terms,

$$g(t) = ag_i(t) + \frac{b}{2}g_i(t - n\pi/\omega_c) + \frac{b}{2}g_i(t + n\pi/\omega_c) \qquad (6\text{-}33)$$

where $g_i(t)$ is the response of the ideal low-pass filter (6-23) to the same input. The last two terms in (6-33) are sometimes referred to as *echoes*. In Fig. 6-10 we have shown the impulse and step response of this filter. We observe that if b is small compared to a, the cosine variation of $A(\omega)$ does not affect the rise of the main portion of $a(t)$; it merely introduces an early build-up to the level $b/2$ and a delayed rise to the final value $a + b$.

C. Binomial filter. Suppose next that $A(\omega)$ is given by

$$A(\omega) = (a + b\omega^{2n})p_{\omega_c}(\omega)$$

(Fig. 6-11); then the corresponding system function $H(\omega)$ can be written in the form

$$H(\omega) = (a + b\omega^{2n})p_{\omega_c}(\omega)e^{-j\omega t_0} = aH_i(\omega) + b\omega^{2n}H_i(\omega) \qquad (6\text{-}34)$$

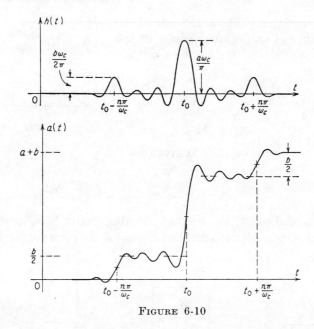

FIGURE 6-10

where $H_i(\omega)$ is the function (6-23). Denoting by $G(\omega)$ the Fourier transform of the output $g(t)$ to an input $f(t)$, we have

$$G(\omega) = F(\omega)H(\omega) = aF(\omega)H_i(\omega) + b\omega^{2n}F(\omega)H_i(\omega) \tag{6-35}$$

But the inverse of $F(\omega)H_i(\omega)$ is the function $g_i(t)$ of (6-24), and the inverse of

$$\omega^{2n}F(\omega)H_i(\omega) = (-1)^n(j\omega)^{2n}F(\omega)H_i(\omega)$$

equals $(-1)^n\, d^{2n}g_i(t)/dt^{2n}$, as we see from (2-41); therefore

$$g(t) = ag_i(t) + (-1)^nb\frac{d^{2n}g_i(t)}{dt^{2n}} \tag{6-36}$$

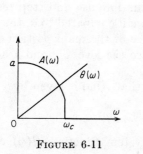

FIGURE 6-11

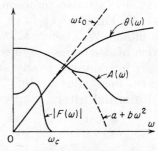

FIGURE 6-12

The above result can be used to evaluate the response of an arbitrary filter to a low-pass narrow-band signal. If $F(\omega)$ has significant values only in a "small" interval $(-\omega_c, \omega_c)$, and in this interval $A(\omega)$ can be approximated by a parabola $A(\omega) \simeq a + b\omega^2$ and $\theta(\omega)$ by a straight line $\theta(\omega) \simeq \omega t_0$ (Fig. 6-12), then $g(t)$ is given by

$$g(t) \simeq af(t - t_0) - b\frac{d^2 f(t - t_0)}{dt^2} \quad (6\text{-}37)$$

as we see from (6-36) and (6-26).

D. General low-pass filter. We shall now consider a filter

$$H(\omega) = A(\omega)e^{-j\omega t_0} \quad (6\text{-}38)$$

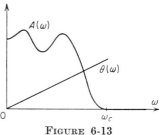

FIGURE 6-13

whose amplitude $A(\omega)$ is arbitrary in the $(-\omega_c, \omega_c)$ interval (Fig. 6-13), and will present two methods for determining its response.

Fourier-series Method. This method is based on a Fourier expansion of $A(\omega)$ and makes use of the results in filter B. We expand the even function $A(\omega)$ into a cosine series in the $(-\omega_c, \omega_c)$ interval

$$A(\omega) = \sum_{n=0}^{\infty} a_n \cos \frac{n\pi\omega}{\omega_c} \qquad |\omega| < \omega_c \quad (6\text{-}39)$$

where

$$a_n = \frac{2}{\omega_c} \int_0^{\omega_c} A(\omega) \cos \frac{n\pi\omega}{\omega_c} \, d\omega$$

and since $A(\omega) = 0$ for $|\omega| > \omega_c$, we conclude from (6-39) and (6-38) that

$$H(\omega) = a_0 H_i(\omega) + H_i(\omega) \sum_{n=1}^{\infty} \frac{a_n}{2} e^{-jn\pi\omega/\omega_c} + H_i(\omega) \sum_{n=1}^{\infty} \frac{a_n}{2} e^{jn\pi\omega/\omega_c} \quad (6\text{-}40)$$

with $H_i(\omega)$ as in (6-23). Thus our filter is equivalent to an infinite array of ideal low-pass filters (Fig. 6-14) with time delays $t_0 \pm n\pi/\omega_c$.

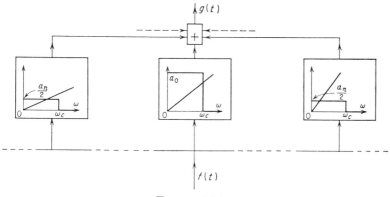

FIGURE 6-14

Its response $g(t)$ to an arbitrary input $f(t)$ can therefore be written as a sum of the function $g_i(t)$ of (6-24) and its echoes

$$g(t) = a_0 g_i(t) + \sum_{n=1}^{\infty} \frac{a_n}{2} g_i\left(t - \frac{n\pi}{\omega_c}\right) + \sum_{n=1}^{\infty} \frac{a_n}{2} g_i\left(t + \frac{n\pi}{\omega_c}\right) \qquad (6\text{-}41)$$

Power-series Method.   We expand $A(\omega)$ into a power series in the $(-\omega_c, \omega_c)$ interval

$$A(\omega) = \sum_{n=0}^{\infty} b_n \omega^{2n} \qquad |\omega| < \omega_c \qquad (6\text{-}42)$$

where
$$b_n = \frac{1}{n!} \frac{d^{2n} A(0)}{d\omega^{2n}}$$

and we obtain as in filter C

$$H(\omega) = \sum_{n=0}^{\infty} b_n \omega^{2n} H_i(\omega) \qquad (6\text{-}43)$$

Therefore

$$g(t) = b_0 g_i(t) - b_1 \frac{d^2 g_i(t)}{dt^2} + b_2 \frac{d^4 g_i(t)}{dt^4} + \cdots \qquad (6\text{-}44)$$

The above two methods are of value only if a small number of terms in (6-39) or (6-42) suffices to give a satisfactory approximation to $A(\omega)$. If the same number of terms is required in both series, then it is preferable to use (6-41) because it involves only shifting of $g_i(t)$, whereas differentiation is necessary in (6-44).

E. Gaussian filter.   The filter

$$H(\omega) = A_0 e^{-\alpha\omega^2} e^{-j\omega t_0} \qquad (6\text{-}45)$$

shown in Fig. 6-15a is called Gaussian.   To determine its impulse response $h(t)$, we use the result in (2-69)

$$h(t) = \frac{A_0}{\pi} \int_0^{\infty} e^{-\alpha\omega^2} \cos \omega(t - t_0)\, d\omega = \frac{A_0}{2\sqrt{\pi\alpha}} e^{-(t-t_0)^2/4\alpha} \qquad (6\text{-}46)$$

The maximum $h_{\max}$ of $h(t)$ and the rise time t_r [see (6-15)] are given by (Fig. 6-15b)

$$h_{\max} = A_0/2\sqrt{\pi\alpha} \qquad t_r = 2\sqrt{\pi\alpha} \qquad (6\text{-}47)$$

The step response is best obtained from the above and (6-14)

$$a(t) = \frac{A_0}{2} + \frac{A_0}{2\sqrt{\pi\alpha}} \int_0^{t-t_0} e^{-\tau^2/4\alpha}\, d\tau \qquad (6\text{-}48)$$

and can be expressed in terms of the tabulated *error function erf x* defined by

$$\text{erf } x = \frac{2}{\sqrt{\pi}} \int_0^x e^{-y^2}\, dy \qquad (6\text{-}49)$$

Inserting into (6-48), we obtain the function

$$a(t) = \frac{A_0}{2}\left(1 + erf\,\frac{t-t_0}{2\sqrt{\alpha}}\right) \qquad (6\text{-}50)$$

shown in Fig. 6-15c.   As is proved in Sec. 4-4, this filter has the property of minimizing the product of the RMS durations of $h(t)$ and $A(\omega)$.

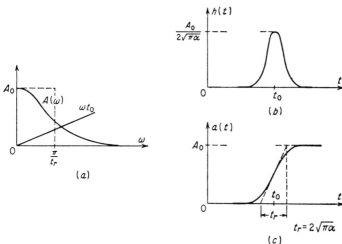

FIGURE 6-15

F. Butterworth filter with linear phase.   A filter with amplitude characteristic

$$A(\omega) = \frac{A_0}{\sqrt{1 + (\omega/\alpha)^{2n}}} \qquad (6\text{-}51)$$

is known as Butterworth.   The phase commonly associated with $A(\omega)$ is so chosen as to make the resulting system function $H(\omega)$ causal. Thus, for $n = 1$, $H(\omega)$ is given by

$$H(\omega) = \frac{A_0}{1 + j\omega/\alpha}$$

However, one can associate to the above amplitude a linear phase shift $\theta(\omega) = \omega t_0$; a noncausal system†

$$H(\omega) = \frac{A_0}{\sqrt{1 + (\omega/\alpha)^2}}\, e^{-j\omega t_0} \qquad (6\text{-}52)$$

† T. Murakami and Murlan S. Corrington, Applications of the Fourier Integral in the Analysis of Color Television Systems, *IRE-PGCT*, September, 1955.

results, without phase distortion (Fig. 6-16). The corresponding impulse response is given by

$$h(t) = \frac{A_0 \alpha}{\pi} \int_0^\infty \frac{\cos \omega(t - t_0)}{\sqrt{\alpha^2 + \omega^2}} \, d\omega \qquad (6\text{-}53)$$

It is known that the above integral can be expressed in terms of the modified Bessel function $K_0(t)$ of the second kind

$$\int_0^\infty \frac{\cos \omega t}{\sqrt{\alpha^2 + \omega^2}} \, d\omega = K_0(\alpha t) \qquad (6\text{-}54)$$

Hence

$$h(t) = \frac{A_0 \alpha}{\pi} K_0[\alpha(t - t_0)] \qquad (6\text{-}55)$$

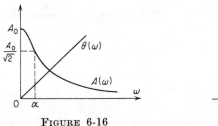

FIGURE 6-16   FIGURE 6-17

From (6-55) and (6-14) we obtain the following expression for the step response $a(t)$:

$$a(t) = \frac{A_0}{2} + \frac{A_0 \alpha}{\pi} \int_0^{t - t_0} K_0(\alpha \tau) \, d\tau \qquad (6\text{-}56)$$

shown in Fig. 6-17.

6-3. Causal Systems with Linear Phase

The systems discussed so far are not physically realizable because their impulse response $h(t)$ is not zero for negative t. We observe, however, that if the delay time t_0 is sufficiently large, then $h(t)$ is negligible for $t < 0$. We would therefore expect that with minor changes in the characteristics of the filters of Sec. 6-2, one might be able to create causal systems; this is indeed the case. To this end we form the causal function

$$h_c(t) = h(t)p_{t_0}(t - t_0) \qquad (6\text{-}57)$$

where $p_{t_0}(t - t_0)$ is a rectangular pulse shown in Fig. 6-18. Since

$$p_{t_0}(t - t_0) \leftrightarrow \frac{2 \sin \omega t_0}{\omega} e^{-j\omega t_0}$$

we conclude that the transform $H_c(\omega)$ of $h_c(t)$ is given by [see (2-74)]

$$H_c(\omega) = \frac{1}{2\pi} H(\omega) * \frac{2 \sin \omega t_0}{\omega} e^{-j\omega t_0} = \left[A(\omega) * \frac{\sin \omega t_0}{\pi \omega} \right] e^{-j\omega t_0} \qquad (6\text{-}58)$$

(The last equality can be easily established.) Thus the causal system function
$H_c(\omega)$ has a linear phase (with possible
discontinuities), and its amplitude $A_c(\omega)$
equals the convolution of $A(\omega)$ with
$(\sin \omega t_0)/\pi\omega$. In Fig. 6-19a we plotted
the spectrum $A(\omega)$ of the ideal low-pass
filter A, and in Fig. 6-19b its causal modification. If $A(\omega)$ is continuous, then,
for large enough t_0, $A_c(\omega) \simeq A(\omega)$; however, this is not true for discontinuous
functions because of the Gibbs' phenomenon. The discussion of Sec. 2-6
suggests an alternative way of creating a
causal system. With $q_{t_0}(t - t_0)$ the triangular pulse of Fig. 6-20, we form
the function

$$h_F(t) = h(t)q_{t_0}(t - t_0) \qquad (6\text{-}59)$$

Its transform is given by [see (2-74) and
(2-58)]

$$H_F(\omega) = \left[A(\omega) * \frac{2 \sin^2 (t_0\omega/2)}{\pi t_0 \omega^2} \right] e^{-j\omega t_0} \qquad (6\text{-}60)$$

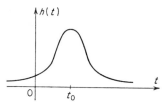

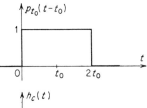

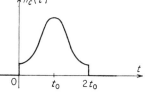

FIGURE 6-18

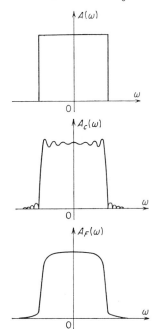

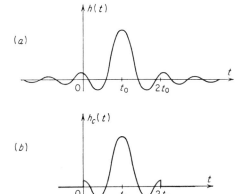

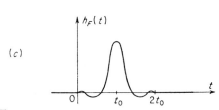

(a)

(b)

(c)

FIGURE 6-19

$H_F(\omega)$ is causal, and its amplitude $A_F(\omega)$ equals the convolution of $A(\omega)$ with the Fejér kernel. Again, $A_F(\omega) \to A(\omega)$ with $t_0 \to \infty$. The functions $h_F(t)$ and $A_F(\omega)$ are sketched in Fig. 6-19c for the ideal low-pass filter.

Paley-Wiener Condition. One might wonder whether it is possible, by allowing phase distortion, to obtain causal systems with the same amplitude characteristic as the filters of Sec. 6-2. As is shown in Chap. 10, this is in general impossible unless $A(\omega)$ satisfies the Paley-Wiener condition

$$\int_{-\infty}^{\infty} \frac{|\ln A(\omega)|}{1 + \omega^2} \, d\omega < \infty \qquad (6\text{-}61)$$

It is easy to see that, among the filters considered so far, only the Butterworth (6-51) satisfies the above condition. Thus if we associate to

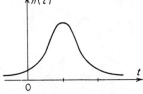

$$A(\omega) = \frac{A_0}{\sqrt{1 + (\omega/\alpha)^2}}$$

the angle $\theta(\omega) = \tan^{-1}(\omega/\alpha)$, then the system function

$$H(\omega) = \frac{A_0}{1 + (j\omega/\alpha)}$$

results, whose inverse $h(t)$, given by [see (2-61)]

$$h(t) = \alpha A_0 e^{-\alpha t} U(t)$$

is indeed causal. The corresponding step response

$$a(t) = A_0(1 - e^{-\alpha t})U(t)$$

is plotted in Fig. 6-21; in the same figure the step response (6-56) of the noncausal filter (6-52) with the same amplitude is also shown.

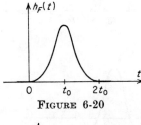

FIGURE 6-20

6-4. Phase Distortion

We now abandon the assumption of linearity of the phase shift $\theta(\omega)$ and assume a general system function

$$H(\omega) = A(\omega)e^{-j\theta(\omega)}$$

To separate the effects of amplitude and phase distortion on the time properties of our system, we introduce the function

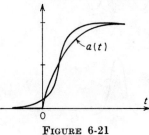

FIGURE 6-21

$$H_0(\omega) = A(\omega)e^{-j\omega t_0} \qquad (6\text{-}62)$$

having the same amplitude $A(\omega)$ and a linear phase, with conveniently chosen slope t_0. We shall compare the impulse response

$$h(t) = \frac{1}{\pi} \int_0^\infty A(\omega) \cos[\omega t - \theta(\omega)]\, d\omega$$

of our system to the inverse transform

$$h_0(t) \longleftrightarrow A(\omega)e^{-j\omega t_0} \qquad (6\text{-}63)$$

of the system (6-62). In Fig. 6-22a we have shown for future reference

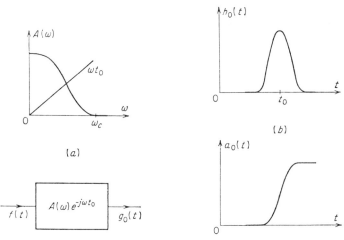

FIGURE 6-22

the system function $H_0(\omega)$ and in Fig. 6-22b its impulse response $h_0(t)$ and step response $a_0(t)$. We remark that, since

$$H(\omega) = H_0(\omega)e^{-j[\theta(\omega) - \omega t_0]} \qquad (6\text{-}64)$$

the phase distortion is due to the presence of the term

$$\Delta\theta(\omega) = \theta(\omega) - \omega t_0$$

in (6-64).

Ideal all-pass filter. As a first example, we shall consider a filter without amplitude distortion and with a phase shift that is linear for $|\omega| < \omega_c$ and constant for $|\omega| > \omega_c$ (Fig. 6-23a):

$$\theta(\omega) = \begin{cases} t_0\omega & \text{for } |\omega| < \omega_c \\ t_0\omega_c & \text{for } \omega > \omega_c \end{cases} \qquad (6\text{-}65)$$

and for simplicity we shall assume

$$t_0 = \frac{n\pi}{\omega_c}$$

[For an n-pole rational all-pass, $\theta(\infty)$ is indeed $n\pi$.]   It can be easily seen from Fig. 6-23a that $H(\omega)$ is given by

$$H(\omega) = A_0 p_{\omega_c}(\omega)e^{-j\omega t_0} + A_0[1 - p_{\omega_c}(\omega)]e^{-jn\pi} \tag{6-66}$$

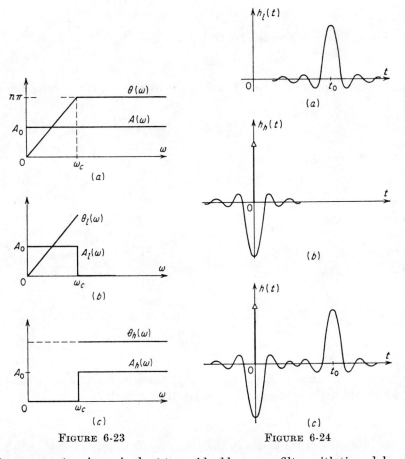

FIGURE 6-23   FIGURE 6-24

Thus our system is equivalent to an ideal low-pass filter with time delay t_0 (Fig. 6-23b)

$$H_l(\omega) = A_0 p_{\omega_c}(\omega)e^{-j\omega t_0} \tag{6-67}$$

and an ideal high-pass filter with zero delay (Fig. 6-23c)

$$H_h(\omega) = (-1)^n A_0[1 - p_{\omega_c}(\omega)] \tag{6-68}$$

With $h_l(t)$ and $h_h(t)$ the inverse transforms of $H_l(\omega)$ and $H_h(\omega)$, we obtain from (6-21) and (6-30)

$$h_l(t) = \frac{A_0 \sin \omega_c(t - t_0)}{\pi(t - t_0)} \tag{6-69}$$

$$h_h(t) = (-1)^n A_0\, \delta(t) - (-1)^n \frac{A_0 \sin \omega_c t}{\pi t} \tag{6-70}$$

The functions $h_l(t)$ and $h_h(t)$ are shown in Fig. 6-24a and b for n even. Since

$$H(\omega) = H_l(\omega) + H_h(\omega)$$

we conclude that the impulse response $h(t)$ of our filter is given by

$$h(t) = h_l(t) + h_h(t)$$

(Fig. 6-24c) and its step response $a(t)$ by

$$a(t) = \frac{A_0}{2}\left\{1 + \frac{2}{\pi}Si[\omega_c(t - t_0)]\right\} + A_0 U(t) - \frac{A_0}{2}\left[1 + \frac{2}{\pi}Si(\omega_c t)\right] \quad (6\text{-}71)$$

[see (6-31) and (6-22)]. This function is shown in Fig. 6-25; in the same figure we have also shown the step response $AU(t - t_0)$ of a delay line with delay equal to t_0.

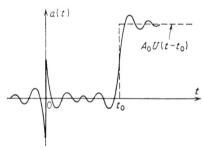

FIGURE 6-25

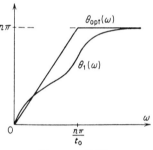

FIGURE 6-26

Comment. The above filter has the following optimal property: consider an arbitrary all-pass system

$$H_1(\omega) = A_0 e^{-j\theta_1(\omega)}$$

such that $\theta_1(\omega)$ increases monotonically to the final value $n\pi$ as in Fig. 6-26. [This is the case, for example, if $H_1(\omega)$ is rational.] We denote by $a_1(\omega)$ its step response. With t_0 a given constant, we want the conditions on $\theta_1(\omega)$ so that the RMS error

$$\int_{-\infty}^{\infty} [A_0 U(t - t_0) - a_1(t)]^2\, dt$$

between the delayed step $A_0 U(t - t_0)$ and $a_1(t)$ will be minimum. The answer is that $\theta_1(\omega)$ should equal the angle $\theta(\omega)$ of (6-65). (For the proof see Prob. 24.)

SMALL PHASE DISTORTION. In the next three examples of low-pass filters, we shall assume that the phase distortion $\Delta\theta(\omega)$ is small compared to one, in the passband $|\omega| < \omega_c$ (Fig. 6-27):

$$\Delta\theta(\omega) = \theta(\omega) - \omega t_0 \ll 1 \qquad |\omega| < \omega_c \qquad (6\text{-}72)$$

Since $e^{-j\,\Delta\theta(\omega)} \simeq 1 - j\,\Delta\theta(\omega)$, we conclude that

$$H(\omega) = A(\omega)e^{-j\theta(\omega)} \simeq A(\omega)e^{-j\omega t_0}[1 - j\Delta\theta(\omega)] \qquad (6\text{-}73)$$

and with $H_0(\omega)$ as in (6-62)

$$H(\omega) \simeq H_0(\omega)[1 - j\Delta\theta(\omega)] \qquad (6\text{-}74)$$

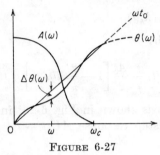

FIGURE 6-27

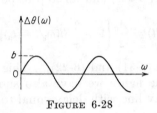

FIGURE 6-28

Sine variation of phase distortion.   If $\Delta\theta(\omega)$ is given by the sine curve of Fig. 6-28

$$\Delta\theta(\omega) = b \sin \frac{n\pi\omega}{\omega_c} \qquad b \ll 1 \qquad (6\text{-}75)$$

then, since

$$j\Delta\theta(\omega) = \frac{b}{2}\left(e^{jn\pi\omega/\omega_c} - e^{-jn\pi\omega/\omega_c}\right)$$

we obtain from (6-73)

$$H(\omega) \simeq A(\omega)e^{-j\omega t_0}\left(1 - \frac{b}{2}e^{jn\pi\omega/\omega_c} + \frac{b}{2}e^{-jn\pi\omega/\omega_c}\right) \qquad (6\text{-}76)$$

With $h_0(t)$ the impulse response (6-63) of the filter without phase distortion (Fig. 6-22b), we see from (6-76) that

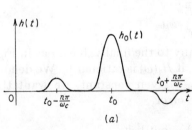

(a)

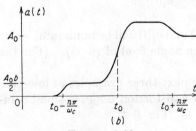

(b)

FIGURE 6-29

$$h(t) \simeq h_0(t) - \frac{b}{2} h_0\left(t + \frac{n\pi}{\omega_c}\right) + \frac{b}{2} h_0\left(t - \frac{n\pi}{\omega_c}\right) \qquad (6\text{-}77)$$

The function $h(t)$ is shown in Fig. 6-29a. Denoting by $g_0(t)$ the response of $H_0(\omega)$ to an arbitrary input $f(t)$ (Fig. 6-22a), we see from (6-76) that the response of our system $H(\omega)$ to $f(t)$ is given by

$$g(t) \simeq g_0(t) - \frac{b}{2} g_0\left(t + \frac{n\pi}{\omega_c}\right) + \frac{b}{2} g_0\left(t - \frac{n\pi}{\omega_c}\right) \qquad (6\text{-}78)$$

In Fig. 6-29b we have shown the step response $a(t)$.

Cosine variation of phase delay. In this example we shall assume that the quantity $\theta(\omega)/\omega$, known as phase delay (see Sec. 7-5), varies as in Fig. 6-30a

$$\frac{\theta(\omega)}{\omega} = t_0 + b \cos \frac{n\pi\omega}{\omega_c} \tag{6-79}$$

i.e., that $\theta(\omega) = \omega t_0 + b\omega \cos (n\pi\omega/\omega_c)$ for $|\omega| < \omega_c$ (Fig. 6-30b). The condition for small distortion is satisfied if

$$b\omega_c \ll 1$$

Since

$$j\Delta\theta(\omega) = j\omega b \cos \frac{n\pi\omega}{\omega_c} = \frac{j\omega b}{2} (e^{jn\pi\omega/\omega_c} + e^{-jn\pi\omega/\omega_c}) \tag{6-80}$$

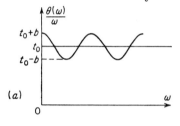

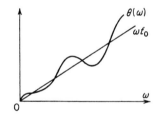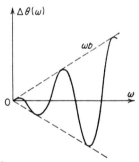

FIGURE 6-30

we see from (6-74) that

$$H(\omega) \simeq H_0(\omega)\left(1 - \frac{j\omega b}{2} e^{jn\pi\omega/\omega_c} - \frac{j\omega b}{2} e^{-jn\pi\omega/\omega_c}\right) \tag{6-81}$$

But the inverse of $j\omega H_0(\omega)$ equals $dh_0(t)/dt$ [see (2-41)]; therefore

$$h(t) \simeq h_0(t) - \frac{b}{2}\frac{dh_0[t - (n\pi/\omega_c)]}{dt} - \frac{b}{2}\frac{dh_0[t + (n\pi/\omega_c)]}{dt} \tag{6-82}$$

This function is shown in Fig. 6-31a. The response $g(t)$ to an arbitrary input $f(t)$ is given by

$$g(t) \simeq g_0(t) - \frac{b}{2}\frac{dg_0[t - (n\pi/\omega_c)]}{dt} - \frac{b}{2}\frac{dg_0[t + (n\pi/\omega_c)]}{dt} \tag{6-83}$$

where again $g_0(t)$ is the response of the same filter, without phase distortion, to $f(t)$. The step response $a(t)$ is shown in Fig. 6-31b.

Arbitrary phase distortion. We shall now consider a low-pass filter with small but arbitrary phase distortion and will present two methods for determining its response. These methods are based on the results of the last two cases.

Fourier Expansion of Phase Distortion. We expand the odd function $\Delta\theta(\omega)$ into a Fourier sine series in the $(-\omega_c,\omega_c)$ interval

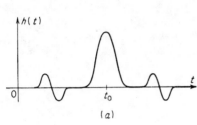

(a)

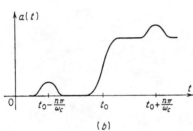

(b)

FIGURE 6-31

$$\Delta\theta(\omega) = \sum_{n=1}^{\infty} b_n \sin \frac{n\pi\omega}{\omega_c}$$

$$b_n = \frac{2}{\omega_c} \int_0^{\omega_c} \Delta\theta(\omega) \sin \frac{n\pi\omega}{\omega_c}\, d\omega \quad (6\text{-}84)$$

Inserting the above into (6-74), we obtain

$$H(\omega) \simeq H_0(\omega)\left(1 - \tfrac{1}{2}\sum_{n=1}^{\infty} b_n e^{j\,n\pi\omega/\omega_c}\right.$$

$$\left. + \tfrac{1}{2}\sum_{n=1}^{\infty} b_n e^{-j\,n\pi\omega/\omega_c}\right) \quad (6\text{-}85)$$

Therefore the response $g(t)$ to an arbitrary input $f(t)$ is given by

$$g(t) \simeq g_0(t) - \tfrac{1}{2}\sum_{n=1}^{\infty} b_n g_0\left(t + \frac{n\pi}{\omega_c}\right)$$

$$+ \tfrac{1}{2}\sum_{n=1}^{\infty} b_n g_0\left(t - \frac{n\pi}{\omega_c}\right) \quad (6\text{-}86)$$

Fourier Expansion of Phase Delay. We expand the even function $\theta(\omega)/\omega$ into a cosine series in the $(-\omega_c,\omega_c)$ interval:

$$\frac{\theta(\omega)}{\omega} = t_0 + \sum_{n=1}^{\infty} b_n \cos \frac{n\pi\omega}{\omega_c} \qquad b_n = \frac{2}{\omega_c} \int_0^{\omega_c} \frac{\theta(\omega)}{\omega} \cos \frac{n\pi\omega}{\omega_c}\, d\omega \quad (6\text{-}87)$$

Thus
$$\Delta\theta(\omega) = \omega \sum_{n=1}^{\infty} b_n \cos \frac{n\pi\omega}{\omega_c}$$

Therefore

$$H(\omega) \simeq H_0(\omega)\left(1 - \frac{j\omega}{2}\sum_{n=1}^{\infty} b_n e^{j\,n\pi\omega/\omega_c} - \frac{j\omega}{2}\sum_{n=1}^{\infty} b_n e^{-j\,n\pi\omega/\omega_c}\right) \quad (6\text{-}88)$$

Reasoning as in (6-83), we obtain

$$g(t) \simeq g_0(t) - \tfrac{1}{2}\sum_{n=1}^{\infty} b_n \frac{dg_0[t + (n\pi/\omega_c)]}{dt} - \tfrac{1}{2}\sum_{n=1}^{\infty} b_n \frac{dg_0[t - (n\pi/\omega_c)]}{dt} \quad (6\text{-}89)$$

In most cases the second method is used because (6-89) converges faster than (6-86).

LARGE PHASE DISTORTION. If the phase distortion $\Delta\theta(\omega)$ is large compared to one, then the above results cannot be used because the approximation (6-73) is no longer valid. To determine the time response, one could use the numerical techniques of Chap. 4. However, a number of special techniques merit some consideration.

Sine variation of phase distortion. Suppose that $\theta(\omega)$ varies sinusoidally with ω:

$$\theta(\omega) = \omega t_0 - b \sin \frac{\pi\omega}{\omega_c}$$

but that b is not small compared to one. To find $g(t)$, we expand the periodic function

$$e^{-j\,\Delta\theta} = e^{jb\,\sin\,(\pi\omega/\omega_c)}$$

into a Fourier exponential series in the $(-\omega_c, \omega_c)$ interval. The coefficients of this series are given by

$$\alpha_n = \frac{1}{2\omega_c} \int_{-\omega_c}^{\omega_c} e^{jb\,\sin\,(\pi\omega/\omega_c)} e^{-jn\pi\omega/\omega_c}\,d\omega = \frac{1}{2\pi} \int_{-\pi}^{\pi} e^{j(b\,\sin\,x - nx)}\,dx \quad (6\text{-}90)$$

It is well known that the last integral equals the Bessel function $J_n(b)$ of the first kind (Lommel's formula),

$$J_n(b) = \frac{1}{2\pi} \int_{-\pi}^{\pi} e^{j(b\,\sin\,x - nx)}\,dx \quad (6\text{-}91)$$

Therefore

$$e^{jb\,\sin\,(\pi\omega/\omega_c)} = J_0(b) + J_1(b)e^{j\pi\omega/\omega_c} + J_{-1}(b)e^{-j\pi\omega/\omega_c} + \cdots \quad (6\text{-}92)$$

With $H_0(\omega)$ as in (6-62), we obtain from (6-92) the following expression for our system function:

$$H(\omega) = H_0(\omega)[J_0(b) + J_1(b)e^{j\pi\omega/\omega_c} + J_{-1}(b)e^{-j\pi\omega/\omega_c} + \cdots] \quad (6\text{-}93)$$

Therefore the response to $f(t)$ is given by

$$g(t) = J_0(b)g_0(t) + J_1(b)g_0\left(t + \frac{\pi}{\omega_c}\right) + J_{-1}(b)g_0\left(t - \frac{\pi}{\omega_c}\right) + \cdots \quad (6\text{-}94)$$

where $g_0(t)$ is the response of the system

$$H_0(\omega) = A(\omega)^{-j\omega t_0}$$

to the same input. It is easily seen that (6-78) is a special case of (6-94).

Polygonal approximation of $\theta(\omega)$. We shall now consider a low-pass filter with constant amplitude $A(\omega) = A_0$ for $|\omega| < \omega_c$ and arbitrary phase $\theta(\omega)$ (Fig. 6-32a), and will determine its impulse response by a polygonal approximation of $\theta(\omega)$. We divide the passband into n intervals (ω_k, ω_{k+1}) and approximate $\theta(\omega)$ by a straight line

$$\theta(\omega) \simeq t_k\omega + b_k \qquad \omega_k < \omega < \omega_{k+1} \quad (6\text{-}95)$$

in each interval. From (6-2) and (6-95) we obtain

$$h(t) = \frac{A_0}{\pi} \int_0^\infty \cos\left[\omega t - \theta(\omega)\right] d\omega = \frac{A_0}{\pi} \sum_{k=0}^{n-1} \int_{\omega_k}^{\omega_{k+1}} \cos\left(\omega t - t_k \omega - b_k\right) d\omega$$

$$= \frac{2A_0}{\pi} \sum_{k=0}^{n-1} \frac{\sin \dfrac{\Delta\omega_k}{2}(t - t_k)}{t - t_k} \sin\left[\bar\omega_k(t - t_k) - b_k\right] \tag{6-96}$$

where $\Delta\omega_k = \omega_{k+1} - \omega_k$ is the length of the kth interval, and

$$\bar\omega_k = (\omega_k + \omega_{k+1})/2$$

its mid-frequency. In Fig. 6-32b we have shown a typical term of the above sum.

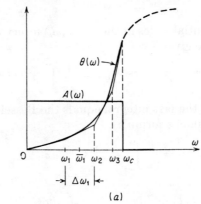

(a)

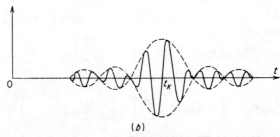

(b)

FIGURE 6-32

Pulse-forming filter. The phase of the filters discussed so far was continuous. A simple example of a filter with discontinuous phase is shown in Fig. 6-33. Its impulse response is a rectangular pulse

$$h(t) = p_T(t - t_0) \tag{6-97}$$

and the corresponding system function is given by [see (2-56)]

$$H(\omega) = \frac{2 \sin \omega T}{\omega} e^{-j\omega t_0} \tag{6-98}$$

It has a linear phase for $\omega \neq n\pi/T$ and, because of the change in the sign of $\sin \omega T$, it jumps by π at the points $\omega = n\pi/T$. The step response is a ramp as in Fig. 6-33.

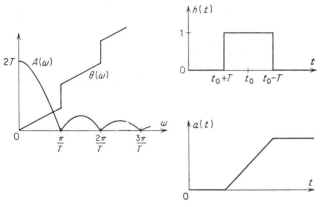

FIGURE 6-33

Rational system functions. We shall now discuss, for later reference, two simple filters with rational system functions. With

$$\left.\begin{array}{r} p_1 \\ p_2 \end{array}\right\} = -\alpha \pm j\beta$$

two complex numbers (see Fig. 6-34), we form the function

$$H(\omega) = \frac{(j\omega + p_1)(j\omega + p_2)}{(j\omega - p_1)(j\omega - p_2)} \tag{6-99}$$

It can easily be seen from Fig. 6-34 that

$$|H(\omega)| = \frac{(OA)(OB)}{(OC)(OD)} = 1 \qquad \theta(\omega) = 2\gamma + 2\delta$$

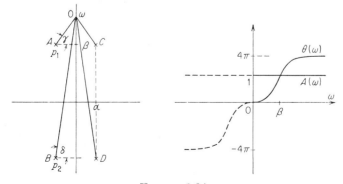

FIGURE 6-34

Hence the filter is all-pass and its phase shift $\theta(\omega)$ increases monotonically.   From (6-99) we obtain with well-known techniques (see also Chap. 9)

$$h(t) = \delta(t) + \frac{4\alpha\sqrt{\alpha^2 + \beta^2}}{\beta}\, U(t)e^{-\alpha t}\sin\left(\beta t - \tan^{-1}\frac{\beta}{\alpha}\right) \qquad (6\text{-}100)$$

and

$$a(t) = U(t)\left(1 - \frac{4\alpha}{\beta}\, e^{-\alpha t}\sin\beta t\right) \qquad (6\text{-}101)$$

The functions $h(t)$ and $a(t)$ are shown in Fig. 6-35.

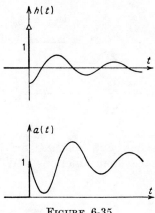

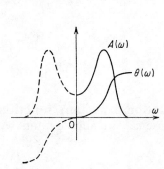

FIGURE 6-35

FIGURE 6-36

Consider next the function

$$H(\omega) = \frac{1}{(j\omega - p_1)(j\omega - p_2)} \qquad (6\text{-}102)$$

with p_1 and p_2 as above.   We easily see that

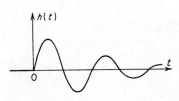

$$A(\omega) = \frac{1}{\sqrt{(\alpha^2 + \beta^2 - \omega^2)^2 + 4\alpha^2\,\omega^2}}$$

$$\theta(\omega) = \tan^{-1}\frac{2\alpha\omega}{\alpha^2 + \beta^2 - \omega^2} \qquad (6\text{-}103)$$

as in Fig. 6-36.   The impulse response of the above filter is given by (Fig. 6-37)

$$h(t) = \frac{U(t)}{\beta}\, e^{-\alpha t}\sin\beta t \qquad (6\text{-}104)$$

and its step response by

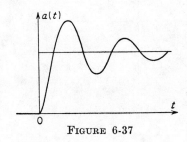

FIGURE 6-37

$$a(t) = U(t)\left[\frac{1}{\alpha^2 + \beta^2} + \frac{1}{\beta\sqrt{\alpha^2 + \beta^2}}\right.$$

$$\left. \times\, e^{-\alpha t}\sin\left(\beta t - \tan^{-1}\frac{\beta}{\alpha}\right)\right] \qquad (6\text{-}105)$$

6-5. Summary

To summarize the previous results, we shall consider the effect of the various parts of a system function

$$H(\omega) = A(\omega)e^{-j\,\Delta\theta(\omega)}e^{-j\omega t_0}$$

on the shape of an input signal $f(t)$ equal to a rectangular pulse.   We shall represent the system as a cascade of three sections with characteristics $A(\omega)$, $e^{-j\,\Delta\theta(\omega)}$, and $e^{-j\omega t_0}$ respectively (Fig. 6-38) and shall

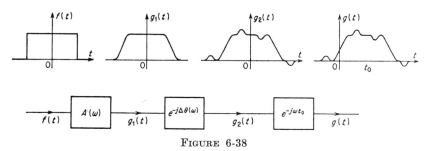

FIGURE 6-38

examine the deformation of $f(t)$ as it passes through each section. The amplitude distortion of the first section causes smoothing of the discontinuities of $f(t)$ and a certain widening.   The phase distortion of the second section introduces advance and retarded signals and a distortion of the main part of the pulse.   Finally, the last section causes a pure delay.

Chapter 7. Bandpass Filters

A bandpass filter is a system whose amplitude characteristic $A(\omega)$ has significant values only in an interval not containing the origin (Fig. 7-1a).   The response $g(t)$ to an arbitrary input $f(t)$ is determined from (5-23).   In a number of important cases, however, special techniques lead to simple results.   It will be shown that the impulse response $h(t)$ of the system is a modulated signal that can be easily found from the response of an appropriately chosen low-pass filter; this result is derived from the low-pass bandpass transformation theorem.   It is further shown that if the spectrum of the input is constant in the bandpass, then $g(t)$ is of the same form as $h(t)$; this is approximately true for arbitrary inputs, provided the band of the filter is sufficiently narrow. The response to signals of the form $f(t) = f_i(t) \cos \omega_0 t$, where ω_0 is in the band of the filter and $f_i(t)$ has a low frequency spectrum, is derived from the response of a low-pass filter to the envelope $f_i(t)$ of $f(t)$.

In Sec. 7-6 we apply the filter ideas to propagation phenomena and discuss the concepts of phase, group, and signal-front velocities, and in Sec. 7-7 we present the stationary-phase method for determining the asymptotic form of these signals.

7-1. Symmetrical Systems

It will be convenient in this analysis to express the system function $H(\omega)$ of a filter as a sum

$$H(\omega) = H_1(\omega) + H_2(\omega) \tag{7-1}$$

of the functions

$$H_1(\omega) = H(\omega)U(\omega) \qquad H_2(\omega) = H(\omega)U(-\omega) \tag{7-2}$$

shown in Fig. 7-1b.   Since $h(t)$ is real, we have [see (2-16)] $\overset{*}{H}(-\omega) = H(\omega)$; therefore

$$H_1(-\omega) = \overset{*}{H}_2(\omega) \tag{7-3}$$

as we can easily see from the definition of these functions.

A bandpass system is called *symmetrical* if

$$H_1(\omega + \omega_0) = \overset{*}{H}_1(\omega_0 - \omega) \qquad (7\text{-}4)$$

or, equivalently, if

$$H_2(\omega - \omega_0) = \overset{*}{H}_2(-\omega_0 - \omega) \qquad (7\text{-}5)$$

where ω_0 is its center frequency.   Thus, with

$$H_1(\omega) = A_1(\omega)e^{-j\theta_1(\omega)} \qquad H_2(\omega) = A_2(\omega)e^{-j\theta_2(\omega)} \qquad (7\text{-}6)$$

we have for a symmetrical filter

$$A_1(\omega + \omega_0) = A_1(\omega_0 - \omega) \qquad \theta_1(\omega + \omega_0) = -\theta_1(\omega_0 - \omega) \qquad (7\text{-}7)$$

i.e., $A_1(\omega)$ is even and $\theta_1(\omega)$ is odd about ω_0.   Similarly

$$A_2(\omega - \omega_0) = A_2(-\omega_0 - \omega) \qquad \theta_2(\omega - \omega_0) = -\theta_2(-\omega_0 - \omega) \qquad (7\text{-}8)$$

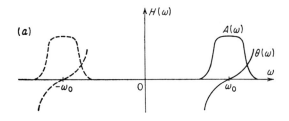

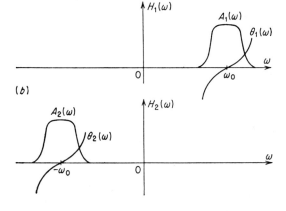

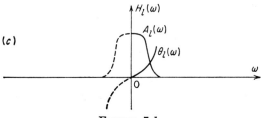

FIGURE 7-1

Equivalent low-pass filter.   We introduce the function $H_l(\omega)$ of Fig. 7-1c obtained by shifting $H_1(\omega)$ to the left by ω_0:

$$H_l(\omega) = H_1(\omega + \omega_0) \qquad H_l(\omega - \omega_0) = H_1(\omega) \qquad (7\text{-}9)$$

and call the corresponding system the *equivalent low-pass filter*. Clearly $H_l(\omega)$ can also be obtained by shifting $H_2(\omega)$ to the right:

$$H_l(\omega) = H_2(\omega - \omega_0) \qquad (7\text{-}10)$$

From the symmetry of our system it follows that

$$H_l(-\omega) = \overset{*}{H_l}(\omega) \qquad (7\text{-}11)$$

i.e., $\qquad\qquad A_l(-\omega) = A_l(\omega) \qquad \theta_l(-\omega) = -\theta_l(\omega) \qquad (7\text{-}12)$

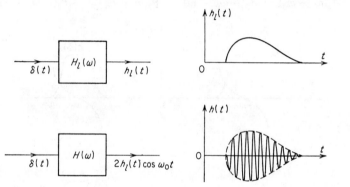

FIGURE 7-2

Therefore, the inverse transform $h_l(t)$ of $H_l(\omega)$

$$h_l(t) \leftrightarrow H_l(\omega) \qquad (7\text{-}13)$$

is a real function.

THEOREM.   The impulse response $h(t)$ of a symmetrical bandpass filter, of center frequency ω_0, is an amplitude-modulated signal

$$h(t) = 2h_l(t) \cos \omega_0 t \qquad (7\text{-}14)$$

with carrier frequency ω_0 and envelope twice the impulse response of the equivalent low-pass filter (Fig. 7-2).

Proof.   Using the shifting theorem (2-37), we obtain from (7-9) and (7-10)

$$h_l(t)e^{j\omega_0 t} \leftrightarrow H_1(\omega)$$
$$h_l(t)e^{-j\omega_0 t} \leftrightarrow H_2(\omega)$$

Hence $\qquad h_l(t)e^{j\omega_0 t} + h_l(t)e^{-j\omega_0 t} \leftrightarrow H_1(\omega) + H_2(\omega)$

and (7-14) follows, because the inverse of $H_1(\omega) + H_2(\omega) = H(\omega)$ is $h(t)$.

Comment. The above result (7-14) can be obtained more directly from

$$h(t) = \text{Re} \frac{1}{\pi} \int_0^\infty H(\omega)e^{j\omega t}\,d\omega$$

[see (2-17)]. Indeed, it follows from the symmetry condition (7-4) that $H(\omega) = 0$ for $\omega > 2\omega_0$; hence, with $\omega - \omega_0 = y$,

$$h(t) = \text{Re} \frac{1}{\pi} e^{j\omega_0 t} \int_{-\omega_0}^{\omega_0} H(\omega_0 + y)e^{jyt}\,dy$$

$$= \text{Re} \frac{1}{\pi} e^{j\omega_0 t} \int_{-\omega_0}^{\omega_0} H_l(y)\,e^{jyt}\,dy$$

and (7-14) follows because the last integral equals the real quantity $2\pi h_l(t)$.

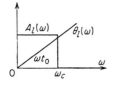 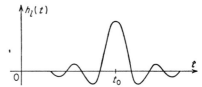

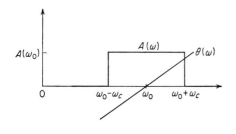 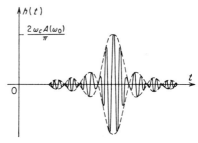

FIGURE 7-3

From the above theorem we can readily find the impulse responses of systems with equivalent low-pass filters as in Chap. 6. Thus, for the ideal bandpass system of Fig. 7-3, of center frequency ω_0 and bandwidth $2\omega_c$, we have

$$h(t) = \frac{2A(\omega_0)}{\pi} \frac{\sin \omega_c(t - t_0)}{t - t_0} \cos \omega_0 t \qquad (7\text{-}15)$$

since [see (6-21)]

$$h_l(t) = \frac{A_l(0)}{\pi} \frac{\sin \omega_c(t - t_0)}{t - t_0}$$

and $A_l(0) = A(\omega_0)$.

The symmetry condition, as formulated in (7-4), is seldom fulfilled by actual systems because of the requirement $\theta_1(\omega_0) = 0$ implicit in (7-7). A more realistic condition is to assume that not $\theta_1(\omega)$, but $\theta_1(\omega) - \theta_1(\omega_0)$ is odd about ω_0

$$\theta_1(\omega + \omega_0) - \theta_1(\omega_0) = \theta_1(\omega_0) - \theta_1(\omega_0 - \omega) \qquad (7\text{-}16)$$

as in Fig. 7-4. This condition can also be stated in the form

$$H_1(\omega + \omega_0)e^{j\theta(\omega_0)} = \overset{*}{H}_1(\omega_0 - \omega)e^{-j\theta(\omega_0)} \qquad (7\text{-}17)$$

and is approximately true for many narrow-band filters. In the following, a filter will be called symmetrical if it satisfies (7-17). We define the equivalent low-pass filter $H_l(\omega)$ of our system by

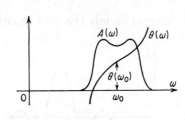

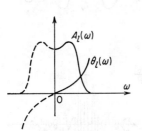

$$H_l(\omega) = H_1(\omega + \omega_0)e^{j\theta(\omega_0)}$$
$$= H_2(\omega - \omega_0)e^{-j\theta(\omega_0)} \qquad (7\text{-}18)$$

and we can easily see that theorem (7-14) still holds with a minor modification. Indeed from (7-17) and (7-18) we conclude that

$$H_l(-\omega) = \overset{*}{H}_l(\omega) \qquad (7\text{-}19)$$

Therefore $H_l(\omega)$ has a real inverse $h_l(t)$

$$h_l(t) \leftrightarrow H_l(\omega)$$

and

$$e^{-j\theta(\omega_0)}h_l(t)e^{j\omega_0 t} \leftrightarrow H_1(\omega)$$
$$e^{j\theta(\omega_0)}h_l(t)e^{-j\omega_0 t} \leftrightarrow H_2(\omega) \qquad (7\text{-}20)$$

Adding, we obtain

$$2h_l(t) \cos[\omega_0 t - \theta(\omega_0)] \leftrightarrow$$
$$H_1(\omega) + H_2(\omega)$$

FIGURE 7-4

Therefore the impulse response $h(t)$ of the system $H_1(\omega) + H_2(\omega) = H(\omega)$ is given by

$$h(t) = 2h_l(t) \cos[\omega_0 t - \theta(\omega_0)] \qquad (7\text{-}21)$$

As an application of (7-21) we shall determine the impulse response of a double-tuned circuit. Its system function is given by†

$$H(\omega) = K \frac{(j\omega)^3}{[(j\omega + \alpha_1)^2 + \beta_1{}^2][(j\omega + \alpha_2)^2 + \beta_2{}^2]} \qquad (7\text{-}22)$$

It has a triple zero at the origin and four poles

$$\left.\begin{array}{c} p_1 \\ \overset{*}{p}_1 \end{array}\right\} = -\alpha_1 \pm j\beta_1 \qquad \left.\begin{array}{c} p_2 \\ \overset{*}{p}_2 \end{array}\right\} = -\alpha_2 \pm j\beta_2$$

† E. J. Angelo, Jr., "Electronic Circuits," McGraw-Hill Book Company, Inc., New York, 1958.

(with respect to the variable $j\omega$) as shown in Fig. 7-5a.   If

$$\alpha_1 = \alpha_2 = \alpha \quad \text{and} \quad \alpha \ll \beta_1 \quad \beta_1 - \beta_2 \ll \beta_1$$

then the frequency response of this system is symmetrical about

$$\omega_0 = \frac{\beta_1 + \beta_2}{2}$$

with $\theta(\omega_0) \simeq -\pi/2$ (Fig. 7-5b).   This can be shown by factoring $H(\omega)$ and observing that the factors $j\omega$, $(j\omega - \overset{*}{p}_1)$, and $(j\omega - \overset{*}{p}_2)$ can be

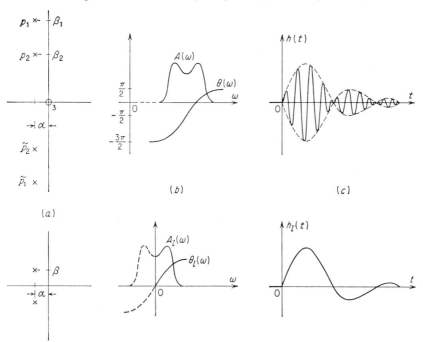

Figure 7-5

considered as constant for ω near ω_0.   From this reasoning it follows also that the equivalent low-pass filter is a rational function, of the form

$$H_l(\omega) = \frac{K_l}{(j\omega + \alpha)^2 + \beta^2}$$

as in (6-102), where

$$\beta = \frac{\beta_1 - \beta_2}{2}$$

and K_l is such that $A_l(0) = A(\omega_0)$.   But

$$A(\omega_0) \simeq \frac{K\omega_0^{3}}{(\alpha^2 + \beta^2)(2\omega_0)^2} \qquad A_l(0) = \frac{K_l}{\alpha^2 + \beta^2}$$

Hence
$$K_l = \frac{K\omega_0}{4}$$

The impulse response $h_l(t)$ is given by [see (6-104)]

$$h_l(t) = \frac{K_l}{\beta} e^{-\alpha t} \sin \beta t \, U(t)$$

and from (7-21) it follows that (Fig. 7-5c)

$$h(t) \simeq \frac{K\omega_0}{2\beta} e^{-\alpha t} \sin \beta t \cos \left(\omega_0 t + \frac{\pi}{2}\right) U(t)$$

Narrow-band systems. A system is called narrow-band if the cutoff frequency ω_c of the equivalent low-pass filter is small compared to ω_0

$$\omega_c \ll \omega_0$$

We shall show that the step response of such a filter is approximately given by the simple expression

$$a(t) \simeq \frac{2h_l(t)}{\omega_0} \sin \left[\omega_0 t - \theta(\omega_0)\right] \tag{7-23}$$

with $h_l(t)$ and $\theta(\omega_0)$ as in (7-21).   Indeed, the Fourier transform of $a(t)$ equals

$$\left[\pi \, \delta(\omega) + \frac{1}{j\omega}\right] H(\omega) = \frac{H(\omega)}{j\omega}$$

because $H(\omega) \, \delta(\omega) = H(0) \, \delta(\omega) = 0$.   For a narrow-band system the variation of $1/j\omega$ in the bandpass can be ignored; hence for $\omega > 0$

$$\frac{H(\omega)}{j\omega} \sim \frac{H(\omega)}{j\omega_0}$$

From the above and (7-21), the desired result (7-23) follows readily.

This result can be generalized to give the response of the filter to an arbitrary real input $f(t)$, whose transform $F(\omega)$ is approximately constant in the bandpass

$$F(\omega) \simeq F(\omega_0) = |F(\omega_0)|e^{j\phi_F(\omega_0)} \qquad \omega_0 - \omega_c < \omega < \omega_0 + \omega_c$$

From
$$G(\omega) = F(\omega)H(\omega) \simeq F(\omega_0)H(\omega) \qquad \omega > 0$$
we obtain
$$g(t) \simeq 2 \, |F(\omega_0)| \, h_l(t) \cos \left[\omega_0 t + \phi_F(\omega_0) - \theta(\omega_0)\right] \tag{7-24}$$

Clearly (7-23) is a special case of (7-24).

7-2. Modulated Input†

In this section we shall determine the response $g(t)$ of a symmetrical bandpass filter to an amplitude-modulated input

$$f(t) = f_l(t) \cos \omega_0 t \qquad (7\text{-}25)$$

of carrier frequency equal to the center frequency ω_0 of the filter.

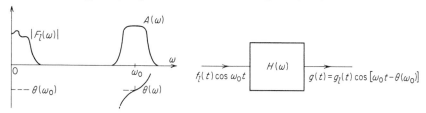

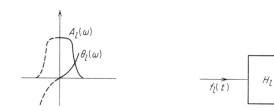

FIGURE 7-6

Narrow-band envelope. We shall first assume that the Fourier transform $F_l(\omega)$ of the envelope $f_l(t)$ is narrow-band in the sense

$$F_l(\omega) = 0 \qquad \text{for } |\omega| > \omega_0 \qquad (7\text{-}26)$$

(see Fig. 7-6). We denote by $g_l(t)$ the response of the equivalent low-pass filter to $f_l(t)$, and we shall prove that

$$g(t) = g_l(t) \cos [\omega_0 t - \theta(\omega_0)] \qquad (7\text{-}27)$$

Proof. The transform $F(\omega)$ of the input $f(t)$ is given by [see (2-38)]

$$F(\omega) = \frac{F_l(\omega + \omega_0) + F_l(\omega - \omega_0)}{2}$$

Hence, with $H(\omega) = H_1(\omega) + H_2(\omega)$ as in (7-1), we have

$$G(\omega) = F(\omega)H(\omega) = \frac{F_l(\omega + \omega_0) + F_l(\omega - \omega_0)}{2} [H_1(\omega) + H_2(\omega)] \qquad (7\text{-}28)$$

† M. Schwartz, "Information Transmission, Modulation, and Noise," McGraw-Hill Book Company, Inc., New York, 1959; S. Goldman, "Frequency Analysis, Modulation, and Noise," McGraw-Hill Book Company, Inc., New York, 1948.

From (7-26) it easily follows that

$$F_i(\omega - \omega_0)H_2(\omega) = 0 \qquad F_i(\omega + \omega_0)H_1(\omega) = 0 \qquad (7\text{-}29)$$

Therefore

$$G(\omega) = \tfrac{1}{2}F_i(\omega - \omega_0)H_1(\omega) + \tfrac{1}{2}F_i(\omega + \omega_0)H_2(\omega) \qquad (7\text{-}30)$$

With $G_i(\omega)$ the transform of $g_i(t)$, we have

$$g_i(t) \leftrightarrow G_i(\omega) = F_i(\omega)H_i(\omega) \qquad (7\text{-}31)$$

Therefore

$$g_i(t)\cos[\omega_0 t - \theta(\omega_0)] \leftrightarrow \tfrac{1}{2}F_i(\omega - \omega_0)H_i(\omega - \omega_0)e^{-j\theta(\omega_0)}$$
$$+ \tfrac{1}{2}F_i(\omega + \omega_0)H_i(\omega + \omega_0)e^{j\theta(\omega_0)} \qquad (7\text{-}32)$$

But [see (7-18)]

$$H_i(\omega - \omega_0)e^{-j\theta(\omega_0)} = H_1(\omega) \qquad H_i(\omega + \omega_0)e^{-j\theta(\omega_0)} = H_2(\omega)$$

Hence

$$g_i(t)[\cos \omega_0 t - \theta(\omega_0)] \leftrightarrow \tfrac{1}{2}F_i(\omega - \omega_0)H_1(\omega) + \tfrac{1}{2}F_i(\omega + \omega_0)H_2(\omega) \qquad (7\text{-}33)$$

and (7-27) follows [see (7-30)].

Comment. The above result (7-27) was based on assumption (7-26). However, this result is true under the more general condition

$$F_i(\omega) = 0 \qquad \text{for } |\omega| > 2\omega_0 - \omega_c \qquad (7\text{-}34)$$

Indeed, (7-34) ensures the validity of (7-29), as we can easily see.

Wide-band envelope. Our result (7-27) was based on the assumption of a narrow-band envelope. If this restriction is not imposed on $f_i(t)$, then $G(\omega)$ is given by

$$G(\omega) = \tfrac{1}{2}F_i(\omega - \omega_0)H_1(\omega) + \tfrac{1}{2}F_i(\omega + \omega_0)H_2(\omega) + \tfrac{1}{2}F_i(\omega + \omega_0)H_1(\omega)$$
$$+ \tfrac{1}{2}F_i(\omega - \omega_0)H_2(\omega) \qquad (7\text{-}35)$$

as we see from (7-28). The inverse of the first two terms above equals $g_i(t)\cos[\omega_0 t - \theta(\omega_0)]$; therefore $g(t)$ can be written in the form

$$g(t) = g_i(t)\cos[\omega_0 t - \theta(\omega_0)] + e(t) \qquad (7\text{-}36)$$

where the *error term* $e(t)$ is determined from

$$e(t) \leftrightarrow \tfrac{1}{2}F_i(\omega + \omega_0)H_1(\omega) + \tfrac{1}{2}F_i(\omega - \omega_0)H_2(\omega) \qquad (7\text{-}37)$$

In many applications, the exact determination of $e(t)$ is not necessary. If it is assumed, as is often the case for narrow-band filters, that the spectrum of $f_i(t)$ is approximately constant in the ω_c vicinity of $2\omega_0$

$$F_i(\omega) \simeq F_i(2\omega_0) = |F_i(2\omega_0)|e^{j\phi_{Fi}(2\omega_0)} \qquad 2\omega_0 - \omega_c < \omega < 2\omega_0 + \omega_c \quad (7\text{-}38)$$

then

$$e(t) \simeq |F_i(2\omega_0)| \, h_i(t)\cos[\omega_0 t + \phi_{Fi}(2\omega_0) - \theta(\omega_0)] \qquad (7\text{-}39)$$

with $h_i(t)$ and $\theta(\omega_0)$ as in (7-21). Thus the error term is proportional to the impulse response $h(t)$ of the filter, with a phase shift in the carrier.

To prove (7-39), we observe from (7-38) that

$$\tfrac{1}{2}F_l(\omega + \omega_0)H_1(\omega) \simeq \tfrac{1}{2}F_l(2\omega_0)H_1(\omega)$$

$$\tfrac{1}{2}F_l(\omega - \omega_0)H_2(\omega) \simeq \tfrac{1}{2}F_l(-2\omega_0)H_2(\omega) \tag{7-40}$$

But [see (7-20)]

$$e^{-j\theta(\omega_0)}h_l(t)e^{j\omega_0 t} \leftrightarrow H_1(\omega) \tag{7-41}$$

$$e^{j\theta(\omega_0)}h_l(t)e^{-j\omega_0 t} \leftrightarrow H_2(\omega) \tag{7-42}$$

Multiplying (7-41) by $F_l(2\omega_0)/2$ and (7-42) by $F_l(-2\omega_0)/2$, and adding, we obtain

$$|F(2\omega_0)|\, h_l(t) \cos [\omega_0 t + \phi_{Fl}(2\omega_0) - \theta(\omega_0)] \leftrightarrow \tfrac{1}{2}F_l(2\omega_0)H_1(\omega)$$

$$+ \tfrac{1}{2}F_l(-2\omega_0)H_2(\omega) \tag{7-43}$$

and (7-39) follows.

As an example of (7-36) we shall determine the response $g(t)$ of an ideal bandpass system (Fig. 7-3) to a step-modulated input

$$f(t) = U(t) \cos \omega_0 t$$

The response $a_l(t)$ of the equivalent low-pass filter to a unit step is given by [see (6-22)]

$$a_l(t) = \frac{A(\omega_0)}{2} \left\{ 1 + \frac{2}{\pi} Si[\omega_c(t - t_0)] \right\}$$

Therefore $$g(t) = \frac{A(\omega_0)}{2} \left\{ 1 + \frac{2}{\pi} Si[\omega_c(t - t_0)] \right\} \cos \omega_0 t + e(t) \tag{7-44}$$

If $\omega_c \ll \omega_0$, then $e(t)$ can be obtained from (7-39). The Fourier transform of the envelope $U(t)$, evaluated at $\omega = 2\omega_0$, equals $1/2j\omega_0$:

$$F(2\omega_0) = \frac{1}{2j\omega_0} = \frac{1}{2\omega_0} e^{-j\pi/2}$$

and since [see (6-21)]

$$h_l(t) = \frac{A(\omega_0)}{\pi} \frac{\sin \omega_c(t - t_0)}{t - t_0}$$

we conclude from (7-39) that

$$e(t) \simeq \frac{A(\omega_0)}{2\pi\omega_0} \frac{\sin \omega_c(t - t_0)}{t - t_0} \sin \omega_0 t$$

Notice that the maximum of $e(t)$ equals

$$\frac{A(\omega_0)}{2\pi} \frac{\omega_c}{\omega_0}$$

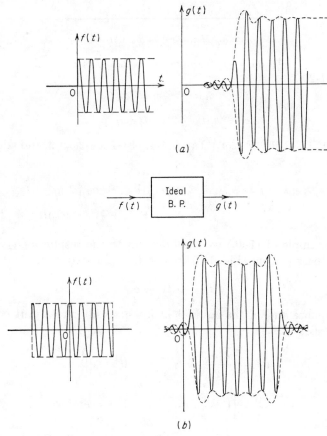

FIGURE 7-7

and is small compared to $A(\omega_0)$ because $\omega_c \ll \omega_0$; hence the term $e(t)$ can be neglected in (7-44). In Fig. 7-7a we have shown the response of the filter to $U(t) \cos \omega_0 t$, and in Fig. 7-7b its response to a pulse-modulated input $p_T(t) \cos \omega_0 t$.

The response of the double-tuned circuit (7-22) to a step-modulated input $f(t) = U(t) \cos \omega_0 t$ with

$$\omega_0 = \frac{\beta_1 + \beta_2}{2}$$

can be similarly found. The response of the equivalent low-pass filter to the envelope $U(t)$ of $f(t)$ is obtained from (6-105)

$$a_l(t) = \frac{K\omega_0}{4} U(t)\left[\frac{1}{\alpha^2 + \beta^2} + \frac{1}{\beta\sqrt{\alpha^2 + \beta^2}} e^{-\alpha t} \sin\left(\beta t - \tan^{-1}\frac{\beta}{\alpha}\right)\right]$$

Hence [see (7-36)]

$$g(t) = a_l(t) \cos \left(\omega_0 t - \frac{\pi}{2} \right) + e(t)$$

where

$$e(t) \simeq \frac{-K}{8\beta} e^{-\alpha t} \sin \beta t \cos \omega_0 t \; U(t)$$

as we can see from (7-39) and (6-104).

7-3. Unsymmetrical Systems

As in the symmetrical case, we write $H(\omega)$ (Fig. 7-8a) as a sum

$$H(\omega) = H_1(\omega) + H_2(\omega)$$

where $H_1(\omega)$ and $H_2(\omega)$ are as in (7-2). If the system has no sharp cutoff points, then there is no clearly defined center frequency ω_0. In such a case we choose ω_0 somewhere in the center of the band and introduce the functions

$$H_{l1}(\omega) = H_1(\omega + \omega_0) \qquad H_{l2}(\omega) = H_2(\omega - \omega_0) \qquad (7\text{-}45)$$

obtained by shifting $H_1(\omega)$ and $H_2(\omega)$ to the left and right respectively, by ω_0. In Fig. 7-8b we have shown $H_{l1}(\omega)$ and $H_{l2}(\omega)$; to simplify the drawing, we assumed $H(\omega)$ real. In the symmetrical case the

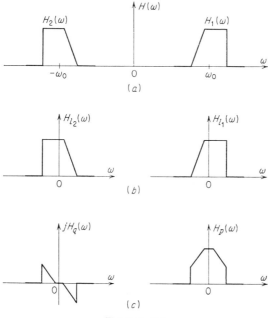

FIGURE 7-8

functions $H_{l1}(\omega)$ and $H_{l2}(\omega)$ were equal; this is no longer true for the unsymmetrical case; however,

$$H_{l1}(-\omega) = \overset{*}{H}_{l2}(\omega) \tag{7-46}$$

This follows from $H_1(\omega_0 - \omega) = \overset{*}{H}_2(\omega - \omega_0)$ [see (7-3) and (7-45)]. We now define the *in-phase* and *quadrature* system functions

$$H_p(\omega) = \frac{H_{l1}(\omega) + H_{l2}(\omega)}{2} \qquad H_q(\omega) = \frac{H_{l2}(\omega) - H_{l1}(\omega)}{2j} \tag{7-47}$$

shown in Fig. 7-8c, and we observe from (7-46) that

$$H_p(-\omega) = \frac{H_{l1}(-\omega) + H_{l2}(-\omega)}{2} = \frac{\overset{*}{H}_{l2}(\omega) + \overset{*}{H}_{l1}(\omega)}{2} = \overset{*}{H}_p(\omega) \tag{7-48}$$

Similarly

$$H_q(-\omega) = \overset{*}{H}_q(\omega) \tag{7-49}$$

Therefore $H_p(\omega)$ and $H_q(\omega)$ are the Fourier transforms of two real functions $h_p(t)$ and $h_q(t)$

$$h_p(t) \leftrightarrow H_p(\omega) \qquad h_q(t) \leftrightarrow H_q(\omega) \tag{7-50}$$

The impulse response $h(t)$ of the given system $H(\omega)$ can be easily expressed in terms of $h_p(t)$ and $h_q(t)$.

THEOREM. $h(t)$ is given by

$$h(t) = 2h_p(t) \cos \omega_0 t + 2h_q(t) \sin \omega_0 t \tag{7-51}$$

Proof.   From (7-47) and (7-50) we easily obtain

$$h_p(t) - jh_q(t) \leftrightarrow H_{l1}(\omega)$$
$$h_p(t) + jh_q(t) \leftrightarrow H_{l2}(\omega) \tag{7-52}$$

But [see (7-45)]

$$H_{l1}(\omega - \omega_0) = H_1(\omega) \qquad H_{l2}(\omega + \omega_0) = H_2(\omega)$$

Therefore

$$e^{j\omega_0 t}[h_p(t) - jh_q(t)] \leftrightarrow H_1(\omega)$$
$$e^{-j\omega_0 t}[h_p(t) + jh_q(t)] \leftrightarrow H_2(\omega)$$

Adding the above two relationships, we obtain

$$2h_p(t) \cos \omega_0 t + 2h_q(t) \sin \omega_0 t \leftrightarrow H_1(\omega) + H_2(\omega) = H(\omega)$$

and (7-51) is proved.

Comment.   Unlike the symmetrical case, the impulse response $h(t)$ is amplitude- and phase-modulated

$$h(t) = 2\sqrt{h_p^2(t) + h_q^2(t)} \cos \left[\omega_0 t - \tan^{-1} \frac{h_q(t)}{h_p(t)} \right] \tag{7-53}$$

as we easily see from (7-51).

7-4. Modulated Input

We shall now determine the response $g(t)$ of an unsymmetrical filter to a modulated input $f(t) = f_i(t) \cos \omega_0 t$. We introduce the in-phase and quadrature filters $H_p(\omega)$ and $H_q(\omega)$ defined by (7-47), where ω_0 is the frequency of the carrier and not necessarily the mid-band frequency of the filter. In fact ω_0 might even be outside the bandpass. The responses of the filters $H_p(\omega)$ and $H_q(\omega)$ to the envelope $f_i(t)$ of $f(t)$ we denote by $g_p(t)$ and $g_q(t)$ respectively (see

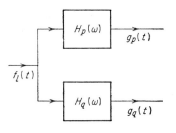

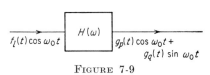

FIGURE 7-9

Fig. 7-9) and will express $g(t)$ in terms of these functions.

Narrow-band envelope. If the Fourier spectrum of the envelope $f_i(t)$ is narrow in the sense of (7-34), then the response $g(t)$ of the unsymmetrical system is given by

$$g(t) = g_p(t) \cos \omega_0 t + g_q(t) \sin \omega_0 t \qquad (7\text{-}54)$$

Notice that $g(t)$ is amplitude- and phase-modulated

$$g(t) = \sqrt{g_p{}^2(t) + g_q{}^2(t)} \cos \left[\omega_0 t - \tan^{-1} \frac{g_q(t)}{g_p(t)} \right] \qquad (7\text{-}55)$$

The proof of (7-54) is similar to the proof of (7-27) and will be omitted.

Wide-band envelope. For a wide-band envelope the response $g(t)$ is given by

$$g(t) = g_p(t) \cos \omega_0 t + g_q(t) \sin \omega_0 t + e(t) \qquad (7\text{-}56)$$

as in (7-36). Assuming that (7-38) holds, we obtain for the error term $e(t)$ the following approximation:

$$e(t) \simeq |F_i(2\omega_0)| \left\{ h_p(t) \cos [\omega_0 t + \phi_{F_i}(2\omega_0)] + h_q(t) \sin [\omega_0 t + \phi_{F_i}(2\omega_0)] \right\} \qquad (7\text{-}57)$$

where $h_p(t)$ and $h_q(t)$ are the impulse responses of the in-phase and quadrature filters. The proof of (7-56) and (7-57) is similar to the proof of the corresponding Eqs. (7-36) and (7-39) of the symmetrical filter.

Comment. The above results can be used to determine the response of a symmetrical filter to a modulated input $f_i(t) \cos \omega_1 t$ whose carrier frequency ω_1 is different from the center frequency ω_0 of the filter. Indeed, with

$$H_{i1}(\omega) = H_1(\omega + \omega_1) \qquad H_{i2}(\omega) = H_2(\omega - \omega_1)$$

and $H_p(\omega)$ and $H_q(\omega)$ as in (7-47), the response $g(t)$ is given by (7-54) and (7-56). The method can also be used to obtain the response of low-pass filters to modulated inputs.

7-5. Group, Phase, and Signal-front Delay

These terms are used to describe the delay of the various parts of a signal $f(t)$ as it passes through a linear system as in Fig. 7-10, and it

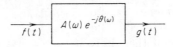

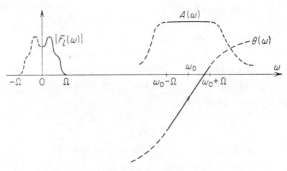

FIGURE 7-10

is the purpose of this section to explain their meaning. With

$$H(\omega) = A(\omega)e^{-j\theta(\omega)}$$

the system function of the given system, we define its group delay t_{gr}, phase delay t_{ph}, and signal-front delay t_{fr} by

$$t_{gr}(\omega) = \frac{d\theta(\omega)}{d\omega} \qquad t_{ph}(\omega) = \frac{\theta(\omega)}{\omega} \qquad t_{fr} = \lim_{\omega \to \infty} \frac{\theta(\omega)}{\omega} \qquad (7\text{-}58)$$

Thus (see Fig. 7-11), t_{gr} is the slope of $\theta(\omega)$ at a given frequency ω, t_{ph}

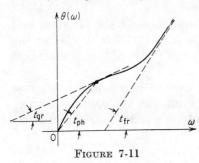

FIGURE 7-11

is the slope of the line from the origin to a point $[\omega, \theta(\omega)]$ of the phase curve, and t_{fr} is the asymptotic slope of $\theta(\omega)$. If $\theta(\omega)$ does not tend to a straight line as ω tends to infinity, then the term signal-front delay has no meaning. The group and phase delays are functions of ω, and the signal-front delay is independent of ω.

Group and phase delay.
These terms are applied to inputs
that are narrow-band in the following sense: suppose that the
input to our system is amplitude-
modulated,

$$f(t) = f_l(t) \cos \omega_0 t \qquad (7\text{-}59)$$

With $F_l(\omega)$ the Fourier transform
of the envelope $f_l(t)$, we assume
that

$$|F_l(\omega)| = 0 \qquad \text{for } |\omega| > \Omega \quad (7\text{-}60)$$

where Ω is a constant, and that in
the $(\omega_0 - \Omega, \omega_0 + \Omega)$ interval, the
amplitude of $H(\omega)$ is constant and
its phase linear (Fig. 7-10):

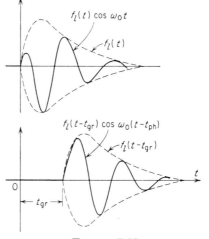

FIGURE 7-12

$$A(\omega) = A(\omega_0) \qquad \omega_0 - \Omega < \omega < \omega_0 + \Omega \tag{7-61}$$
$$\theta(\omega) = \theta(\omega_0) + \theta'(\omega_0)(\omega - \omega_0) = \omega_0 t_{ph}(\omega_0) + (\omega - \omega_0)t_{gr}(\omega_0)$$

Thus the Fourier integral $F(\omega) = \frac{1}{2}[F_l(\omega + \omega_0) + F_l(\omega - \omega_0)]$ of the
input $f(t)$ is zero outside the band $(\omega_0 - \Omega, \omega_0 + \Omega)$ and its image, and
in this band the system acts as an ideal symmetrical bandpass filter.
Therefore [see (6-26) and (7-27)] the response $g(t)$ is given by

$$g(t) = f_l(t - t_{gr}) \cos \omega_0(t - t_{ph}) \tag{7-62}$$

where for simplicity we assumed $A(\omega_0) = 1$. From (7-62) we see that
the group delay t_{gr}, evaluated at the carrier frequency $\omega = \omega_0$, equals
the delay of the envelope $f_l(t)$ of the input (Fig. 7-12); the phase delay
t_{ph} equals the delay of the carrier.

If (7-61) is not true, then the output $g(t)$ is no longer given by (7-62).
The variation of $A(\omega)$ and the nonlinear part of $\theta(\omega)$ cause a distortion
and a broadening of the envelope $f_l(t)$ as in Fig. 6-38, with a possible
phase modulation due to the lack of symmetry of the system. However, if these effects are small, then the envelope of the input is still
recognizable, and t_{gr} gives the displacement of its center of gravity,
since it equals the slope of $\theta_l(\omega)$ at $\omega = 0$ [see (2-55)]. For wide-band
systems, the terms group and phase delay have no meaning. From the
above discussion it follows that the narrow-band requirement, necessary for the validity of (7-62), depends not only on the signal but also
on the shape of $A(\omega)$ and $\theta(\omega)$ in the band of $F(\omega)$.

Signal-front delay. With t_{fr} the asymptotic slope of $\theta(\omega)$ [see
(7-58)], it will be shown in Sec. 9-5 that the impulse response $h(t)$ of

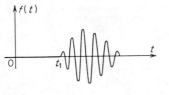

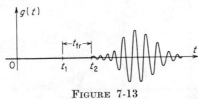

FIGURE 7-13

the given system equals zero for $t < t_{fr}$:

$$h(t) = 0 \qquad t < t_{fr}$$

From this property of $h(t)$, we easily conclude that if $f(t)$ starts at $t = t_1$, i.e., if:

$$f(t) = 0 \qquad \text{for } t < t_1 \quad (7\text{-}63)$$

then the response $g(t)$ starts at $t = t_1 + t_{fr}$ (Fig. 7-13):

$$g(t) = 0 \qquad \text{for } t < t_1 + t_{fr} \quad (7\text{-}64)$$

Thus the signal-front delay t_{fr} is the delay of the beginning, or *front*, of a signal, as it passes through a linear device.

One might raise the point that if the signal $f(t)$ is narrow-band in the above sense, then there is no distortion in the envelope; therefore the delay of its front should equal the delay of the center of gravity, although t_{fr} and t_{gr} might be different. The answer to this apparent contradiction, is the following: if $F_i(\omega)$ satisfies (7-60), then $f_i(t)$ cannot have a front; i.e., it cannot be zero in an interval extending to $-\infty$, and the concept of signal-front delay cannot be applied to it. The reason is that (7-60) is incompatible with the Paley-Wiener condition (6-61) satisfied by all signals that do not extend to $-\infty$.

7-6. Group, Phase and Signal-front Velocity

These terms characterize the propagation velocities of the various parts of a traveling wave $f(x,t)$, and they are related to the various delays discussed in the preceding section. The system concepts can be applied to propagation phenomena if one considers the medium as a linear device (Fig. 7-14)

$$H(\omega) = e^{-\alpha(\omega)x} e^{-j\beta(\omega)x} \tag{7-65}$$

with input $f(0,t)$ and output $f(x,t)$. We shall justify (7-65) for two special cases of some interest.

Plane waves. Consider an electromagnetic plane wave traveling in a uniform isotropic medium. We assume that x is the direction of propagation and that $\bar{E}$ has only a y component and $\bar{B}$ a z component (Fig. 7-15):

$$E_y = e(x,t) \qquad B_z = b(x,t)$$

Maxwell's equations

$$\nabla \times \bar{E} = -\frac{\partial \bar{B}}{\partial t} \qquad \nabla \times \bar{H} = \sigma \bar{E} + \frac{\partial \bar{D}}{\partial t}$$

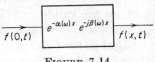

FIGURE 7-14

take the form

$$\frac{\partial e}{\partial x} = -\frac{\partial b}{\partial t} \qquad -\frac{1}{\mu}\frac{\partial b}{\partial x} = \sigma e + \epsilon \frac{\partial e}{\partial t} \tag{7-66}$$

where σ, ϵ, μ are the familiar constants of the medium. Differentiating the first equation in (7-66) with respect to x and the second with respect to t and eliminating b, we obtain the wave equation

$$\frac{\partial^2 e}{\partial x^2} - \sigma\mu \frac{\partial e}{\partial t} - \mu\epsilon \frac{\partial^2 e}{\partial t^2} = 0 \tag{7-67}$$

With $E(x,\omega)$ the Fourier transform of $e(x,t)$ we have

$$E(x,\omega) = \int_{-\infty}^{\infty} e(x,t)e^{-j\omega t}\, dt \tag{7-68}$$

$$e(x,t) = \frac{1}{2\pi}\int_{-\infty}^{\infty} E(x,\omega)e^{j\omega t}\, d\omega \tag{7-69}$$

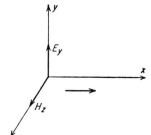

Figure 7-15

The above equations are the same as (2-1) and (2-3), with x appearing as a parameter. Differentiating (7-69) with respect to x and t, we obtain the pairs

$$\frac{\partial e(x,t)}{\partial x} \leftrightarrow \frac{\partial E(x,\omega)}{\partial x} \tag{7-70}$$

$$\frac{\partial e(x,t)}{\partial t} \leftrightarrow j\omega E(x,\omega) \tag{7-71}$$

$$\frac{\partial^2 e(x,t)}{\partial t^2} \leftrightarrow (j\omega)^2 E(x,\omega) \tag{7-72}$$

We now take the Fourier integral of both sides of (7-67); i.e., we multiply by $e^{j\omega t}$ and integrate from $-\infty$ to $+\infty$. Because of (7-70) to (7-72), the following equation results:

$$\frac{\partial^2 E(x,\omega)}{\partial x^2} = j\omega\mu(\sigma + j\omega\epsilon)E(x,\omega) \tag{7-73}$$

or with

$$\gamma^2(\omega) = j\omega\mu(\sigma + j\omega\epsilon) \tag{7-74}$$

$$\frac{\partial^2 E}{\partial x^2} - \gamma^2 E = 0 \tag{7-75}$$

For a wave traveling in the $+x$ direction, the solution of (7-75) is

$$E(x,\omega) = K(\omega)e^{-\gamma(\omega)x} \tag{7-76}$$

Clearly $K(\omega) = E(0,\omega)$; hence the medium from 0 to x can be considered as a linear system with system function $e^{-\gamma(\omega)x}$. The constant

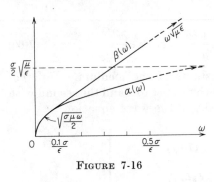

FIGURE 7-16

$\gamma(\omega)$ is complex

$$\gamma(\omega) = \alpha(\omega) + j\beta(\omega) \quad (7\text{-}77)$$

where, as we can easily find from (7-74),

$$\beta(\omega) = \omega\left[\frac{\mu\epsilon}{2}\left(\sqrt{1 + \frac{\sigma^2}{\omega^2\epsilon^2}} + 1\right)\right]^{\frac{1}{2}} \quad (7\text{-}78)$$

$$\alpha(\omega) = \omega\left[\frac{\mu\epsilon}{2}\left(\sqrt{1 + \frac{\sigma^2}{\omega^2\epsilon^2}} - 1\right)\right]^{\frac{1}{2}}$$

In Fig. 7-16 we have shown the attenuation $\alpha(\omega)$ and phase shift $\beta(\omega)$ of our system.

Transmission lines.† Similar results can be obtained for an infinite, uniform transmission line (Fig. 7-17). Denoting by $v(x,t)$ the voltage difference between the wires at a distance x from the beginning

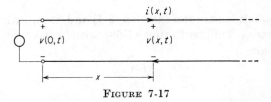

FIGURE 7-17

and by $i(x,t)$ the current along the wire, we obtain from Kirchhoff's laws

$$-\frac{\partial v}{\partial x} = Ri + L\frac{\partial i}{\partial t} \qquad -\frac{\partial i}{\partial x} = Gv + C\frac{\partial v}{\partial t} \quad (7\text{-}79)$$

where R, L, G, C are the familiar constants of the line. From (7-79) we have

$$\frac{\partial^2 v}{\partial x^2} = RGv + (LG + RC)\frac{\partial v}{\partial t} + LC\frac{\partial^2 v}{\partial t^2} \quad (7\text{-}80)$$

and with $V(x,\omega)$ the Fourier transform of $v(x,t)$

$$\frac{\partial^2 V(x,\omega)}{\partial x^2} = \gamma^2(\omega)V(x,\omega) \quad (7\text{-}81)$$

where

$$\gamma^2(\omega) = (R + j\omega L)(G + j\omega C) = [\alpha(\omega) + j\beta(\omega)]^2 \quad (7\text{-}82)$$

Thus

$$V(x,\omega) = K(\omega)e^{-\gamma(\omega)x} \quad (7\text{-}83)$$

---

† E. Weber, "Linear Transient Analysis," John Wiley & Sons, Inc., New York, 1956.

and (7-65) is again established.   Clearly

$$K(\omega) = V(0,\omega)$$

is the Fourier transform of the input voltage $v(0,t)$.
We now return to the system function

$$H(\omega) = e^{-\alpha(\omega)x} e^{-j\beta(\omega)x}$$

The front of the wave $f(0,t)$ reaches the output x with a time delay t_{fr}
[see (7-58)]:

$$t_{fr} = \frac{\beta(\omega)x}{\omega}\bigg|_{\omega=\infty}$$

Hence it travels with a velocity

$$v_{fr} = \frac{x}{t_{fr}} = \frac{1}{\beta(\omega)/\omega}\bigg|_{\omega=\infty} \tag{7-84}$$

known as signal-front velocity: v_{fr} equals the inverse asymptotic slope
of $\beta(\omega)$.   We similarly define the group velocity

$$v_{gr} = \frac{1}{d\beta(\omega)/d\omega} = \frac{x}{t_{gr}} \tag{7-85}$$

and phase velocity

$$v_{ph} = \frac{1}{\beta(\omega)/\omega} = \frac{x}{t_{ph}} \tag{7-86}$$

v_{gr} is the velocity of propagation of the envelope of $f(x,t)$ and v_{ph} the
velocity of propagation of the carrier.

7-7. The Principle of Stationary Phase

From the discussion of the preceding section we see that the propagation of time signals is determined by integrals of the form

$$f(x,t) = \frac{1}{2\pi} \int_{-\infty}^{\infty} e^{-\alpha(\omega)x} e^{j[\omega t - x\beta(\omega)]}\, d\omega \tag{7-87}$$

involving the parameters x and t, and their exact evaluation is in
general complicated.   In many cases one is interested in the value
of $f(x,t)$ for large x; since the group velocity is finite, for large x
$f(x,t)$ has significant values when t is also large.   With

$$A(\omega) = e^{-\alpha(\omega)x}$$

and

$$\omega - \frac{x}{t}\beta(\omega) = \mu(\omega) \tag{7-88}$$

we conclude that the asymptotic form of $f(x,t)$ involves the determination of the integral

$$I = \int_{-\infty}^{\infty} A(\omega)e^{jt\mu(\omega)}\, d\omega \tag{7-89}$$

for large values of t.   We shall show that I can be very simply found asymptotically by a method known as *the principle of stationary phase*. We plan to justify the principle using a heuristic argument; its exact formulation is rather involved and will not be given.†

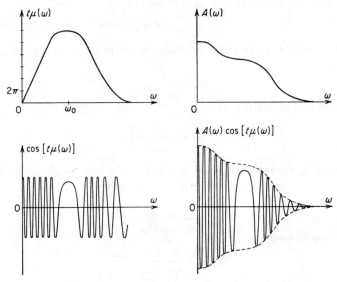

FIGURE 7-18

We shall assume first that the slope of $\mu'(\omega)$ equals zero at only one point ω_0 (Fig. 7-18)

$$\mu'(\omega_0) = 0$$

and at this point, its second derivative $\mu''(\omega)$ is positive

$$\mu''(\omega_0) > 0$$

and we shall show that

$$\int_{-\infty}^{\infty} A(\omega)e^{jt\mu(\omega)}\, d\omega \sim A(\omega_0) \sqrt{\frac{2\pi}{t\mu''(\omega_0)}}\, e^{j[t\mu(\omega_0) + \pi/4]} \tag{7-90}$$

where by the symbol $y_1(t) \sim y_2(t)$ we shall mean that the functions $y_1(t)$ and $y_2(t)$ are approximately equal for large t, i.e., that

$$y_1(t) - y_2(t) \ll y_1(t) \qquad t \to \infty$$

Proof. I is a sum of integrals involving the functions $A(\omega)\cos[t\mu(\omega)]$ and $A(\omega)\sin[t\mu(\omega)]$.   For large values of t, the function $\cos[t\mu(\omega)]$ oscillates rapidly everywhere except near ω_0 (Fig. 7-18).   In an interval $[k\pi/t, (k+1)\pi/t]$ away from ω_0, the function $\mu(\omega)$ can be

† E. T. Copson, "The Asymptotic Expansion of a Function Defined by a Definite Integral or Contour Integral," The Admiralty, London, S.R. E/ACS 106, 1946.

considered linear and $A(\omega)$ constant; therefore the integral over that interval equals zero. The same is true for the sine term. Thus the value of I is determined by the behavior of the integrand only in the vicinity of the point ω_0. Consider now the integral

$$I_1 = \int_{-\infty}^{\infty} A(\omega_0) e^{jt[\mu(\omega_0) + \mu''(\omega_0)(\omega - \omega_0)^2/2]} \, d\omega$$

Reasoning similarly, we conclude that I_1 depends also only on the value of the integrand near ω_0. But for ω close to ω_0 we have

$$\mu(\omega) \simeq \mu(\omega_0) + \frac{\mu''(\omega_0)}{2}(\omega - \omega_0)^2 \qquad A(\omega) \simeq A(\omega_0) \qquad (7\text{-}91)$$

Thus the integrands of I and I_1 are equal in the ω_0 vicinity: hence

$$I \sim I_1$$

The integral

$$I_1 = A(\omega_0) e^{jt\mu(\omega_0)} \int_{-\infty}^{\infty} e^{jt\mu''(\omega_0)(\omega - \omega_0)^2/2} \, d\omega$$

can be easily evaluated; with

$$\frac{t\mu''(\omega_0)(\omega - \omega_0)^2}{2} = y^2$$

we have

$$I_1 = A(\omega_0) e^{jt\mu(\omega_0)} \sqrt{\frac{2}{t\mu''(\omega_0)}} \int_{-\infty}^{\infty} e^{jy^2} \, dy$$

But the last integral equals $\sqrt{\pi j}$ (see Prob. 30), and (7-90) follows.

If
$$\mu''(\omega_0) < 0$$

we obtain by a similar argument

$$\int_{-\infty}^{\infty} A(\omega) e^{jt\mu(\omega)} \, d\omega \sim A(\omega_0) \sqrt{\frac{2\pi}{t\,|\mu''(\omega_0)|}} \, e^{j[t\mu(\omega_0) - \pi/4]} \qquad (7\text{-}92)$$

Returning to (7-88), we see that

$$\mu'(\omega) = 1 - \frac{x}{t}\beta'(\omega) \qquad \mu''(\omega) = -\frac{x}{t}\beta''(\omega)$$

Therefore the point ω_0 of stationary phase is determined from

$$\frac{1}{\beta'(\omega_0)} = \frac{x}{t} \qquad (7\text{-}93)$$

Thus at that point the group velocity $v_{gr}(\omega_0) = 1/\beta'(\omega_0)$ equals x/t. We also have [see (7-87) and (7-92)]

$$f(x,t) \sim e^{-\alpha(\omega_0)x} \sqrt{\frac{1}{2\pi x \beta''(\omega_0)}} \, e^{j[\omega_0 t - x\beta(\omega_0) - \pi/4]} \qquad (7\text{-}94)$$

Hence $f(x,t)$ is a modulated signal with carrier frequency ω_0.

Suppose now that $\mu(\omega)$ has n stationary points $\omega_1, \omega_2, \ldots, \omega_n$

$$\mu'(\omega_i) = 0 \qquad i = 1, 2, \ldots, n$$

By dividing the ω axis into n intervals, each of which contains only one zero of $\mu'(\omega)$, and reasoning as in (7-90), we easily obtain

$$\int_{-\infty}^{\infty} A(\omega)e^{jt\mu(\omega)}\, d\omega \sim \sum_{i=1}^{n} A(\omega_i)\sqrt{\frac{2\pi}{t\mu''(\omega_i)}}\; e^{j[t\mu(\omega_i)+\pi/4]} \qquad (7\text{-}95)$$

Thus, if $\mu(\omega)$ is odd and ω_0 is the only zero of $\mu'(\omega)$ for $\omega > 0$, then $\mu'(-\omega_0)$ is also zero; hence the summation in (7-95) contains two conjugate terms and

$$\int_{-\infty}^{\infty} A(\omega)e^{jt\mu(\omega)}\, d\omega \sim 2A(\omega_0)\sqrt{\frac{2\pi}{t\mu''(\omega_0)}}\; \cos[t\mu(\omega_0)+\pi/4] \qquad (7\text{-}96)$$

Example 7-1. Using (7-96), we shall determine the asymptotic form of the Bessel function $J_0(t)$ of the first kind, for large values of t. It is well known that $J_0(t)$ can be written as an integral [see (6-91)]

$$J_0(t) = \frac{1}{2\pi}\int_{-\pi}^{\pi} e^{jt\sin\omega}\, d\omega \qquad (7\text{-}97)$$

This integral is of the form (7-89) with

$$\mu(\omega) = \sin\omega$$

The limits of integration can be made infinite by assuming the integrand to equal zero for $|\omega| > \pi$. The odd function $\sin\omega$ has only one zero in the $(0, \pi)$ interval

$$\omega_0 = \frac{\pi}{2}$$

and $\qquad\qquad \sin\omega_0 = 1 \qquad d^2\sin\omega_0/d\omega^2 = -1$

Inserting into (7-96), we obtain

$$J_0(t) \sim \sqrt{\frac{2}{\pi t}}\cos\left(t - \frac{\pi}{4}\right) \qquad (7\text{-}98)$$

One can similarly find

$$J_n(t) = \frac{1}{2\pi}\int_{-\pi}^{\pi} e^{j(t\sin\omega - \omega n)}\, d\omega \sim \sqrt{\frac{2}{\pi t}}\cos\left(t - \frac{2n+1}{4}\pi\right) \qquad (7\text{-}99)$$

where $J_n(t)$ is the nth-order Bessel function (see Prob. 28).

If $\mu'(\omega)$ has one double zero at $\omega = \omega_0$ and no other zeros

$$\mu'(\omega_0) = \mu''(\omega_0) = 0 \qquad \mu'''(\omega_0) \neq 0$$

then $\qquad \displaystyle\int_{-\infty}^{\infty} A(\omega)e^{jt\mu(\omega)}\, d\omega \sim \frac{\Gamma(\frac{1}{3})}{\sqrt{3}}\left[\frac{6}{t\,|\mu'''(\omega_0)|}\right]^{\frac{1}{3}}A(\omega_0)e^{jt\mu(\omega_0)} \qquad (7\text{-}100)$

where $\Gamma(\frac{1}{3})$ is the gamma function; the proof is left as a problem.

Example 7-2. We shall now find the asymptotic form of $J_n(t)$ for large values of t and n. Assuming $t > n$ we introduce the quantity β defined by

$$n = t \cos \beta \qquad 0 < \beta < \frac{\pi}{2} \tag{7-101}$$

and obtain from (7-99)

$$J_n(t) = \frac{1}{2\pi} \int_{-\pi}^{\pi} e^{jt(\sin \omega - \omega \cos \beta)} \, d\omega$$

With $\mu(\omega) = \sin \omega - \omega \cos \beta$, we have $\mu'(\omega) = \cos \omega - \cos \beta$; hence, for $0 < \omega < \pi$ the point of stationary phase is $\omega_0 = \beta$. Since $-\beta$ is also a stationary point, we conclude, as in (7-96), that

$$J_n(t) \sim \frac{2}{2\pi} \sqrt{\frac{2\pi}{t \sin \beta}} \cos \left[t(\sin \beta - \beta \cos \beta) - \frac{\pi}{4} \right] \tag{7-102}$$

where β is given by (7-101).

If $n = t$ then $\beta = 0$, hence, the above result does not hold because $\mu''(\omega_0) = 0$. In this case we must use (7-100); since $\mu(\omega_0) = 0$ and $\mu'''(\omega_0) = -1$, we obtain

$$J_n(n) \sim \frac{1}{2\pi} \frac{\Gamma(1/3)}{\sqrt{3}} \left(\frac{6}{n} \right)^{1/3} \tag{7-103}$$

If n is not an integer then $J_n(t)$ is no longer given by (7-99); however, it can be shown that the estimates (7-102) and (7-103) are still valid (see E. T. Copson, *op. cit.*).

Chapter 8. Spectrum Analyzers

A spectrum analyzer is a physical system measuring the Fourier spectrum of an arbitrary signal $f(t)$. If $f(t)$ is given analytically or is recorded for every t, then it is possible to evaluate its Fourier transform $F(\omega)$, i.e., the integral (2-1), by an analogue or digital computer. The computer is then, in effect, a device measuring a definite integral and not an analyzer in the usual sense of the word. The term analyzer is reserved for systems as in Fig. 8-1, where their input is $f(t)$ and their one or more outputs $g(t)$ are functions of the time of observation and are supposed to measure, somehow, the desired transform $F(\omega)$. Since these systems are causal, their output $g(t)$ at a given time $t = t_0$ depends only on the values of $f(t)$ prior to $t = t_0$, whereas $F(\omega)$ depends on the entire history of $f(t)$ for t from $-\infty$ to $+\infty$. Therefore $g(t_0)$ cannot possibly measure $F(\omega)$ unless $f(t)$ equals zero for $t > t_0$ or its future is uniquely determined from its past, as is the case for periodic signals. Our object in this chapter is to establish the relationship between $g(t)$ and the unknown transform $F(\omega)$ and to examine the conditions for a satisfactory operation. We do not intend to present the various methods of realizing an analyzer and the ways of sensing its response.

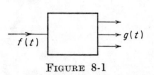

FIGURE 8-1

Spectrum analyzers are generally divided into two classes. In the first class the system has many outputs (Fig. 8-2), and at a given time $t = t_0$ each output measures one frequency component of the input. The desired spectrum is thus read *simultaneously*. In the second class (Fig. 8-8) there is only one output measuring the various frequency components at consecutive times. The spectrum is determined *sequentially*.

The choice of the analyzer and the interpretation of the results depends on the nature of the input signal $f(t)$. We shall consider signals that are either periodic, or aperiodic with finite energy. Aperiodic signals with finite nonzero power have no Fourier transforms and

144

are characterized by their power spectra (Sec. 12-3). The discussion of this chapter can be useful in the measurement of power spectra; for an intelligent interpretation of the results, however, probabilistic considerations are necessary, but outside the scope of this book.† ‡

8-1. Simultaneous Spectral Analysis

A simultaneous spectrum analyzer works on the principle of multiple *filtering*. The signal $f(t)$ is fed into an array of n bandpass filters as in Fig. 8-2, and the output of each filter is related to the Fourier

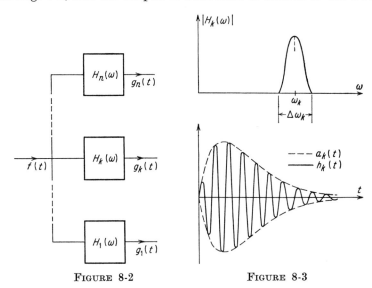

FIGURE 8-2   FIGURE 8-3

transform $F(\omega)$ of $f(t)$ evaluated at the center frequency of that filter. We denote by $H_k(\omega)$ the system function of the kth filter and by $h_k(t)$ its impulse response. $h_k(t)$ can be written in the form [see (7-51)]

$$h_k(t) = a_k(t) \cos \omega_k t + b_k(t) \sin \omega_k t \qquad (8\text{-}1)$$

For symmetrical filters the sine term above is zero [see (7-21)] with, possibly, a constant phase added to the cosine term. In this case $a_k(t)$ equals twice the impulse response of the equivalent low-pass filter. To simplify the notations, we shall assume that

$$h_k(t) = a_k(t) \cos \omega_k t \qquad (8\text{-}2)$$

as in Fig. 8-3; the general case (8-1) can be treated similarly. At a

† R. B. Blackman and J. W. Tukey, "The Measurement of Power Spectra," Dover Publications, New York, 1958.

‡ A. A. Kharkevich, "Spectra and Analysis" (translated from Russian), Consultants Bureau Enterprises, New York, 1960.

given time $t = t_0$, the response $g_k(t_0)$ of the kth filter is given by [see (5-10)]

$$g_k(t_0) = \int_{-\infty}^{\infty} f(t)h_k(t_0 - t)\,dt = \int_{-\infty}^{\infty} f(t)a_k(t_0 - t)\cos \omega_k(t_0 - t)\,dt \quad (8\text{-}3)$$

(for causal systems the integrand is zero for $t > t_0$). Expanding the cosine term in (8-3), we obtain

$$g_k(t_0) = \cos \omega_k t_0 \int_{-\infty}^{\infty} f(t)a_k(t_0 - t) \cos \omega_k t\,dt$$

$$+ \sin \omega_k t_0 \int_{-\infty}^{\infty} f(t)a_k(t_0 - t) \sin \omega_k t\,dt \quad (8\text{-}4)$$

Denoting by

$$F_{t0}(\omega) = R_{t0}(\omega) + jX_{t0}(\omega) = A_{t0}(\omega)e^{j\phi_{t0}(\omega)} \quad (8\text{-}5)$$

the Fourier transform of the function $f(t)a_k(t_0 - t)$, we have

$$F_{t0}(\omega) = R_{t0}(\omega) + jX_{t0}(\omega) = \int_{-\infty}^{\infty} f(t)a_k(t_0 - t)e^{-j\omega t}\,dt \quad (8\text{-}6)$$

and (8-4) can be written in the form

$$g_k(t_0) = R_{t0}(\omega_k) \cos \omega_k t_0 - X_{t0}(\omega_k) \sin \omega_k t_0$$

$$= A_{t0}(\omega_k) \cos [\omega_k t_0 + \phi_{t0}(\omega_k)] \quad (8\text{-}7)$$

The above result (8-7) can be stated as follows. The output $g_k(t)$ of the kth filter is an amplitude- and phase-modulated signal. At a given time $t = t_0$, its envelope $A_{t0}(\omega_k)$ equals the Fourier spectrum of the function $f(t)a_k(t_0 - t)$, evaluated at the frequency ω_k. The function $f(t)a_k(t_0 - t)$ is obtained by multiplying the given signal $f(t)$ by the envelope of the impulse response of the kth filter, shifted and mirrored about t_0 as in Fig. 8-4.

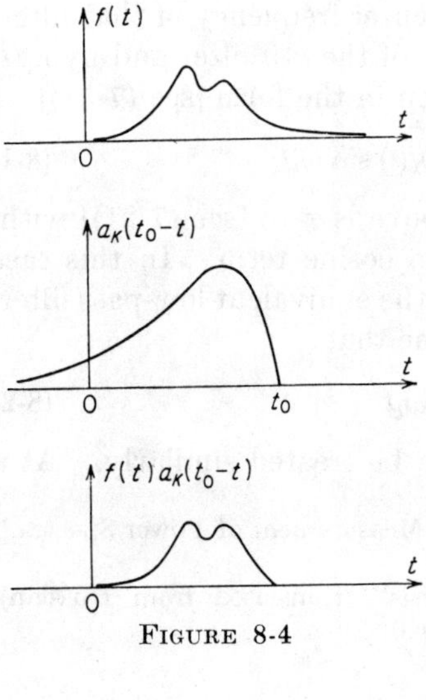

The Fourier transform $F_{t0}(\omega)$

$$f(t)a_k(t_0 - t) \leftrightarrow F_{t0}(\omega) \quad (8\text{-}8)$$

can be expressed in terms of $F(\omega)$, and the Fourier transform $a_k(\omega)$ of $a_k(t)$

$$a_k(t) \leftrightarrow a_k(\omega)$$

Indeed, from $a_k(t - t_0) \leftrightarrow e^{-j\omega t_0} a_k(\omega)$ we have

$$a_k(t_0 - t) \leftrightarrow e^{j\omega t_0} \overset{*}{a}_k(\omega) \quad (8\text{-}9)$$

and from the frequency convolution theorem (2-74) and (8-8) we obtain

$$F_{t0}(\omega) = \frac{1}{2\pi} F(\omega) * [e^{j\omega t_0} \overset{*}{a}_k(\omega)] \quad (8\text{-}10)$$

FIGURE 8-4

If the kth filter is symmetrical, then $a_k(\omega)$ is the system function of its
equivalent low-pass.

Time-limited signals. From the above we see that an analyzer
measures not $F(\omega)$ but the spectrum of $f(t)a_k(t_0 - t)$. However, if $f(t)$
is of finite duration, then, by properly selecting $a_k(t)$, we can determine
$F(\omega)$. Indeed, suppose that the given signal $f(t)$ starts at a time
$t = t_1$ and terminates at $t = t_2$ as in Fig. 8-5. We select the kth filter

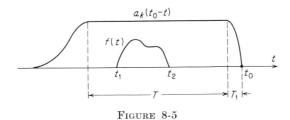

FIGURE 8-5

in such a way that its envelope $a_k(t)$ is constant in an interval T large
compared to $t_2 - t_1$

$$a_k(t) \simeq c_k = \text{constant} \qquad \text{for } T_1 < t < T_1 + T \qquad (8\text{-}11)$$

We then have

$$f(t)a_k(t_0 - t) \simeq c_k f(t) \qquad \text{for } t_2 + T_1 < t_0 < t_1 + T_1 + T \qquad (8\text{-}12)$$

Since the envelope of the response $g_k(t)$ equals the spectrum of
$f(t)a_k(t_0 - t)$, we conclude from (8-12) that in the time interval

$$t_2 + T_1 < t < t_1 + T_1 + T$$

this envelope is constant and equals $c_k A(\omega_k)$, where $A(\omega)$ is the spectrum of $f(t)$. Clearly, the characteristics of the various filters forming
the analyzer can differ from each other as long as their envelopes $a_k(t)$
remain constant over a sufficiently long time. The value of c_k must,
of course, be taken into account in the final recording of the spectrum.

Energy Considerations. We shall now show that, if the spectrum
$A(\omega)$ of $f(t)$ is approximately constant in the band $(\omega_k - \Delta\omega_k/2, \omega_k + \Delta\omega_k/2)$ of the kth filter (see Fig. 8-6)

$$A(\omega) \simeq A(\omega_k) \qquad \omega_k - \frac{\Delta\omega_k}{2} < \omega < \omega_k + \frac{\Delta\omega_k}{2} \qquad (8\text{-}13)$$

then the total energy of its output $g_k(t)$ is proportional to $A^2(\omega_k)$. The
determination of $A(\omega_k)$ can then be accomplished by a measurement,
say thermal, of that energy. The above requirement (8-13), loosely
phrased in the time domain, says that the duration of $f(t)$ should be
small compared to the duration of $a_k(t)$.

To prove our statement, we observe from (8-13) that

$$A(\omega)H_k(\omega) \simeq A(\omega_k)H_k(\omega) \tag{8-14}$$

Therefore, with $G_k(\omega)$ the transform of the output $g_k(t)$, we have

$$|G_k(\omega)| = A(\omega) |H_k(\omega)| \simeq A(\omega_k) |H_k(\omega)| \tag{8-15}$$

From Parseval's formula (2-75) we have

$$\int_{-\infty}^{\infty} g_k^2(t)\, dt = \frac{1}{2\pi} \int_{-\infty}^{\infty} |G_k(\omega)|^2\, d\omega \simeq \frac{A^2(\omega_k)}{2\pi} \int_{-\infty}^{\infty} |H_k(\omega)|^2\, d\omega = E_k A^2(\omega_k) \tag{8-16}$$

where

$$E_k = \frac{1}{2\pi} \int_{-\infty}^{\infty} |H_k(\omega)|^2\, d\omega$$

is the energy of $h_k(t)$ and is independent of $f(t)$. Thus the output energy is, indeed, proportional to the energy spectrum $A^2(\omega_k)$ of the input.

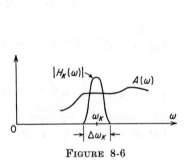

FIGURE 8-6

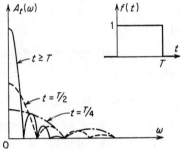

FIGURE 8-7

Example 8-1. As an illustration, suppose that

$$h_k(t) = e^{-\alpha t} \cos \omega_k t\, U(t) \tag{8-17}$$

A filter with the above response appears often in applications. It has a pair of complex poles and can be simply realized by a tuned circuit. If the duration of the input $f(t)$ is small compared to the time constant $1/\alpha$, then condition (8-13), necessary for the validity of (8-16), is satisfied. We easily find

$$\int_{-\infty}^{\infty} g_k^2(t)\, dt \simeq \frac{2\alpha^2 + \omega_k^2}{4\alpha(\alpha^2 + \omega_k^2)} A^2(\omega_k)$$

Running spectra. A useful concept in this discussion is the *running transform* $F_t(\omega)$ of $f(t)$ defined by

$$F_t(\omega) = \int_{-\infty}^{t} f(\tau)e^{-j\omega\tau}\, d\tau \tag{8-18}$$

Its magnitude $A_t(\omega) = |F_t(\omega)|$ is known as the *running spectrum* of $f(t)$. The above integral (8-18) agrees with (8-6) only if $a_k(t)$ is a unit step. It will not be confusing, however, to use the same notation. The function $F_t(\omega)$ is the Fourier transform of $f(\tau)U(t - \tau)$, if τ is the variable of integration in (2-1) and t a parameter

$$f(\tau)U(t - \tau) \longleftrightarrow F_t(\omega) \qquad (8\text{-}19)$$

We clearly have $F_\infty(\omega) = F(\omega)$, and since

$$U(t - \tau) \longleftrightarrow e^{j\omega t}\left[\pi\,\delta(\omega) - \frac{1}{j\omega}\right]$$

we conclude from (8-19) and (2-74) that

$$F_t(\omega) = \frac{1}{2\pi}\,F(\omega) * e^{j\omega t}\left[\pi\,\delta(\omega) - \frac{1}{j\omega}\right] \qquad (8\text{-}20)$$

Example 8-2. We shall determine the running spectrum of a rectangular pulse

$$f(t) = U(t) - U(t - T) \qquad (8\text{-}21)$$

From (8-18) we readily obtain

$$A_t(\omega) = \begin{cases} 0 & t < 0 \\[2mm] \dfrac{2\sin(\omega t/2)}{\omega} & 0 < t < T \\[2mm] \dfrac{2\sin(\omega T/2)}{\omega} & t > T \end{cases} \qquad (8\text{-}22)$$

The above spectrum is shown in Fig. 8-7 for $t = T/4,\ T/2,\ T$.

We return to the analyzer. If the impulse response $h_k(t)$ is step-modulated

$$h_k(t) = U(t)\cos\omega_k t \qquad (8\text{-}23)$$

then [see (8-6) and (8-18)] the instantaneous value of the envelope of its response equals the running spectrum $A_t(\omega_k)$ of the input $f(t)$. The analyzer (8-23) is in a sense ideal.

Sometimes one is interested in the spectrum of only a portion of $f(t)$ from $t - T$ to t, i.e., in the quantity

$$\int_{t-T}^{t} f(\tau)e^{-j\omega\tau}\,d\tau = F_{t-T}(\omega) - F_t(\omega) \qquad (8\text{-}24)$$

and the determination of its dependence on t. In this case the envelope $a_k(t)$ of all analyzer filters should be constant in the $(0,T)$ interval and zero outside this interval:

$$h_k(t) = [U(t) - U(t - T)]\cos\omega_k t \qquad (8\text{-}25)$$

Instantaneous Power Spectrum.† If we differentiate the *running energy spectrum* $A_t^{\,2}(\omega)$ of a function $f(t)$ with respect to t, we obtain a quantity

$$\rho_t(\omega) = \frac{\partial}{\partial t}\,A_t^{\,2}(\omega) \qquad (8\text{-}26)$$

† C. H. Page, Instantaneous Power Spectra, *J. Appl. Phys.*, pp. 103–106, 1952.

known as the *instantaneous power spectrum*. This spectrum has the property†

$$\frac{1}{2\pi} \int_{-\infty}^{\infty} \rho_t(\omega) \, d\omega = f^2(t) \tag{8-27}$$

Indeed, from (8-19) and Parseval's formula (2-75) we have

$$\frac{1}{2\pi} \int_{-\infty}^{\infty} A_t^2(\omega) \, d\omega = \int_{-\infty}^{\infty} f^2(\tau) U^2(t-\tau) \, d\tau = \int_{-\infty}^{t} f^2(\tau) \, d\tau \tag{8-28}$$

Differentiating with respect to t, we obtain (8-27). It is easy to see from (8-26) that, if an analyzer is ideal in the sense (8-23) and a sensing device measures the square of the envelope $A_t(\omega_k)$ of its response $g_k(t)$, then the rate of change of this measurement equals $\rho_t(\omega_k)$.

8-2. Sequential Spectral Analysis

A sequential analyzer either consists of a single filter whose frequency characteristics vary with time, or it works on the principle of *modulation*. The first scheme involves time-varying elements, and a general analysis cannot be given unless the actual configuration of the filter is known. We shall, therefore, discuss only the second scheme. The given signal $f(t)$ and a frequency-modulated signal $s(t)$ are inserted into a multiplier M as in Fig. 8-8, and the output $f(t) s(t)$ is fed into a filter $H(\omega)$. This filter is either low-pass or bandpass. The analysis is similar for both cases and will be given only for the low-pass system. We shall assume that the modulated signal $s(t)$ is of the form

$$s(t) = \cos \phi(t) = \cos \alpha t^2 \tag{8-29}$$

Its instantaneous frequency $\omega(t)$, given by

$$\omega(t) = \frac{d\phi(t)}{dt} = 2\alpha t \tag{8-30}$$

† One can similarly derive the more general result

$$f(t)f(t-|\tau|) = \frac{1}{2\pi} \int_{-\infty}^{\infty} \rho_t(\omega)e^{j\omega\tau} \, d\omega$$

From the above it readily follows that the average of $\rho_t(\omega)$ equals the power, spectrum $S(\omega)$ of $f(t)$ (see Sec. 12.3). Using probabilistic concepts, we can phrase this result as follows: if $f(t)$ is a stationary random process then, with E for expected value, we have

$$E\{f(t)f(t-\tau)\} = \frac{1}{2\pi} \int_{-\infty}^{\infty} E\{\rho_t(\omega)\}e^{j\omega\tau} \, d\omega$$

But the expected value of $f(t)f(t-\tau)$ equals the autocorrelation $R(\tau)$ of $f(t)$ whose Fourier transform equals $S(\omega)$; hence

$$E\{\rho_t(\omega)\} = S(\omega)$$

varies linearly with time as in Fig. 8-9a. The exact determination of
the output $g(t)$ of the filter is in general complicated; however, if the
slope 2α of the $\omega(t)$ line is small, then an approximate analysis can be
developed.

With $h(t)$ the impulse response of the filter, its response $g(t_0)$ to the
input $f(t) \cos \phi(t)$ is given by

$$g(t_0) = \int_{-\infty}^{\infty} f(t) \cos \phi(t) h(t_0 - t)\, dt \qquad (8\text{-}31)$$

If the duration of $h(t)$ is T, then (8-31) takes the form

$$g(t_0) = \int_{t_0-T}^{t_0} f(t) \cos \phi(t) h(t_0 - t)\, dt \qquad (8\text{-}32)$$

Thus, in the above integral, the variable of integration t takes values
only in the $(t_0 - T, t_0)$ interval. We now assume α so small that in

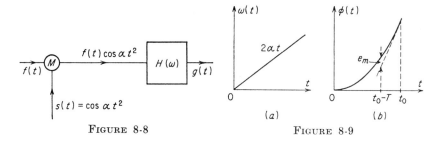

FIGURE 8-8   FIGURE 8-9

this interval the curve $\phi(t)$ can be approximated by its tangent $\phi_0(t)$
at the point $t = t_0$ (Fig. 8-9b)

$$\phi(t) \simeq 2\alpha t_0 t - \alpha t_0{}^2 = \phi_0(t) \qquad t_0 - T < t < t_0 \qquad (8\text{-}33)$$

In this approximation, the maximum error e_m occurs at $t = t_0 - T$
and is given by

$$e_m = \phi(t_0 - T) - [2\alpha t_0(t_0 - T) - \alpha t_0{}^2] = \alpha T^2 \qquad (8\text{-}34)$$

Thus, if $$\alpha T^2 \ll 1 \qquad (8\text{-}35)$$

then the effect of the error on the value of $\cos \phi(t)$ in the integral (8-32)
can be neglected. With this assumption we obtain, inserting (8-33)
into (8-32),

$$g(t_0) \simeq \int_{t_0-T}^{t_0} f(t) \cos (\alpha t_0{}^2 - 2\alpha t_0 t) h(t_0 - t)\, dt$$

$$= \cos \alpha t_0{}^2 \int_{t_0-T}^{t_0} f(t) h(t_0 - t) \cos 2\alpha t_0 t\, dt$$

$$+ \sin \alpha t_0{}^2 \int_{t_0-T}^{t_0} f(t) h(t_0 - t) \sin 2\alpha t_0 t\, dt \qquad (8\text{-}36)$$

The last two integrals above are equal to the real and imaginary parts of the Fourier transform of $f(t)h(t_0 - t)$, evaluated at $\omega = 2\alpha t_0$. Since $2\alpha t_0$ is the instantaneous frequency of the modulator, we conclude, reasoning as in (8-4), that the output $g(t)$ of the analyzer is a modulated signal whose envelope, at a given time $t = t_0$, equals the Fourier spectrum of the function $f(t)h(t_0 - t)$, evaluated at the instantaneous frequency $2\alpha t_0$ of the carrier.

If the filter is a pulse-forming network $h(t) = U(t) - U(t - T)$, then the envelope of the output equals the magnitude of

$$F_{t0}(2\alpha t_0) - F_{t0-T}(2\alpha t_0)$$

where $F_{t0}(\omega)$ is the running transform (8-17) of $f(t)$.

The above result was obtained under assumption (8-35). This assumption can also be phrased as follows: since, in a time interval equal to the duration T of $h(t)$, the instantaneous frequency of $s(t)$ increases by $\Delta\omega = 2\alpha T$ [see (8-30)], (8-35) is equivalent to

$$\Delta\omega T \ll 1$$

But $1/T$ is of the order of magnitude of the bandwidth of the output filter (see Sec. 4-4); therefore during the time T the increase in the frequency of the modulator should be small compared to this bandwidth.

Error Correction. In deriving (8-36) from (8-32), we replaced $\phi(t)$ by its tangent $\phi_0(t)$ as in (8-33), introducing an error

$$e(t) = \phi(t) - \phi_0(t) = \alpha(t_0 - t)^2 \tag{8-37}$$

In the following we shall give a first-order correction to the resulting error $\Delta g(t)$ in the evaluation of $g(t)$. For this purpose we shall use the approximation

$$\cos e(t) \simeq 1 \qquad \sin e(t) \simeq e(t)$$

or the equivalent

$$\cos \phi(t) \simeq \cos \phi_0(t) - e(t) \sin \phi_0(t) \tag{8-38}$$

From (8-38) and (8-32) we see that the first-order correction is given by

$$
\begin{aligned}
\Delta g(t) &\simeq - \int_{t_0-T}^{t_0} \sin \phi_0(t)\alpha(t - t_0)^2 f(t)h(t_0 - t)\, dt \\
&= - \cos \alpha t_0^2 \int_{t_0-T}^{t_0} \alpha(t_0 - t)^2 f(t)h(t_0 - t) \sin 2\alpha t_0 t \, dt \\
&\quad - \sin \alpha t_0^2 \int_{t_0-T}^{t_0} \alpha(t_0 - t)^2 f(t)h(t_0 - t) \cos 2\alpha t_0 t \, dt
\end{aligned}
\tag{8-39}
$$

It is a modulated signal like (8-36), with envelope equal to the Fourier spectrum of the function $\alpha(t_0 - t)^2 f(t)h(t_0 - t)$, evaluated at the instantaneous carrier frequency $2\alpha t_0$.

8-3. Periodic Signals

Periodic signals are spectrally analyzed in one of the above schemes; the interpretation of the results, however, requires special considerations. We shall consider two cases depending on the available measurement time. In the first case we shall assume that the transients have died down and steady state is reached. This means that the interval of time, from the moment the input $f(t)$ is applied to the moment of observation of the output, is larger than the duration of the impulse response of the analyzer, so that the observed output is periodic. In the second case the transient effects will be taken into consideration. In the simultaneous analysis the steady-state condition is, usually, fulfilled; in the sequential analysis, however, this is not true, because the time of observation of each frequency component is limited by the design of the analyzer.

Steady state. We expand the periodic function $f(t)$ into a Fourier series

$$f(t) = \sum_{n=-\infty}^{\infty} \alpha_n e^{j n \omega_0 t} \qquad \omega_0 = \frac{2\pi}{T} \qquad (8\text{-}40)$$

where T is its period. The steady-state response $g_k(t)$ of the output of the kth filter of Fig. 8-2 is given by

$$g_k(t) = \sum_{n=-\infty}^{\infty} \alpha_k H_k(n\omega_0) e^{j n \omega_0 t} \qquad (8\text{-}41)$$

If the bandwidth $\Delta\omega_k$ of each filter is smaller than the fundamental ω_0 of the input

$$\Delta\omega_k \leq \omega_0 \qquad (8\text{-}42)$$

then $g_k(t)$ will contain at most one harmonic of $f(t)$ if such harmonic is in the band $(\omega_k - \Delta\omega_k/2,\ \omega_k + \Delta\omega_k/2)$. Suppose that each analyzer filter is ideal, as in Fig. 8-10. If

$$g_k(t) = \alpha_m e^{j\omega_m t} \qquad (8\text{-}43)$$

then $f(t)$ contains a frequency component ω_m in the band of the kth filter

$$\omega_k - \frac{\Delta\omega_k}{2} < \omega_m < \omega_k + \frac{\Delta\omega_k}{2} \qquad (8\text{-}44)$$

of amplitude $|\alpha_m|$. Thus, there is a $\Delta\omega_k$ indeterminancy in the value of ω_m. A more exact determination of ω_m is possible if the characteristics of adjacent bands overlap as in Fig. 8-11, where we assumed triangular forms. From the responses

$$g_k(t) = \beta_k e^{j\omega_m t} \qquad g_{k-1}(t) = \beta_{k-1} e^{j\omega_m t} \qquad (8\text{-}45)$$

of the kth and $(k - 1)$th filter, we conclude that

$$\omega_m = \omega_{k-1} + \frac{\Delta\omega}{2}\frac{|\beta_k|}{|\beta_{k-1}| + |\beta_k|} \tag{8-46}$$

We shall now consider a sequential analyzer consisting of a single time-varying filter. We assume that its characteristics vary slowly

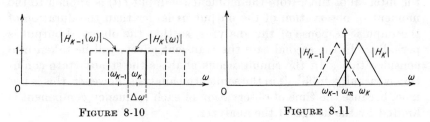

FIGURE 8-10   FIGURE 8-11

enough (quasi-static) so that $H(\omega)$ can be given the static interpretation. We denote by ω_r the frequency for which $|H(\omega)|$ is maximum (Fig. 8-12a), and we assume that the time dependence of $H(\omega)$ is such that, with increasing time, the entire curve moves to the right, keeping its shape constant. If ω_r varies slowly, then the steady-state response to $f(t) = e^{j\omega_m t}$ is given by

$$g(t) = \beta(\omega_r)e^{j\omega_m t} \tag{8-47}$$

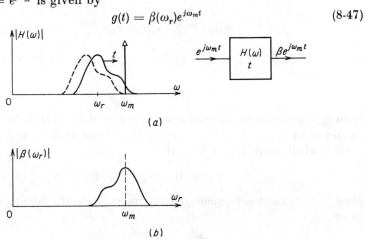

FIGURE 8-12

where β is a function of time; hence of ω_r. The plot of $|\beta(\omega_r)|$ is shown in Fig. 8-12b. It is obtained by moving $H(\omega)$ so that $\omega_r = \omega_m$, and mirroring about a vertical axis through ω_m. Thus, from the location of ω_r at maximum output and the corresponding amplitude of β, one can determine the amplitude and frequency of the input.

If instead of a variable filter we use a modulator as in Fig. 8-8, then the results are similar. Indeed, the input to the low-pass filter $H(\omega)$ is

$$e^{j\omega_m t} \cos \alpha t^2 = \frac{e^{j(\omega_m t + \alpha t^2)} + e^{j(\omega_m t - \alpha t^2)}}{2}$$

For slowly varying α, only the second term above contributes to the steady-state value of the response

$$g(t) = \beta(t)e^{j\omega_m t}$$

and the filter responds only when the instantaneous frequency

$$\omega_m - 2\alpha t$$

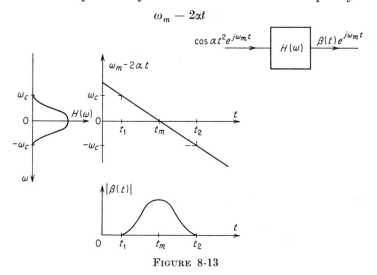

FIGURE 8-13

lies in its band $(-\omega_c, \omega_c)$, i.e., for

$$t_1 = \frac{\omega_m - \omega_c}{2\alpha} < t < \frac{\omega_m + \omega_c}{2\alpha} = t_2$$

(Fig. 8-13). The amplitude $\beta(t)$ of the response is a curve similar to $H(\omega)$

$$\beta(t) = \tfrac{1}{2}H(\omega_m - 2\alpha t)$$

It attains its maximum for $t = t_m = \omega_m/2\alpha$, when the instantaneous frequency $2\alpha t$ of the modulator equals ω_m.

Suppose now that the bandwidth of our filters is not smaller than the fundamental ω_0 of the input $f(t)$. In such a case the output $g(t)$ will contain more than one frequency component. To examine the effect of neighboring components on the selectivity of each filter, we shall assume that $f(t)$ contains two components contributing to the output

$$f(t) = \alpha_m e^{j\omega_m t} + \alpha_s e^{j\omega_s t} \tag{8-48}$$

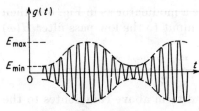

FIGURE 8-14

The steady-state value of $g(t)$ is given by

$$g(t) = \alpha_m H(\omega_m)e^{j\omega_m t} + \alpha_s H(\omega_s)e^{j\omega_s t}$$

$$= e^{j\omega_m t}[\alpha_m H(\omega_m)$$

$$+ \alpha_s H(\omega_s)e^{j(\omega_s - \omega_m)t}] \qquad (8\text{-}49)$$

Thus $g(t)$ is a modulated signal with carrier frequency ω_m and with envelope the bracket in (8-49). $g(t)$ is shown in Fig. 8-14 for

$$\omega_s - \omega_m \ll \omega_m \quad \text{(beats)}$$

When the two terms in the above bracket are in phase or $180°$ out of phase, the envelope $E(t)$ is maximum or minimum:

$$\begin{aligned} E_{\max} &= |\alpha_m H(\omega_m)| + |\alpha_s H(\omega_s)| \\ E_{\min} &= |\alpha_m H(\omega_m)| - |\alpha_s H(\omega_s)| \end{aligned} \qquad (8\text{-}50)$$

Example 8-3. Suppose that the filter is a tuned circuit

$$H(\omega) = \frac{1}{(j\omega - p_1)(j\omega - p_2)} \qquad \left.\begin{aligned} p_1 \\ p_2 \end{aligned}\right\} = -\alpha \pm j\omega_r \qquad (8\text{-}51)$$

If we assume that $\alpha \ll \omega_r$ (high-Q case), then $H(\omega)$ takes significant values only for ω near $\pm\omega_r$. We thus have for $\omega > 0$

$$j\omega - p_2 \simeq 2j\omega_r \qquad H(\omega) \simeq \frac{1}{2j\omega_r[\alpha + j(\omega - \omega_r)]} \qquad (8\text{-}52)$$

and $|H(\omega)|$ is given by the familiar resonance curve

$$|H(\omega)| \simeq \frac{1}{2\alpha\omega_r\sqrt{1 + (\omega - \omega_r)^2/\alpha^2}} \qquad (8\text{-}53)$$

Inserting $|H(\omega)|$ into (8-50), we obtain

$$2\alpha\omega_r E_{\max} = \frac{|\alpha_m|}{\sqrt{1 + (\omega_m - \omega_r)^2/\alpha^2}} + \frac{|\alpha_s|}{\sqrt{1 + (\omega_s - \omega_r)^2/\alpha^2}} \qquad (8\text{-}54)$$

If the filter is time-variable as in Fig. 8-12, then the maximum $E_{\max}$ of the envelope of the response is a function of the location of ω_r relative to ω_m and ω_s. For $\omega_s - \omega_m \gg \alpha$ the maxima of $E_{\max}$ are obtained for $\omega_r \simeq \omega_s$ and $\omega_r \simeq \omega_m$. Thus from the location of these maxima one can determine the frequencies of the input. However, if $\omega_s - \omega_m$ is of the order of α, then this is no longer true. The maxima shift closer to each other, and for

$$\omega_s - \omega_m < A\alpha$$

the curve has only a single maximum. The constant A depends on $|\alpha_s/\alpha_m|$. For $|\alpha_s| = |\alpha_m|$ its value equals $\sqrt{2}$. In Fig. 8-15 we have shown

E_{max} for $\omega_s - \omega_m = 3\alpha$ and $|\alpha_s| = |a_m|, |\alpha_s| = 3|\alpha_m|$.   For $\omega_s - \omega_m < \alpha\sqrt{2}$
the analyzer can no longer discriminate between ω_s and ω_m.   If α_s and α_m
differ considerably from each other, the analyzer cannot discriminate even for
large values of $\omega_s - \omega_m$, as we see from Fig. 8-15.   We should point out that
even if the E_{max} curve has a single maximum, one can still determine the
two frequency components ω_s and ω_m.   However, this can be done only with
a measurement of the entire $E_{max}(\omega_r)$ curve and a proper interpretation of the
result.

Comment.   The results of this section were based on the assumption
that $f(t)$ was periodic.   This fact should be known and does not follow
from the periodicity of the response $g(t)$.   It is possible, although un-
likely, for a nonperiodic input to cause a periodic output.   Indeed,

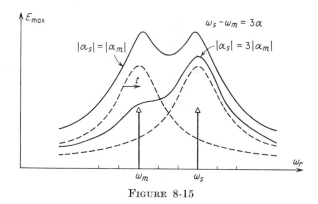

FIGURE 8-15

suppose first that $f(t) = e^t$ and that the impulse response of the filter
is given by

$$h(t) = (3e^{-2t} - 1)U(t)$$

Then
$$g(t) = \int_{-\infty}^{t} e^\tau h(t - \tau)\, d\tau = 0$$

Therefore, the response to the nonperiodic input $f(t) = e^t + e^{j\omega_0 t}$ is
periodic,

$$g(t) = H(\omega_0)e^{j\omega_0 t}$$

where $H(\omega)$ is the system function.   Thus, from the periodicity of $g(t)$
it is false to deduce the periodicity of $f(t)$.   It is true that in our example $f(t)$ tends to infinity for $t \to \infty$; however, one can always trun-
cate it above $t = t_0$.   The response $g(t)$ will then be periodic for $t < t_0$.

Transients.   In the preceding discussion it was assumed that the
response $g(t)$ reached steady state.   If the measurement time is limited,
this assumption is no longer true, and the operation of the analyzer is
not determined from the static characteristic $H(\omega)$.   In the following

we shall investigate the influence of time on the measurement of periodic signals. Suppose that the signal

$$f(t) = e^{j\omega_m t}$$

is applied at $t = 0$. The response $g(t)$ of a filter is given by

$$g(t) = \int_0^t e^{j\omega_m(t-\tau)} h(\tau)\, d\tau = e^{j\omega_m t} \int_0^t h(\tau) e^{-j\omega_m \tau}\, d\tau \qquad (8\text{-}55)$$

It is a modulated signal with envelope

$$H_t(\omega_m) = \int_0^t h(\tau) e^{-j\omega_m t}\, d\tau \qquad (8\text{-}56)$$

Since $h(t) = 0$ for $t < 0$, $H_t(\omega)$ is the running transform of $h(t)$ [see (8-18)]. For a given t, the selectivity of the analyzer depends not on $H(\omega)$ but on the variation of $H_t(\omega)$ with ω. In general, this curve broadens with decreasing t, so that the discrimination power of the instrument is reduced. In the next example we shall consider a tuned circuit and will examine in some detail the frequency dependence of $H_t(\omega)$ and the design considerations.

Example 8-4. We assume that the filter of the analyzer is a high-Q-tuned circuit with impulse response

$$h(t) = e^{-\alpha t} \cos \omega_r t U(t) \qquad \alpha \ll \omega_r$$

Since $\alpha \ll \omega_r$, this filter has essentially the same characteristic as (8-51). Its running transform is given by

$$
\begin{aligned}
H_t(\omega) &= \int_0^t e^{-\alpha \tau} \cos \omega_r \tau e^{-j\omega \tau}\, d\tau \\
&= \frac{e^{-\alpha \tau} e^{j(\omega - \omega_r)\tau}}{2[-\alpha - j(\omega - \omega_r)]} \bigg|_0^t + \frac{e^{-\alpha \tau} e^{j(\omega + \omega_r)\tau}}{2[-\alpha - j(\omega + \omega_r)]} \bigg|_0^t
\end{aligned}
$$

For positive ω, only the first fraction above takes significant values; hence for $\omega > 0$

$$|H_t(\omega)| = A_t(\omega) \simeq \frac{1}{2} \sqrt{\frac{1 + e^{-2\alpha t} - 2e^{-\alpha t} \cos(\omega - \omega_r)t}{\alpha^2 + (\omega - \omega_r)^2}} \qquad (8\text{-}57)$$

To normalize, we introduce the variables

$$\frac{\omega - \omega_r}{\alpha} = \delta \qquad \alpha t = \tau \qquad (8\text{-}58)$$

and with

$$S_\tau(\delta) = \sqrt{\frac{1 + e^{-2\tau} - 2e^{-\tau} \cos \delta\tau}{1 + \delta^2}} \qquad (8\text{-}59)$$

the dynamic response of the filter is given by

$$A_t(\omega) \simeq \frac{1}{2\alpha} S_{\alpha t}\left(\frac{\omega - \omega_r}{\alpha}\right) \qquad (8\text{-}60)$$

The function $S_\tau(\delta)$ is a one-parameter family of curves and is plotted in Fig. 8-16 for $\tau = .2, .4, .8, \infty$. We observe that it broadens with decreasing τ. The circles in this figure indicate the half-power points, i.e., the values $\bar{\delta}_\tau$ for which

$$S_\tau(\bar{\delta}_\tau) = \frac{S_\tau(0)}{\sqrt{2}} \qquad (8\text{-}61)$$

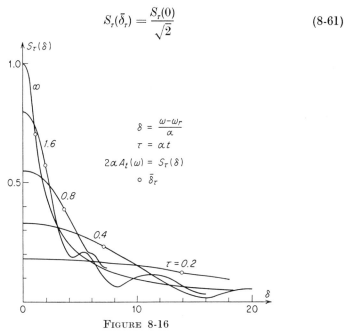

FIGURE 8-16

In Fig. 8-17 we have plotted the product $\tau \bar{\delta}_\tau$ as a function of τ. It is interesting to observe that this product is approximately constant for $\tau < 3$:

$$\tau \bar{\delta}_\tau \simeq 3 \qquad \tau < 3 \qquad (8\text{-}62)$$

We now define the dynamic bandwidth $\Delta\omega_t$ of $A_t(\omega)$ as the interval between the two half-power points ω_1 and ω_2

$$A_t(\omega_1) = A_t(\omega_2) = \frac{A_t(0)}{\sqrt{2}} \qquad \Delta\omega_t = \omega_2 - \omega_1$$

With a proper change in scales, the curve of Fig. 8-17 gives also the product $\Delta\omega_t t$ as a function of t. From (8-62) and (8-58) we conclude that, since $(\omega - \omega_r)t = \delta\tau$, we have

$$\Delta\omega_t t \simeq 6 \qquad \text{for } \alpha t < 3$$

Thus the dynamic bandwidth $\Delta\omega_t$ is inversely proportional to the measurement time, independently of the value of α. For a given t, the bandwidth $\Delta\omega_t$ of the filter is about $6/t$, no matter how small α

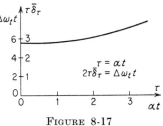

FIGURE 8-17

is chosen. The value of α introduces only a scale factor. Similarly, in order to obtain a certain dynamic bandwidth $\Delta\omega_t$, the measurement time must be about $6/\Delta\omega_t$. The above results hold, of course, for a dynamic high Q, i.e., for

$$\Delta\omega_t\omega_r \ll 1$$

In Eq. (8-55) and in Example 8-4 we interpreted the output of the filter as a modulated wave with carrier frequency ω_m and envelope $| H_t(\omega_m) |$. As in the steady-state case, this envelope depends on the frequency ω_m of the input, and its shape determines the power of resolution of the analyzer at a given time. If we had applied the analysis in Sec. 8-1, we would have interpreted the output $g(t)$ as a modulated wave with carrier frequency equal to the resonance frequency ω_r of the filter, and with envelope the quantity in Eq. (8-6), i.e., the running spectrum of the input, properly weighted. Although the result would agree with Eq. (8-55), its meaning would not be as clear as the meaning of the running spectrum $H_t(\omega)$.

Problems

15. Show that a system with the following properties:

(*a*) it is linear

(*b*) if the input $f(t)$ is specified for $t < t_0$, then the output $g(t)$ is uniquely determined for $t < t_0$

is causal.

16.† Show that if $f(t)$ is causal and the real part $R(\omega)$ of its Fourier transform decreases monotonically to zero

$$R(\omega) \geq 0 \qquad \frac{dR}{d\omega} \leq 0 \qquad R(\infty) = 0$$

then
$$|f(t)| \leq \frac{2R(0)}{\pi t}$$

17.† Prove that if a real linear system is causal and the imaginary part $X(\omega)$ of its system function $H(\omega) = R(\omega) + jX(\omega)$ is negative, then its step response $a(t)$ satisfies the following inequality:

$$a(0) \leq a(t) \leq 2R(0) - a(0) \tag{i}$$

18. The input to an ideal low-pass filter

$$H(\omega) = \begin{cases} A_0 \, e^{-j\omega t_0} & |\omega| < \omega_c \\ 0 & |\omega| > \omega_c \end{cases}$$

is a train of pulses

$$f^*(t) = Tf(t) \sum_{n=-\infty}^{\infty} \delta(t - nT)$$

whose envelope $f(t)$ has a band-limited spectrum

$$|F(\omega)| = 0 \qquad \text{for } |\omega| > \omega_c$$

Prove that if $T < \pi/\omega_c$, then the response $g(t)$ of the filter is given by

$$g(t) = A_0 f(t - t_0)$$

19. *Band-limited interpolation.* Given an arbitrary function $f(t)$ and a constant T, find a function $g(t)$ with a band-limited spectrum $[G(\omega) = 0$ for $|\omega| > \pi/T = \omega_c]$ such that $g(nT) = f(nT)$ for all integer n.

---

† A. H. Zemanian, Bounds Existing on the Time and Frequency Responses of Various Types of Networks, *Proc. IRE*, vol. 42, pp. 835–839, May, 1954.

Answer.

$$g(t) = \sum_{n=-\infty}^{\infty} f(nT) \frac{\sin(\omega_c t - n\pi)}{\omega_c t - n\pi} \tag{i}$$

A *sampler* S is a linear (but not time-invariant) device such that its output to a given input $f(t)$ is given by

$$f^*(t) = T \sum_{n=-\infty}^{\infty} f(nT)\, \delta(t - nT)$$

Show that the above function $g(t)$ can be obtained by cascading a sampler with an ideal low-pass filter as in Fig. P-19. If the information contained in $f(t)$ is the sequence of numbers $f(nT)$, the above operation preserves this information and *shrinks* the spectrum of the signal.

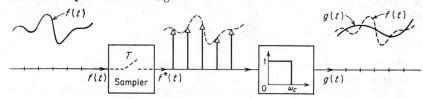

FIGURE P-19

20. *Frequency shifting.* The input to the system of Fig. P-19 is an exponential $f(t) = e^{j\omega_1 t}$ where $(2m - 1)\omega_c < \omega_1 < (2m + 1)\omega_c$ and m is an integer. Show that the output is given by $g(t) = e^{j\omega_0 t}$ where $\omega_0 = \omega_1 - 2m\omega_c$. Thus, a band-limited interpolation of an exponential is an exponential of lower frequency.

21. (a) *Rectification.* A full-wave rectifier is a device such that its response $g(t)$ to an input $f(t)$ is given by $g(t) = |f(t)|$. Show that if $f(t)$ is amplitude-modulated $f(t) = \phi(t) \sin \omega_0 t$, with positive envelope $\phi(t) \geq 0$, then the transform $G(\omega)$ of the output is given by

$$G(\omega) = -\frac{2}{\pi} \sum_{n=-\infty}^{\infty} \frac{\Phi(\omega - 2n\omega_0)}{4n^2 - 1} \tag{i}$$

where $\Phi(\omega)$ is the transform of $\phi(t)$.

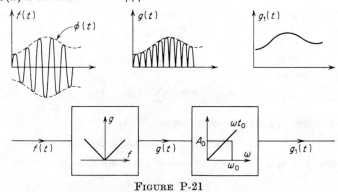

FIGURE P-21

(b) *Detection.* The output $g(t)$ of the above rectifier is inserted into an ideal low-pass filter of cutoff frequency ω_0 (Fig. P-21). Show that if $\Phi(\omega) = 0$ for $|\omega| > \omega_0$, then the output $g_1(t)$ of the filter is of the same form as $\phi(t)$

$$g_1(t) = \frac{2A_0}{\pi} \phi(t - t_0) \tag{ii}$$

22. The input to an ideal low-pass filter $H(\omega) = A_0 e^{-j\omega t_0} p_{\omega c}(\omega)$ is a step-modulated signal $f(t) = E \cos \omega_0 t U(t)$. Show that its response is given by

$$g(t) = \frac{EA_0}{2\pi}[\pi + Si(\omega_c - \omega_0)(t - t_0) + Si(\omega_c + \omega_0)(t - t_0)] \cos \omega_0(t - t_0)$$

$$+ \frac{EA_0}{2\pi}[Ci(\omega_c - \omega_0)(t - t_0) - Ci(\omega_c + \omega_0)(t - t_0)] \sin \omega_0(t - t_0) \qquad \text{(i)}$$

where
$$Si\, x = \int_0^x \frac{\sin y}{y}\, dy \qquad Ci\, x = -\int_x^\infty \frac{\cos y}{y}\, dy$$

Show that

$$\lim_{t \to \infty} g(t) = \begin{cases} EA_0 \cos \omega_0(t - t_0) & \omega_0 < \omega_c \\ 0 & \omega_0 > \omega_c \end{cases} \qquad \text{(ii)}$$

23. The system function of a linear filter is given by

$$H(\omega) = \begin{cases} A_0 e^{-j\theta_0} & \omega > 0 \\ A_0 e^{j\theta_0} & \omega < 0 . \end{cases}$$

as in Fig. P-23a. Determine its impulse response $h(t)$.

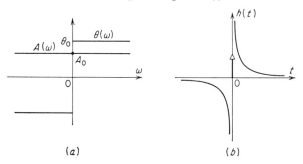

(a) (b)

FIGURE P-23

Answer.

$$h(t) = A_0 \cos \theta_0\, \delta(t) + A_0 \frac{\sin \theta_0}{\pi t} \qquad \text{(i)}$$

as in Fig. P-23b.

24. Given an all-pass filter $|H(j\omega)| = 1$ with monotone-increasing phase lag as in Fig. P-24a. The input $f(t)$ is a signal whose Fourier spectrum decreases with increasing ω

$$\frac{d\, |F(\omega)|}{d\omega} \le 0 \qquad \text{(i)}$$

t_0 is a given constant. Show that the RMS error E between the output $g(t)$ and the delayed input $f(t - t_0)$

$$E = \int_{-\infty}^{\infty} [g(t) - f(t - t_0)]^2\, dt \qquad \text{(ii)}$$

is minimum if $\theta(\omega)$ increases linearly to its final value $\theta(\infty)$ as in (6-65)

$$\theta(\omega) = \begin{cases} \omega t_0 & |\omega| < \omega_c \\ \theta(\infty) & \omega > \omega_c \end{cases} \qquad \omega_c = \frac{\theta(\infty)}{t_0} \qquad \text{(iii)}$$

Special case.† $f(t) = U(t)$; the ideal all-pass (iii) minimizes the error

$$\int_{-\infty}^{\infty} [U(t - t_0) - a(t)]^2\, dt$$

where $a(t)$ is its step response.

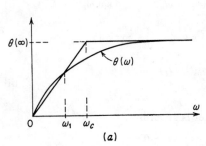 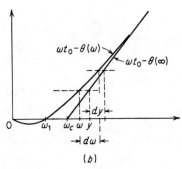

(a)   (b)

FIGURE P-24

25. Show that if the frequency characteristics of a low-pass filter are (Fig. P-25)

$$A(\omega) = \begin{cases} A_0 & |\omega| < \omega_c \\ 0 & |\omega| > \omega_c \end{cases} \qquad \begin{aligned} \theta(\omega) &= t_0\omega + \alpha\omega^2 & \omega > 0 \\ \theta(-\omega) &= -\theta(\omega) \end{aligned}$$

then its impulse response $h(t)$ is given by

$$h(t) = \frac{A_0}{\sqrt{2\pi\alpha}} \Big\{ [C(\omega_c \sqrt{2\alpha/\pi} - \tau) + C(\tau)] \cos \frac{\pi}{2} \tau^2$$

$$+ [S(\omega_c \sqrt{2\alpha/\pi} - \tau) + S(\tau)] \sin \frac{\pi}{2} \tau^2 \Big\}$$

where $\qquad \tau = \dfrac{t - t_0}{\sqrt{2\pi\alpha}}$

FIGURE P-25

and $\qquad C(\tau) = \displaystyle\int_0^\tau \cos \frac{\pi}{2} y^2\, dy \qquad S(\tau) = \displaystyle\int_0^\tau \sin \frac{\pi}{2} y^2\, dy$

are the *Fresnel* integrals.

26. The phase distortion of the filter in Prob. 25 is small

$$\alpha\omega_c^2 \ll 1$$

Show that its step response $a(t)$ is given by

$$a(t) \simeq a_0(t) + \frac{\alpha A_0}{\pi} \left[\frac{2 \sin^2 [\omega_c(t - t_0)/2]}{(t - t_0)^2} - \frac{\omega_c \sin \omega_c(t - t_0)}{t - t_0} \right] \tag{i}$$

where $a_0(t)$ is the response of the same filter without phase distortion.

† G. Wunsch, Bemerkungen und Ergänzungen zur Küpfmüllerschen Ein-schwingformel, *Hochfrequenztechnik und Elektroakustik*, vol. 69, no. 1, February, 1960.

27. The signal

$$f(t) = \begin{cases} \cos \phi(t) & 0 \le t \le T \\ 0 & \text{elsewhere} \end{cases} \qquad \phi(t) = at^2 + bt + c$$

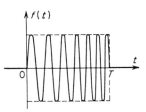

has a pulse envelope and a linearly increasing
frequency from ω_1 to ω_2 (Fig. P-27)

$$\frac{d\phi(t)}{dt} = 2at + b \qquad b = \omega_1 \qquad a = \frac{\omega_2 - \omega_1}{2T}$$

Show that its Fourier transform $F(\omega)$ is given by

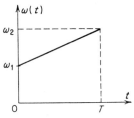

$$2\sqrt{\frac{\omega_2 - \omega_1}{\pi T}}\, F(\omega)$$

$$= e^{j\delta_1} \int_{y_1}^{y_3} e^{-j\pi y^2/2}\, dy + e^{-j\delta_2}\int_{y_2}^{y_4} e^{j\pi y^2/2}\, dy$$

$$= e^{j\delta_1}\{[C(y_3) - C(y_1)] - j[S(y_3) - S(y_1)]\}$$

$$+ e^{-j\delta_2}\{[C(y_4) - C(y_2)] + j[S(y_4) - S(y_2)]\}$$

FIGURE P-27

where

$$\delta_1 = \frac{T(\omega_1 + \omega)^2}{2(\omega_2 - \omega_1)} - c \qquad \delta_2 = \frac{T(\omega_1 - \omega)^2}{2(\omega_2 - \omega_1)} - c$$

$$y_1 = (\omega_1 + \omega)\sqrt{\frac{2T}{\pi(\omega_2 - \omega_1)}} \qquad y_2 = (\omega_1 - \omega)\sqrt{\frac{2T}{\pi(\omega_2 - \omega_1)}}$$

$$y_3 = (\omega_2 + \omega)\sqrt{\frac{2T}{\pi(\omega_2 - \omega_1)}} \qquad y_4 = (\omega_2 - \omega)\sqrt{\frac{2T}{\pi(\omega_2 - \omega_1)}}$$

and $C(y)$ and $S(y)$ are the *Fresnel* integrals (see Prob. 25).

28. Using the principle of stationary phase, show that for large t the Bessel
function $J_n(t)$ is asymptotically given by

$$J_n(t) \sim \sqrt{\frac{2}{\pi t}} \cos\left(t - \frac{2n + 1}{4}\,\pi\right)$$

Solutions

15. Suppose that $f(t) = 0$ for $t < t_0$. From the linearity of the system it
follows that the response to $2f(t)$ is $2g(t)$. For $t < t_0$ we have $f(t) = 0 = 2f(t)$
and from (b) we conclude that for $t < t_0$, $g(t) = 2g(t)$, which is possible only if
$g(t) = 0$.

16. From (4-10) we have

$$tf(t) = -\frac{2}{\pi}\int_0^\infty R'(\omega)\sin \omega t\, d\omega$$

and, since $R'(\omega)$ is negative,

$$|tf(t)| \le -\frac{2}{\pi}\int_0^\infty R'(\omega)\, d\omega = \frac{2R(0)}{\pi}$$

17. From (5-32) we have

$$a(t) = R(0) + \frac{2}{\pi}\int_0^\infty \frac{X(\omega)}{\omega}\cos \omega t\, d\omega$$

and, since $X(\omega) \le 0$,

$$|a(t) - R(0)| \le -\frac{2}{\pi}\int_0^\infty \frac{X(\omega)}{\omega}\, d\omega = R(0) - a(0)$$

and (i) follows easily.

18. Since [see (3-64)] the Fourier transform of the input is given by

$$F^*(\omega) = \sum_{n=-\infty}^{\infty} F\left(\omega + \frac{2\pi n}{T}\right)$$

we have for $T < \dfrac{\pi}{\omega_c}$

$$F^*(\omega)H(\omega) = F(\omega)H(\omega) = F(\omega)A_0 e^{-j\omega t_0}$$

Therefore

$$g(t) = A_0 f(t - t_0)$$

19. With $g(t)$ as in (i) we easily see that $g(nT) = f(nT)$ and [see (2-57)]

$$G(\omega) = T p_{\omega c}(\omega) \sum_{n=-\infty}^{\infty} f(nT) e^{-jnT\omega}$$

Hence $G(\omega) = 0$ for $|\omega| > \omega_c$. Since the response of the low-pass filter to $\delta(t - nT)$ equals $\sin \omega_c(t - nT)/\pi(t - nT)$, we conclude that the output of the system of Fig. P-19 equals $g(t)$. Notice that if $F(\omega) = 0$ for $|\omega| > \omega_c$ then $g(t) = f(t)$ (see Prob. 18).

20. Since $F(\omega) = 2\pi\delta(\omega - \omega_1)$, we have [see (3-64)]

$$F^*(\omega) = 2\pi \sum_{n=-\infty}^{\infty} \delta(\omega - \omega_1 + 2n\omega_c)$$

the only component of the above sum that lies in the band of the filter is $\delta(\omega - \omega_1 + 2m\omega_c)$; hence $g(t) = e^{j(\omega_1 - 2m\omega_c)}$.

21. (a) Expanding $|\sin \omega_0 t|$ into a Fourier series, we have

$$|\sin \omega_0 t| = -\frac{2}{\pi} \sum_{n=-\infty}^{\infty} \frac{e^{j2n\omega_0 t}}{4n^2 - 1}$$

and since $g(t) = \phi(t) |\sin \omega_0 t|$, (i) follows easily [see (2-37)].

(b) From $\Phi(\omega) = 0$ for $|\omega| > \omega_0$ and (i), we obtain $G_1(\omega) = G(\omega)H(\omega) = (2A_0/\pi)\Phi(\omega)e^{-j\omega t_0}$ and (ii) is proved.

22. Since $f(t) = 0$ for $t < 0$ and

$$h(t) = \frac{A_0 \sin \omega_c(t - t_0)}{\pi(t - t_0)}$$

we obtain from (5-10) neglecting the t_0 delay,

$$\frac{\pi g(t)}{EA_0} = \int_{-\infty}^{t} \cos \omega_0(t - \tau) \frac{\sin \omega_c \tau}{\tau} d\tau$$

$$= \cos \omega_0 t \int_{-\infty}^{t} \frac{\sin(\omega_c - \omega_0)\tau + \sin(\omega_c + \omega_0)\tau}{2\tau} d\tau$$

$$+ \sin \omega_0 t \int_{-\infty}^{t} \frac{\cos(\omega_c - \omega_0)\tau - \cos(\omega_c + \omega_0)\tau}{2\tau} d\tau$$

and since $Ci(\infty) = 0$ and the integral of $\sin y/y$ from $-\infty$ to 0 equals $\pi/2$, we easily find (i). For $t \to \infty$, the function $Si\, ax$ tends to $\pi/2$ if $a > 0$ and to $-\pi/2$ if $a > 0$, and (ii) follows from (i).

23. $H(\omega)$ can be written in the form

$$H(\omega) = A_0(\cos \theta_0 - j \sin \theta_0 \operatorname{sgn} \omega)$$

From $\operatorname{sgn} t \longleftrightarrow 2/j\omega$ [see (3-9)] and (2-34) we have

$$2/jt \longleftrightarrow 2\pi \operatorname{sgn} (-\omega) = -2\pi \operatorname{sgn} \omega$$

and since $\cos \theta_0 \delta(t) \longleftrightarrow \cos \theta_0$, (i) follows. Notice that the response of the filter to $\cos \omega_1 t$ equals $A_0 \cos (\omega_1 t - \theta_0)$.

24. The Fourier transform of the error $g(t) - f(t - t_0)$ equals $F(\omega)[e^{-j\theta(\omega)} - e^{-j\omega t_0}]$; therefore, to prove (ii), it suffices to show that the integral

$$\int_{-\infty}^{\infty} |F(\omega)|^2 |e^{j[\theta(\omega) - \omega t_0]} - 1|^2 \, d\omega$$

is minimum for $\theta(\omega)$ as in (iii), i.e., that

$$\int_0^{\infty} |F(\omega)|^2 |e^{j[\theta(\omega) - \omega t_0]} - 1|^2 \, d\omega > \int_{\omega_c}^{\infty} |F(y)|^2 |e^{j[\theta(\infty) - y t_0]} - 1|^2 \, dy \qquad \text{(iv)}$$

To a given y we find the corresponding ω such that $\omega t_0 - \theta(\omega) = y t_0 - \theta(\infty)$. It is easy to see that (see Fig. P-24b) $\omega < y$ and, with $d\omega$ and dy the corresponding increments, we have $d\omega > dy$ because $[t_0 - \theta'(\omega)]\, d\omega = t_0 \, dy$ and $\theta'(\omega) \geq 0$. And since $|F(\omega)| > |F(y)|$ [see (i)], we conclude that

$$|F(\omega)|^2 |e^{j[\theta(\omega) - \omega t_0]} - 1|^2 \, d\omega > |F(y)|^2 |e^{j[\theta(\infty) - y t_0]} - 1|^2 \, dy$$

and (iv) is proved.

25.

$$h(t) = \frac{A_0}{\pi} \int_0^{\omega_c} \cos \left[\omega(t - t_0) - \alpha \omega^2 \right] d\omega$$

$$= \frac{A_0}{\pi} \int_0^{\omega_c} \cos \left[\left(\omega \sqrt{\alpha} - \frac{t - t_0}{2\sqrt{\alpha}} \right)^2 - \left(\frac{t - t_0}{2\sqrt{\alpha}} \right)^2 \right] d\omega$$

and with

$$\tau = \frac{t - t_0}{\sqrt{2\pi\alpha}} \qquad \text{and} \qquad \omega\sqrt{\alpha} - \sqrt{\pi/2}\,\tau = \sqrt{\pi/2}\,x$$

we obtain

$$h(t) = \frac{A_0}{\sqrt{2\pi\alpha}} \int_{-\tau}^{\omega_c\sqrt{2\alpha/\pi}\,-\tau} \cos \frac{\pi}{2}\,(x^2 - \tau^2)\, dx$$

and the desired result follows easily.

26. The Fourier transform of $a(t) - a_0(t)$ is given by [see (6-74)]

$$\left[\pi\delta(\omega) + \frac{1}{j\omega} \right] [H_0(\omega)\,(-j)\,\Delta\theta(\omega)] = -\frac{\Delta\theta(\omega)}{\omega}\, H_0(\omega)$$

because $\Delta\theta(\omega)\delta(\omega) = 0$, and since

$$\Delta\theta(\omega) = \begin{cases} \alpha\omega^2 & \omega > 0 \\ -\alpha\omega^2 & \omega < 0 \end{cases}$$

is odd, we obtain

$$a(t) - a_0(t) = -\frac{\alpha A_0}{\pi} \int_0^{\omega_c} \omega \cos \omega(t - t_0) \, d\omega$$

and (i) follows by integration by parts.

27. With

$$y = \left(\sqrt{a}\, t + \frac{b \pm \omega}{2\sqrt{a}} \right) \sqrt{\frac{2}{\pi}}$$

we obtain as in the solution to Prob. 5

$$F(\omega) = \int_0^T \cos \phi(t)\, e^{-j\omega t}\, dt = \tfrac{1}{2} \int_0^T e^{j[\phi(t) - \omega t]}\, dt + \tfrac{1}{2} \int_0^T e^{-j[\phi(t) + \omega t]} dt$$

$$= \tfrac{1}{2} \sqrt{\frac{\pi}{2a}} \int_{y_2}^{y_4} e^{j(\pi y^2/2 - \delta_2)}\, dy + \tfrac{1}{2} \sqrt{\frac{\pi}{2a}} \int_{y_1}^{y_3} e^{-j(\pi y^2/2 - \delta_1)}\, dy$$

and the desired result follows.

28. From Lommel's formula

$$J_n(t) = \frac{1}{2\pi} \int_{-\pi}^{\pi} e^{-jn\omega}\, e^{jt \sin \omega}\, d\omega$$

and with $A(\omega) = e^{-jn\omega}$, $\mu(\omega) = \sin \omega$, we obtain the stationary-phase points $\omega_0 = \pm\pi/2$, $\mu(\omega_0) = \pm 1$, $\mu''(\omega_0) = \mp 1$; hence [see (7-92)]

$$J_n(t) \sim \frac{e^{-jn\pi/2}}{2\pi} \sqrt{\frac{2\pi}{t}}\, e^{j(t - \pi/4)} + \frac{e^{jn\pi/2}}{2\pi} \sqrt{\frac{2\pi}{t}}\, e^{j(t + \pi/4)}$$

$$= \sqrt{\frac{2}{\pi t}} \cos\left(t - \frac{2n + 1}{4}\pi \right)$$

PART THREE

Chapter 9. The Laplace Transform

In the preceding portion of this book we treated the Fourier integral $F(\omega)$ as a function of the real variable ω. In the next two chapters we shall extend the definition to complex values of ω, for the purpose of examining the analytic properties of $F(\omega)$. This is essential in a number of applications; we mention the evaluation of the inversion formula by contour integration, the characterization of certain linear systems by the nature of their system function, the conditions on $F(\omega)$ for causality of $f(t)$, and its determination in terms of $R(\omega)$ or $A(\omega)$. In order to follow the usual convention in the technical literature, we shall limit ω to real values and will introduce a new variable p such that, when p is purely imaginary, it equals $j\omega$. The resulting function $F(p/j)$ is, of course, the familiar Laplace transform of $f(t)$; in this chapter we shall present certain aspects of the theory of Laplace transforms related to the Fourier integral.

9-1. The Unilateral Laplace Transform†‡

Given a function $f(t)$, defined for every $t \geq 0$, we form the integral

$$\int_0^\infty e^{-pt} f(t)\, dt$$

For the values

$$p = \alpha + j\omega$$

of the complex variable p, for which it converges, it defines a function $F_I(p)$, known as the *unilateral Laplace transform* of $f(t)$. It can be shown that, if the above integral exists for some value $p_0 = \alpha_0 + j\omega_0$ of p, then it exists for every p such that

$$\operatorname{Re} p > \alpha_0$$

† R. V. Churchill, "Modern Operational Mathematics in Engineering," McGraw-Hill Book Company, Inc., New York, 1944.

‡ G. Doetsch, "Theorie und Anwendung der Laplace-Transformation," Dover Publications, New York, 1943.

169

From the above it follows easily (Dedekind cut) that, to every function $f(t)$, a real constant γ can be associated such that $F_\mathrm{I}(p)$ converges for $\mathrm{Re}\, p > \gamma$ and it does not converge for $\mathrm{Re}\, p < \gamma$; for $\mathrm{Re}\, p = \gamma$ no general conclusion can be drawn. Thus the region of existence of

$$F_\mathrm{I}(p) = \int_0^\infty e^{-pt} f(t)\, dt \qquad (9\text{-}1)$$

is a half plane to the right of the vertical line $\mathrm{Re}\, p = \gamma$ (Fig. 9-1). With $p = \alpha + j\omega$, we shall write $F_\mathrm{I}(p)$ in the form

$$F_\mathrm{I}(p) = R(\alpha,\omega) + jX(\alpha,\omega) \qquad (9\text{-}2)$$

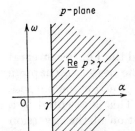

FIGURE 9-1

where $R(\alpha,\omega)$ and $X(\alpha,\omega)$ are its real and imaginary parts. As an example, suppose that k is real and

$$f(t) = e^{-kt} U(t)$$

then we easily see from (9-1) that for $\mathrm{Re}\, p > -k$

$$F_\mathrm{I}(p) = \frac{1}{p + k} = \frac{1}{\alpha + j\omega + k}$$

and

$$R(\alpha,\omega) = \frac{\alpha + k}{(\alpha + k)^2 + \omega^2} \qquad X(\alpha,\omega) = \frac{-\omega}{(\alpha + k)^2 + \omega^2}$$

The following theorem establishes the analytic properties of $F_\mathrm{I}(p)$.

Fundamental theorem. The Laplace transform $F_\mathrm{I}(p)$, defined by (9-1), is an analytic function of p in the region $\mathrm{Re}\, p > \gamma$ of its existence, and its derivative is given by

$$\frac{dF_I(p)}{dp} = \int_0^\infty e^{-pt}(-t)f(t)\, dt \qquad (9\text{-}3)$$

To prove this theorem, it suffices to establish (9-3), i.e., to show that the integral in (9-1) has a derivative, obtained by differentiating under the integral sign.

This is done in two stages; we first prove that the function

$$F_T(p) = \int_0^T e^{-pt} f(t)\, dt$$

is analytic for every p, and its derivative is given by

$$\frac{dF_T(p)}{dp} = \int_0^T e^{-pt}(-t)f(t)\, dt$$

This is a well-known result in general calculus and is easily established by proving that

$$\lim_{\Delta p \to 0} \left[\frac{F_T(p + \Delta p) - F_T(p)}{\Delta p} - \int_0^T e^{-pt}(-t)f(t)\, dt \right]$$

$$= \lim_{\Delta p \to 0} \int_0^T e^{-pt}\left(\frac{e^{-t\Delta p} - 1}{\Delta p} + t \right) f(t)\, dt = 0$$

One then shows that for Re $p > \gamma$, the limit of $dF_T(p)/dp$, as T tends to infinity, exists and equals the integral in (9-3); the details of the proof will be omitted.

We should point out that the converse of the above theorem is not true; from the analyticity of a function $F_I(p)$ is some half plane, it does not follow that $F_I(p)$ is the Laplace transform of a certain function $f(t)$. Additional requirements must be imposed on $F_I(p)$ for this to be true (see Sec. 10-5).

The function dF_I/dp in (9-3) is the Laplace transform of $(-t)f(t)$; therefore it also has a derivative. Thus repeated application of (9-3) shows that the transform of $(-t)^n f(t)$ equals $d^n F_I(p)/dp^n$:

$$\frac{d^n F_I(p)}{dp^n} = \int_0^\infty e^{-pt}(-t)^n f(t)\, dt \tag{9-4}$$

For example, with $f(t) = e^{-kt}U(t)$, we conclude that the Laplace transform of $t^n e^{-kt}U(t)$ is given by

$$(-1)^n \frac{d^n(1/(p + k))}{dp^n} = \frac{n!}{(p + k)^{n+1}}$$

Thus the unilateral Laplace transform of t^n equals $n!/p^{n+1}$. For later reference we shall develop the transform of t^α where $\alpha > -1$. With the change of variable $pt = x$, we obtain from (9-1)

$$\int_0^\infty e^{-pt}t^\alpha\, dt = \frac{1}{p^{\alpha+1}} \int_0^\infty e^{-x}x^\alpha\, dx$$

The last integral is a quantity known as the *gamma function* $\Gamma(\alpha + 1)$. Thus

$$\int_0^\infty e^{-pt}t^\alpha\, dt = \frac{\Gamma(\alpha + 1)}{p^{\alpha+1}} \tag{9-5}$$

It is easy to see from the above that if α is an integer, then $\Gamma(\alpha + 1) = \alpha!$

9-2. Relationship between the Fourier Integral of a Causal Function and the Unilateral Laplace Transform

We shall now relate the Fourier integral

$$F(\omega) = \int_0^\infty e^{-j\omega t} f(t)\, dt = R(\omega) + jX(\omega) \tag{9-6}$$

of a causal function $f(t)$ to its Laplace transform $F_I(p)$ given by (9-1). Specifically, we shall derive $F(\omega)$ from $F_I(p)$ and will show how $F_I(p)$ can be found if $F(\omega)$ is known for real ω.

$F(\omega)$ **from** $F_I(p)$. There are three cases to consider:

If the region of convergence of $F_I(p)$ contains the $j\omega$ axis in its interior, i.e., if $\gamma < 0$, then the integral (9-6) is a special case of (9-1), with $p = j\omega$, and $F(\omega)$ is given by

$$F(\omega) = F_I(j\omega) \tag{9-7}$$

Hence [see (9-2) and (9-6)]

$$R(\omega) = R(0,\omega) \qquad X(\omega) = X(0,\omega)$$

Thus if $F_I(p) = 1/(p + k)$ and $k > 0$, then $F(\omega) = 1/(j\omega + k)$.

If the $j\omega$ axis is outside the region of convergence of $F_I(p)$, i.e., if $\gamma > 0$, then $F(\omega)$ does not exist; the function $f(t)$ has no Fourier transform.

The last case is $\gamma = 0$; the function $F_I(p)$ is analytic for Re $p > 0$, but at least one of its singular points lies on the $j\omega$ axis. Suppose first that

$$F_I(p) = \frac{1}{p - j\omega_0}$$

This function is the Laplace transform of

$$f(t) = e^{j\omega_0 t} U(t)$$

as it is easy to see from (9-1). And since the Fourier transform of $e^{j\omega_0 t} U(t)$ is given by [see (3-14)]

$$\pi\delta(\omega - \omega_0) + \frac{1}{j\omega - j\omega_0}$$

the correspondence

$$F_I(p) = \frac{1}{p - j\omega_0} \qquad F(\omega) = \pi\delta(\omega - \omega_0) + F_I(j\omega) \tag{9-8}$$

results. Suppose, next, that

$$F_I(p) = \frac{1}{(p - j\omega_0)^n}$$

It is easy to see from (9-1) that, for Re $p > 0$, the above function is the Laplace transform of

$$f(t) = \frac{t^{n-1}}{(n-1)!} e^{j\omega_0 t} U(t)$$

Reasoning as in (3-36), we conclude that (see also comment on page 185)

$$F(\omega) = \frac{\pi j^{n-1}}{(n-1)!} \delta^{(n-1)}(\omega - \omega_0) + \frac{1}{(j\omega - j\omega_0)^n}$$

Thus,

$$F_I(p) = \frac{1}{(p - j\omega_0)^n} \qquad F(\omega) = \frac{\pi j^{n-1}}{(n-1)!} \delta^{(n-1)}(\omega - \omega_0) + F_I(j\omega) \qquad (9\text{-}9)$$

Consider now a function $F_I(p)$ with n simple poles $j\omega_1, j\omega_2, \ldots, j\omega_n$ and no other singularities in the half plane Re $p \geq 0$. This function can be written in the form

$$F_I(p) = G(p) + \sum_{m=1}^{n} \frac{a_m}{p - j\omega_m}$$

where $G(p)$ is free of singularities for Re $p \geq 0$. From (9-7) it follows that the Fourier transform corresponding to $G(p)$ is given by $G(j\omega)$; therefore

$$F(\omega) = F_I(j\omega) + \pi \sum_{m=1}^{n} a_m \delta(\omega - \omega_m) \qquad (9\text{-}10)$$

as we can easily see from (9-8). Multiple poles can be similarly handled with the help of (9-9).

As an example, consider the function

$$F_I(p) = \frac{A}{p^2 + \omega_0^2} = \frac{A/2j\omega_0}{p - j\omega_0} - \frac{A/2j\omega_0}{p + j\omega_0}$$

From (9-10) it follows that

$$F(\omega) = \frac{A}{\omega_0^2 - \omega^2} + \frac{A\pi}{2j\omega_0} [\delta(\omega - \omega_0) - \delta(\omega + \omega_0)]$$

$F_I(p)$ **from** $F(\omega)$. From the existence of $F(\omega)$ we conclude that $F_I(p)$ is analytic for Re $p > 0$. If $F_I(p)$ exists also for Re $p = 0$ and $F(\omega)$ is given by an analytic function of ω, then

$$F_I(p) = F\left(\frac{p}{j}\right) \qquad \text{Re } p \geq 0 \qquad (9\text{-}11)$$

Often $F(\omega)$ is given only graphically or by analytic pieces. To determine $F_I(p)$, we write it in the form

$$F_I(\alpha + j\omega) = \int_0^\infty e^{-\alpha t} f(t) e^{-j\omega t} \, dt \qquad \alpha > 0$$

Thus $F_{\rm I}(p)$, considered as a function of ω with $\alpha > 0$ a real constant, is the Fourier transform of the function $e^{-\alpha t}f(t)U(t)$:

$$e^{-\alpha t}f(t)U(t) \leftrightarrow F_{\rm I}(\alpha + j\omega) \qquad \alpha > 0 \qquad (9\text{-}12)$$

Since $e^{-\alpha t}f(t)U(t)$ is the product of $f(t)$ and $e^{-\alpha t}U(t)$, we conclude from

$$f(t) \leftrightarrow F(\omega) \qquad e^{-\alpha t}U(t) \leftrightarrow \frac{1}{\alpha + j\omega}$$

and (2-74) that

$$F_{\rm I}(\alpha + j\omega) = \frac{1}{2\pi} F(\omega) * \frac{1}{\alpha + j\omega} = \frac{1}{2\pi} \int_{-\infty}^{\infty} \frac{F(y)}{\alpha + j\omega - jy}\, dy$$

Thus

$$F_{\rm I}(p) = \frac{1}{2\pi} \int_{-\infty}^{\infty} \frac{F(y)}{p - jy}\, dy \qquad {\rm Re}\, p > 0 \qquad (9\text{-}13)$$

If $f(t)$ is causal, then $F_{\rm I}(p)$ can be expressed in terms of the real or imaginary parts of $F(\omega) = R(\omega) + jX(\omega)$. Indeed, for a causal function we have $f(t)U(t) = 2f_e(t)U(t)$, where $f_e(t) \leftrightarrow R(\omega)$ is the even part of $f(t)$. Hence, $e^{-\alpha t}f(t)U(t) = 2f_e(t)e^{-\alpha t}U(t)$; applying again the convolution theorem (2-74), we obtain for $\alpha > 0$

$$F_{\rm I}(\alpha + j\omega) = \frac{1}{\pi} R(\omega) * \frac{1}{\alpha + j\omega}$$

$$= \frac{1}{\pi} \int_{-\infty}^{\infty} \frac{R(y)}{\alpha + j(\omega - y)}\, dy = \frac{1}{\pi} \int_{-\infty}^{\infty} \frac{R(y)}{p - jy}\, dy \qquad (9\text{-}14)$$

or the equivalent

$$R(\alpha,\omega) = \frac{1}{\pi} R(\omega) * \frac{\alpha}{\alpha^2 + \omega^2}$$

$$X(\alpha,\omega) = -\frac{1}{\pi} R(\omega) * \frac{\omega}{\alpha^2 + \omega^2} \qquad (9\text{-}15)$$

From the evenness of $R(\omega)$ and (9-14) it easily follows that

$$F_{\rm I}(p) = \frac{2p}{\pi} \int_{0}^{\infty} \frac{R(y)}{p^2 + y^2}\, dy \qquad \alpha > 0 \qquad (9\text{-}16)$$

If $f(t)$ does not contain an impulse at the origin, it equals $2f_0(t)$ for $t \geq 0$, where $f_0(t) \leftrightarrow jX(\omega)$ is its odd part. Reasoning as in (9-14), we obtain

$$F_{\rm I}(p) = \frac{j}{\pi} \int_{-\infty}^{\infty} \frac{X(y)}{p - jy}\, dy = \frac{-2}{\pi} \int_{0}^{\infty} \frac{yX(y)}{p^2 + y^2}\, dy \qquad \alpha > 0$$

If $f(t)$ contains an impulse at the origin, then $R(\infty)$ must be added in the above expression [see (10-33)]. The previous results can also be derived with a contour integration as in Prob. 37.

To illustrate (9-16), we shall determine the Laplace transform of $f(t) = \cos \omega_0 t\, U(t)$. We have $R(\omega) = (\pi/2)[\delta(\omega - \omega_0) + \delta(\omega + \omega_0)]$; hence

$$F_I(p) = \frac{2p}{\pi} \int_0^\infty \frac{\pi\delta(y - \omega_0)}{2(p^2 + y^2)}\, dy = \frac{p}{p^2 + \omega_0^2}$$

9-3. The Inversion Formula

Using (2-3), we shall express the causal function $f(t)$ in terms of its Laplace transform. From (9-12) we see that $F_I(\alpha + j\omega)$ is the Fourier integral of $e^{-\alpha t}f(t)U(t)$. Applying (2-3) to this pair, we obtain

$$e^{-\alpha t}f(t)U(t) = \frac{1}{2\pi} \int_{-\infty}^{\infty} e^{j\omega t}F_I(\alpha + j\omega)\, d\omega$$

or the equivalent

$$f(t)U(t) = \frac{1}{2\pi} \int_{-\infty}^{\infty} e^{(\alpha + j\omega)t}F_I(\alpha + j\omega)\, d\omega \quad (9\text{-}17)$$

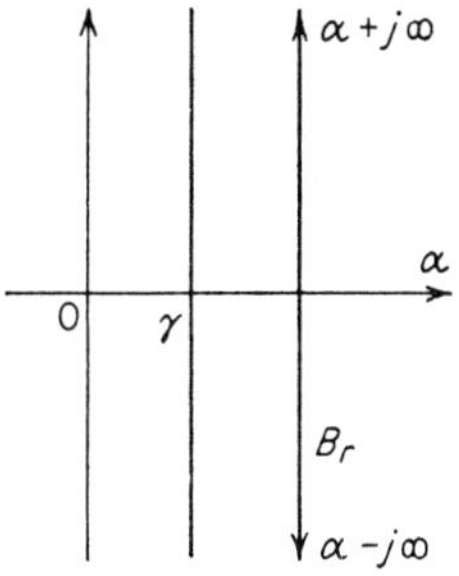

This result can be written as a line integral; with

$$\alpha + j\omega = p \qquad j\, d\omega = dp$$

we observe that, as ω varies from $-\infty$ to $+\infty$, p takes values along the vertical line $\operatorname{Re} p = \alpha$ of Fig. 9-2. Thus (9-17) can be written in the form

$$f(t)U(t) = \frac{1}{2\pi j} \int_{\alpha - j\infty}^{\alpha + j\infty} e^{pt}F_I(p)\, dp \quad (9\text{-}18)$$

FIGURE 9-2

known as the inversion formula of the Laplace transform. For the existence of $F_I(p)$, it is, of course, necessary that $\alpha > \gamma$; therefore, the line of integration must lie in the region $\operatorname{Re} p > \gamma$ of convergence of $F_I(p)$. This line is known as the Bromwich path and is denoted by Br.

By a formal differentiation of (9-18) we obtain

$$\frac{d[f(t)U(t)]}{dt} = \frac{1}{2\pi j} \int_{Br} pF_I(p)e^{pt}\, dp \quad (9\text{-}19)$$

Thus, with $f_i'(t)$ the derivative of $f(t)$ for $t > 0$, we conclude that the transform of

$$\frac{d[f(t)U(t)]}{dt} = f_i'(t)U(t) + f(0^+)\,\delta(t) \quad (9\text{-}20)$$

is given by $pF_I(p)$. For the validity of (9-19) it is necessary to allow the time function in (9-1) to contain impulses at the origin. This can be done if the lower limit of integration is taken as 0^-; it then follows that

$$1 = \int_{0^-}^{\infty} \delta(t) e^{-pt}\, dt$$

and, from the above, one can easily see that the Laplace transform of $f_i{}'(t)$ equals

$$pF(p) - f(0^+)$$

9-4. Evaluation of $f(t)$

The evaluation of $f(t)$ by a direct integration along the Bromwich path (9-18) is in general complicated; simple results can often be obtained, however, by a suitable modification of the path of integration. In doing this, we shall have to consider the function $F_I(p)$ for values of p outside the region Re $p > \gamma$ of convergence of (9-1); this is no problem, since $F_I(p)$ in (9-18) will be treated as an analytic function that might exist for Re $p < \gamma$. Thus the function $1/(p + 2)$, say, is the transform of $e^{-2t}U(t)$ only for Re $p > -2$, however, it exists for every $p \neq -2$.

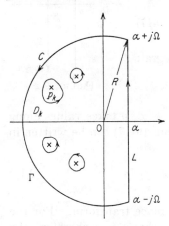

FIGURE 9-3

It will be convenient in this analysis to classify $F_I(p)$ into three categories.

Finitely many poles. We shall first assume that $F_I(p)$ has finitely many poles $p_1, p_2, \ldots, p_n$, and no other singularities; this case includes the important class of rational transforms. In order to apply Cauchy's integral theorem, we must express $f(t)$ as an integral along a closed path. This is done as follows. Consider the closed path C of Fig. 9-3 consisting of a vertical line L from $\alpha - j\Omega$ to $\alpha + j\Omega$ and an arc Γ of circle of radius R. We know from Jordan's lemma (II-56) that if

$$F(p) \rightarrow 0 \text{ as } |p| \rightarrow \infty \text{ on } \Gamma \tag{9-21}$$

then for $t > 0$

$$\int_{\Gamma} e^{pt} F_I(p)\, dp \xrightarrow[R \rightarrow \infty]{} 0 \qquad t > 0$$

and since $\displaystyle\int_{\alpha - j\Omega}^{\alpha + j\Omega} e^{pt} F_I(p)\, dp \rightarrow \int_{Br} e^{pt} F_I(p)\, dp \qquad \Omega \rightarrow \infty \tag{9-22}$

we conclude that $f(t)$ can be written as a limit,

$$f(t) \xrightarrow[R \rightarrow \infty]{} \frac{1}{2\pi j} \int_{C} e^{pt} F_I(p)\, dp \tag{9-23}$$

of an integral along the closed path C. If R is large enough to contain all poles of $F_{\rm I}(p)$, i.e., if R is greater than the maximum $|p_k|$, then the integral along C is independent of R; hence

$$f(t) = \frac{1}{2\pi j} \int_C e^{pt} F_{\rm I}(p)\, dp \qquad R > \max |p_k| \qquad (9\text{-}24)$$

We repeat, (9-24) is true only if (9-21) holds and t is positive. To evaluate (9-24), we enclose each pole p_k by a closed curve D_k having no other poles in its interior (see also page 296). From Cauchy's theorem (II-22) it follows that

$$\int_C e^{pt} F_{\rm I}(p)\, dp = \sum_{k=1}^{n} \int_{D_k} e^{pt} F_{\rm I}(p)\, dp \qquad (9\text{-}25)$$

Each integral in the above summation can be very simply found from Cauchy's formula (II-32). Indeed, if the pole p_k is simple, then the function

$$F_k(p) = F_{\rm I}(p)(p - p_k) \qquad (9\text{-}26)$$

is analytic in the interior of D_k; hence

$$\int_{D_k} e^{pt} F_{\rm I}(p)\, dp = \int_{D_k} \frac{e^{pt} F_k(p)}{p - p_k}\, dp = 2\pi j e^{p_k t} F_k(p_k) \qquad (9\text{-}27)$$

It is easy to see that the constant $F_k(p_k)$, known as the residue of $F_{\rm I}(p)$ at $p = p_k$, is given by

$$F_k(p_k) = F_{\rm I}(p)(p - p_k)\Big|_{p=p_k} \qquad (9\text{-}28)$$

Thus, if all the poles are simple, then $f(t)$ is given by

$$f(t) = \sum_{k=1}^{n} F_k(p_k) e^{p_k t} \qquad t > 0 \qquad (9\text{-}29)$$

If the multiplicity of the pole p_k is $m + 1$, then with

$$F_k(p) = F_{\rm I}(p)(p - p_k)^{m+1} \qquad (9\text{-}30)$$

we have from (II-34)

$$\int_{D_k} e^{pt} F_{\rm I}(p)\, dp = \int_{D_k} \frac{e^{pt} F_k(p)}{(p - p_k)^{m+1}}\, dp$$

$$= \frac{2\pi j}{m!} \frac{d^m [e^{pt} F_k(p)]}{dp^m}\Big|_{p=p_k} \qquad (9\text{-}31)$$

We shall now evaluate the integral (9-18) for $t < 0$; if $F_{\rm I}(p)$ satisfies (9-21), then the integral along the arc Γ_r of the circle of Fig. 9-4 tends to zero as the radius tends to infinity [see (II-54)]. Reasoning as in (9-24), we conclude that

$$f(t)U(t) = \frac{1}{2\pi j} \int_{C_r} e^{pt} F_{\rm I}(p)\, dp \qquad (9\text{-}32)$$

where C_r is the closed curve consisting of the segment L and the arc Γ_r. But $F_\mathrm{I}(p)$ is analytic in the interior of C_r; hence

$$f(t)U(t) = 0 \qquad t < 0 \tag{9-33}$$

as it should be.

Example 9-1

$$F_\mathrm{I}(p) = \frac{p}{(p+1)(p+2)}$$

This function obviously satisfies (9-21); its poles are given by

$$p_1 = -1 \qquad p_2 = -2$$

Hence $\quad F_1(p_1) = \dfrac{p}{p+2}\bigg|_{p=-1} = -1 \qquad F_2(p_2) = \dfrac{p}{p+1}\bigg|_{p=-2} = 2$

Therefore [see (9-29)]

$$f(t) = -e^{-t} + 2e^{-2t} \qquad t > 0$$

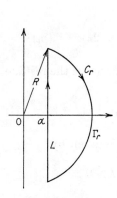

FIGURE 9-4

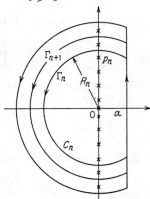

FIGURE 9-5

Example 9-2

$$F_\mathrm{I}(p) = \frac{1}{p^2(p+1)}$$

Its poles are

$$p_1 = 0 \qquad p_2 = -1$$

The first pole is of multiplicity $m = 2$; hence with

$$F_1(p) = \frac{1}{p+1}$$

the value of the integral along the first circle D_1 is obtained from

$$\frac{d}{dp}\left(\frac{e^{pt}}{p+1}\right)_{p=0} = \left[\frac{te^{pt}}{p+1} - \frac{e^{pt}}{(p+1)^2}\right]_{p=0} = t - 1$$

and since
$$F_2(p) = \frac{1}{p^2} \qquad F_2(p_2) = 1$$

we have
$$f(t) = t - 1 + e^{-t} \qquad t > 0$$

Example 9-3

$$F_I(p) = \frac{e^{-pT}}{p}$$

It is no longer true that for $t > 0$ the function $F_I(p)$ tends to zero as p tends to infinity; therefore we cannot use (9-29). To evaluate $f(t)$, we write the inversion integral in the form

$$f(t) = \frac{1}{2\pi j} \int_{B_r} \frac{e^{p(t-T)}}{p} \, dp$$

Since $1/p \to 0$ with $p \to \infty$, we conclude that if $t > T$, then from Jordan's lemma it follows that the integral along the arc Γ of Fig. 9-3 tends to zero with $R \to \infty$; hence, reasoning as in (9-24), we obtain

$$f(t) = \frac{1}{2\pi j} \int_C \frac{e^{p(t-T)}}{p} \, dp \qquad t > T$$

The only pole of the above is the point $p = 0$; hence

$$f(t) = e^{p(t-T)} \big|_{p=0} = 1 \qquad t > T$$

For $t < T$ we apply (9-32) and obtain $f(t) = 0$; thus

$$f(t) = U(t - T)$$

Infinitely many poles. Suppose now that $F_I(p)$ has infinitely many poles, as in Fig. 9-5. This case requires special consideration because it is no longer true that $F_I(p)$ tends to zero on the arc Γ of Fig. 9-3, as R tends to infinity; in fact, it might be infinite if one of its poles is on Γ. In this case we shall assume that a sequence of circular arcs $\Gamma_1, \Gamma_2, \ldots, \Gamma_n, \ldots$, with radii tending to infinity, can be found (Fig. 9-5) such that

$$F_I(p) \to 0 \quad \text{as} \quad p \to \infty \text{ on } \Gamma_n \tag{9-34}$$

Applying Jordan's lemma to the integrals along these arcs, we have

$$\int_{\Gamma_n} e^{pt} F_I(p) \, dp \xrightarrow[n \to \infty]{} 0 \qquad t > 0 \tag{9-35}$$

and with C_n the closed curve, consisting of Γ_n and the vertical line $\operatorname{Re} p = \alpha$, we obtain as in (9-24)

$$f(t) = \lim_{n \to \infty} \frac{1}{2\pi j} \int_{C_n} e^{pt} F_I(p) \, dp \qquad t > 0 \tag{9-36}$$

From the above it follows that, if the poles $p_1, p_2, \ldots, p_n, \ldots$, of $F_{\mathrm{I}}(p)$ are simple, then, with $F_k(p) = F_{\mathrm{I}}(p)(p - p_k)$, $f(t)$ is given by

$$f(t) = \sum_{k=1}^{\infty} F_k(p_k) e^{p_k t} \qquad (9\text{-}37)$$

Multiple poles can be handled as in (9-31).

Example 9-4

$$F_{\mathrm{I}}(p) = \frac{1}{p \cosh ap} \qquad a > 0 \qquad (9\text{-}38)$$

The poles of this function are given by

$$p_0 = 0 \qquad p_k = \pm j \frac{(2k - 1)\pi}{2a} \qquad k = 1, 2, \ldots$$

as it is easy to see (Fig. 9-5). We select the arcs Γ_n such that their radii are given by

$$R_n = jn\pi$$

It can be shown that for p on Γ_n, the function $1/\cosh ap$ remains bounded; hence

$$\frac{1}{p \cosh ap} \to 0 \quad \text{as} \quad p \to \infty \text{ on } \Gamma_n$$

and $f(t)$ is given by (9-36). Since

$$pF_{\mathrm{I}}(p)\big|_{p=0} = 1 \qquad (p - p_k)F_{\mathrm{I}}(p)\big|_{p=p_k} = \frac{(-1)^k 2}{(2k - 1)\pi}$$

we conclude from (9-37) that

$$f(t) = 1 + \frac{2}{\pi} \sum_{k=1}^{\infty} \frac{(-1)^k}{2k - 1} e^{p_k t} + \frac{2}{\pi} \sum_{k=1}^{\infty} \frac{(-1)^k}{2k - 1} e^{-p_k t}$$

$$= 1 + \frac{4}{\pi} \sum_{k=1}^{\infty} \frac{(-1)^k}{2k - 1} \cos \frac{(2k - 1)\pi t}{2a} \qquad (9\text{-}39)$$

The above result shows that $f(t)$ is periodic with period $T = 4a$, and (9-39) is its Fourier series expansion; however, it is not easy to recognize its shape from (9-39). The following discussion will permit a more direct evaluation of $f(t)$.

Periodic Functions. Suppose that a function $f(t)$ is periodic for $t > 0$ and its period equals T. It is easy to see from (9-1) that its transform is given by

$$F_{\mathrm{I}}(p) = \frac{\displaystyle\int_0^T e^{-pt} f(t)\, dt}{1 - e^{-pT}} \qquad (9\text{-}40)$$

We form the function (Fig. 9-6)

$$f_0(t) = \begin{cases} f(t) & 0 < t < T \\ 0 & t > T \end{cases}$$

and observe from (9-40) that the numerator of $F_I(p)$ equals the Laplace transform $F_0(p)$ of $f_0(t)$:

$$F_0(p) = \int_0^\infty e^{-pt} f_0(t)\, dt = \int_0^T e^{-pt} f(t)\, dt \qquad (9\text{-}41)$$

Hence

$$F_0(p) = (1 - e^{-pT}) F_I(p) \qquad (9\text{-}42)$$

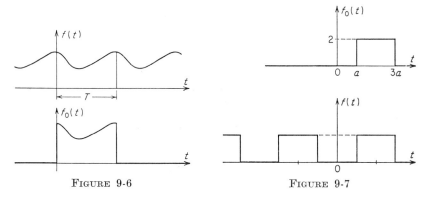

FIGURE 9-6   FIGURE 9-7

Applying (9-42) to the function $F_I(p)$ in (9-38), we obtain with $T = 4a$

$$F_0(p) = \frac{1 - e^{-4ap}}{p \cosh ap} = 2\, \frac{e^{-ap} - e^{-3ap}}{p}$$

Hence (see Example 9-3)

$$f_0(t) = 2U(t - a) - 2U(t - 3a)$$

The function $f_0(t)$ is shown in Fig. 9-7; $f(t)$ is its periodic repetition.

Steady State. The above suggests a method for determining the asymptotic form of a function $f(t)$ as t tends to infinity (steady state). Suppose that certain poles p_n of $F_I(p)$ are on the imaginary axis

$$p_n = jn\omega_0$$

and the remaining poles p'_n lie in the left-hand plane. Then

$$f(t) = \sum a_n e^{jn\omega_0 t} + \sum a'_n e^{p'_n t} = f_{ss}(t) + f_{tr}(t)$$

This is the case, for example, if $f(t)$ is the response of a stable system to a periodic input. Since Re $p'_n < 0$, the second summation above tends to zero as t tends to infinity; hence

$$f(t) \xrightarrow[t \to \infty]{} \sum a_n e^{jn\omega_0 t} = f_{ss}(t)$$

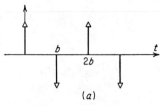

(a)

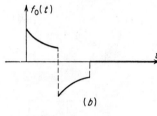

(b)

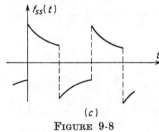

(c)

FIGURE 9-8

Thus $f(t)$ tends to a periodic function which can be obtained with the use of (9-42) if the term $f_{tr}(t)$ is first removed from it. We shall illustrate the method with a simple example.

Example 9-5

$$F_{\mathrm{I}}(p) = \frac{1}{(p + \alpha)(1 + e^{-bp})}$$

[This is the response of a first-order system $H(p) = 1/(p + \alpha)$ to the periodic input of Fig. 9-8a.] The imaginary poles of $F_{\mathrm{I}}(p)$ are given by

$$p_n = \frac{j(2n + 1)\pi}{b}$$

Hence the steady state is periodic with period $T = 2b$. From

$$F_{\mathrm{I}}(p) = \frac{A_1}{p + \alpha} + F_{ss}(p) \qquad A_1 = \frac{1}{1 + e^{\alpha b}}$$

we have $$F_{ss}(p) = \frac{1 - A_1(1 + e^{-bp})}{(p + \alpha)(1 + e^{-bp})}$$

and [see (9-42)]

$$F_0(p) = (1 - e^{-2bp})\frac{1 - A_1(1 + e^{-bp})}{(p + \alpha)(1 + e^{-bp})} = \frac{(1 - e^{-bp}) - A_1(1 - e^{-2bp})}{p + \alpha}$$

Therefore

$$f_0(t) = (1 - A_1)e^{-\alpha t} - e^{-\alpha(t-b)}U(t - b) \qquad 0 < t < 2b$$

and $f_0(t) = 0$ for $t > 2b$ (Fig. 9-8b). The steady-state part of $f(t)$ is the periodic repetition of $f_0(t)$ as in Fig. 9-8c.

Branch points. The method of determining the inverse transform of a function with branch points will be demonstrated in two examples.

Example 9-6

$$F_{\mathrm{I}}(p) = \frac{e^{-a\sqrt{p}}}{\sqrt{p}} \qquad a > 0$$

We know that $F_{\mathrm{I}}(p)$ is analytic to the right of some vertical line $\mathrm{Re}\, p = \gamma$. Therefore the branch cut that makes $\sqrt{p}$ continuous must lie in the left-hand plane (see page 287). We choose for convenience the negative real axis for this cut. From the two possible values of $\sqrt{p}$ we select the one that makes it positive for p positive real, in order to ensure the vanishing of $F_{\mathrm{I}}(p)$ as

$p \to \infty$ in the region Re $p > \gamma$ We thus have completed the definition of $\sqrt{p}$:

$$p = \rho e^{j\theta} \qquad -\pi < \theta \leq \pi \qquad \sqrt{p} = \sqrt{\rho}\, e^{j\theta/2} \qquad (9\text{-}43)$$

We now form the closed curve C of Fig. 9-9, consisting of the vertical line L from $\alpha - j\Omega$ to $\alpha + j\Omega$, the two arcs Γ_1, Γ_2 of radius R, the two line segments AB and DE, and the circle Γ_3 of radius r.   The function $F_{\mathrm{I}}(p)$ is analytic in the interior of C; therefore

$$\int_C e^{pt} F_{\mathrm{I}}(p)\, dp = 0 \qquad (9\text{-}44)$$

It is easy to see that, for p on Γ_1 and Γ_2, the given function tends to zero as R tends to infinity; hence [see (II-56)]

$$\int_{\Gamma_1 + \Gamma_2} e^{pt}\, \frac{e^{-a\sqrt{p}}}{\sqrt{p}}\, dp \xrightarrow[R \to \infty]{} 0 \qquad (9\text{-}45)$$

For p on the circle Γ_3 we have

$$\left| \frac{e^{pt} e^{-a\sqrt{p}}}{\sqrt{p}} \right| \leq \frac{e^{rt}}{\sqrt{r}}$$

Therefore [see (II-17)], for a fixed $t > 0$,

$$\left| \int_{\Gamma_3} e^{pt}\, \frac{e^{-a\sqrt{p}}}{\sqrt{p}}\, dp \right| \leq 2\pi r\, \frac{e^{rt}}{\sqrt{r}} \xrightarrow[r \to 0]{} 0 \qquad (9\text{-}46)$$

We now come to the evaluation of the integrals along AB and DE:

On AB: $p = -x \qquad \sqrt{p} = j\sqrt{x}$

On DE: $p = -x \qquad \sqrt{p} = -j\sqrt{x}$

[see (9-43)]; hence

$$\int_{AB+DE} e^{pt}\, \frac{e^{-a\sqrt{p}}}{\sqrt{p}}\, dp \xrightarrow[R \to \infty]{r \to 0} -\int_{\infty}^{0} e^{-xt}\, \frac{e^{ja\sqrt{x}}}{j\sqrt{x}}\, dx - \int_{0}^{\infty} e^{-xt}\, \frac{e^{-ja\sqrt{x}}}{-j\sqrt{x}}\, dx \quad (9\text{-}47)$$

And since [see (9-18)]

$$\int_L e^{pt}\, \frac{e^{-a\sqrt{p}}}{\sqrt{p}}\, dp \xrightarrow[\Omega \to \infty]{} 2\pi j f(t) \qquad (9\text{-}48)$$

we conclude from (9-44) to (9-48) that

$$f(t) + \frac{1}{2\pi j} \int_0^{\infty} e^{-xt}\, \frac{e^{ja\sqrt{x}} + e^{-ja\sqrt{x}}}{j\sqrt{x}}\, dx = 0 \qquad (9\text{-}49)$$

With $x = y^2$ we have

$$\int_0^{\infty} e^{-xt}\, \frac{\cos a\sqrt{x}}{\sqrt{x}}\, dx = 2 \int_0^{\infty} e^{-y^2 t} \cos ay\, dy$$

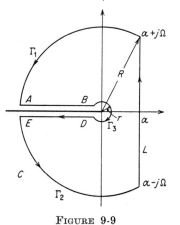

Figure 9-9

But [see (2-69)]

$$2 \int_0^\infty e^{-v^2 t} \cos ay \, dy = \sqrt{\frac{\pi}{t}} \, e^{-a^2/4t}$$

Therefore $$f(t) = \frac{1}{\sqrt{\pi t}} e^{-a^2/4t} \qquad (9\text{-}50)$$

Example 9-7

$$F_I(p) = \frac{1}{\sqrt{p^2 + 1}}$$

This function has two branch points:

$$p = \pm j$$

The branch cuts must again lie outside the region $\operatorname{Re} p > \gamma$; this can be effected either with the vertical cut N from $+j$ to $-j$ as in Fig. 9-10a, or

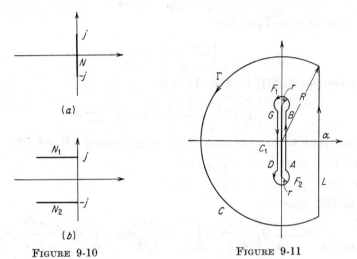

(a)

(b)

FIGURE 9-10   FIGURE 9-11

with the two cuts N_1, N_2 as in Fig. 9-10b. We shall use the cut of Fig. 9-10a. Consider the closed curve C of Fig. 9-11, consisting of the vertical line L (see also Fig. 9-3) and the arc Γ of radius R. Since

$$\frac{1}{\sqrt{p^2 + 1}} \xrightarrow[|p| \to \infty]{} 0$$

we conclude that, for $t > 0$, the integral of $e^{pt}/\sqrt{p^2 + 1}$ along Γ tends to zero as R tends to infinity; hence, as in (9-24), the function $f(t)$ is given by

$$f(t) = \frac{1}{2\pi j} \int_C \frac{e^{pt}}{\sqrt{p^2 + 1}} \, dp \qquad (9\text{-}51)$$

We now form the closed curve C_1 consisting of the two vertical segments AB and GD of Fig. 9-11 and the two circles F_1 and F_2 with radii equal to r.

Since $1/\sqrt{p^2 + 1}$ is analytic in the region between C and C_1, we conclude that $f(t)$ is given by the integral in (9-51) taken along C_1. But

$$\left| \frac{e^{pt}}{\sqrt{p^2 + 1}} \right| < \frac{e^{rt}}{\sqrt{r}\sqrt{2 - r}} \qquad \text{for } p \text{ on } F_1$$

Hence, for a fixed $t > 0$,

$$\left| \int_{F_1} \frac{e^{pt}}{\sqrt{p^2 + 1}} \, dp \right| < \frac{2\pi\sqrt{r}e^{rt}}{\sqrt{2 - r}} \xrightarrow[r \to 0]{} 0$$

Similarly the integral along F_2 tends to zero with r. But [see (II-15)]

On AB: $p = j\omega$ $\sqrt{1 + p^2} = \sqrt{1 - \omega^2}$

On GD: $p = j\omega$ $\sqrt{1 + p^2} = -\sqrt{1 - \omega^2}$

Therefore, for $t > 0$,

$$f(t) = \frac{j}{2\pi j} \int_{-1}^{1} \frac{e^{j\omega t}}{\sqrt{1 - \omega^2}} \, d\omega + \frac{j}{2\pi j} \int_{1}^{-1} \frac{e^{j\omega t}}{-\sqrt{1 - \omega^2}} \, d\omega$$

$$= \frac{1}{\pi} \int_{-1}^{1} \frac{\cos \omega t}{\sqrt{1 - \omega^2}} \, d\omega \qquad\qquad (9\text{-}52)$$

and with $\omega = \sin \theta$

$$f(t) = \frac{1}{\pi} \int_{-\pi/2}^{\pi/2} \cos (t \sin \theta) \, d\theta = J_0(t) \qquad (9\text{-}53)$$

where $J_0(t)$ is the Bessel function of the first kind [see (6-91)].

Comment. The relationship between $F(\omega)$ and $F_I(j\omega)$ (see Sec. 9-2) can be reestablished with the help of the inversion formula (9-18). Suppose that the vertical boundary of analyticity of $F_I(p)$ is the imaginary axis. Its inverse $f(t)$ is given by (9-18), with $\alpha > 0$. Using Cauchy's theorem, we observe that the integration on the line Re $p = \alpha > 0$ is equivalent to the integration along the imaginary axis, indented by semicircles γ_i, bypassing the imaginary singularities of $F_I(p)$ as in Fig. 10-3. The Fourier inversion formula, on the other hand, involves integration along the real ω axis. Thus, $F(\omega)$ equals $F_I(j\omega)$ only if the integration along the semicircles γ_i tends to zero with their radius and if the integral (2-3) is a Cauchy principal value (see footnote, page 10). As an example, suppose that $F_I(p) = 1/\sqrt{p}$. It is easy to see, as in (9-46), that the integration along the semicircle enclosing the origin tends to zero with r; hence, $F(\omega) = 1/\sqrt{j\omega}$. If $F_I(p)$ has imaginary poles $j\omega_i$ with residues A_i, then the integration along γ_i tends to $j\pi A_i$. In this case, suitable impulses, as in (9-8), must be added to $F_I(j\omega)$ to obtain $F(\omega)$.

9-5. Initial-value Theorem

The initial-value theorem, loosely phrased, states that the values of the Laplace transform $F_I(p)$ of a function $f(t)$, for large p, depends on the behavior of $f(t)$ near the origin $t = 0$. To make this statement plausible, we observe that, if p is large, then the term e^{-pt} in

$$\int_0^\infty e^{-pt} f(t)\, dt$$

takes significant values only if t is small; therefore the dominant part of this integral depends on the values of $f(t)$ near the origin. Suppose that

$$f(t) \sim M t^\lambda \qquad \text{for } t \to 0,\ \lambda > -1 \tag{9-54}$$

Then
$$F_I(p) \sim \int_0^\infty e^{-pt} M t^\lambda\, dt = \frac{M\Gamma(\lambda + 1)}{p^{\lambda+1}} \qquad \text{for } p \to \infty \tag{9-55}$$

[see (9-5)]. This result is a special form of an important class of theorems on the asymptotic form of $F_I(p)$; its rigorous proof is not trivial.

We should point out that the term e^{-pt} tends to zero as $p \to \infty$ only if the real part of p tends also to infinity; this is not true, for example, if $p \to \infty$ along a vertical line. Thus the condition $p \to \infty$ in (9-55) should read

$$p \to \infty \qquad |\arg p| \leq \frac{\pi}{2} - \epsilon \tag{9-56}$$

i.e., Re $p \to \infty$. If the asymptotic behavior of $F_I(p)$ has the same form independently of the arguement of p, as is the case for rational functions, for example, then the above restriction (9-56) is not necessary. In this case, assuming that $f(t)$ has a Fourier transform

$$F(\omega) = e^{-\alpha(\omega)} e^{-j\theta(\omega)}$$

we have: if
$$f(t) \sim M t^\lambda \qquad \text{for } t \to 0,\ \lambda > -1$$

then
$$F(\omega) \sim \frac{M\Gamma(\lambda + 1)}{(j\omega)^{\lambda+1}} \qquad \text{for } \omega \to \infty \tag{9-57}$$

or th: equivalent
$$\alpha(\omega) \sim (\lambda + 1) \ln \omega \qquad \theta(\omega) \sim (\lambda + 1)\frac{\pi}{2} \qquad \text{for } \omega \to \infty \tag{9-58}$$

(see Fig. 9-12).

The above theorem is usually stated in a more special form. If $f(t)$ is differentiable at the origin and its first nonzero derivative is of order

$n \geq 0$, then, since

$$f(t) \sim \frac{f^{(n)}(0)}{n!} t^n \qquad t \to 0$$

we have $$F_\mathrm{I}(p) \sim \frac{f^{(n)}(0)}{p^{n+1}} \qquad p \to \infty$$

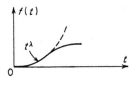

Thus

$$f(0) = f'(0) = \cdots = f^{(n-1)}(0) = 0$$

$$f^{(n)}(0) = \lim_{p \to \infty} p^{n+1} F_\mathrm{I}(p) \qquad (9\text{-}59)$$

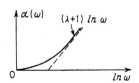

Signal-front delay. We shall now justify the statement (7-64) of Sec. 7-5. Suppose that in the region of convergence of (9-1) the function $e^{ap} F_\mathrm{I}(p)$ tends to zero as $p \to \infty$, where $a > 0$:

$$e^{ap} F_\mathrm{I}(p) \xrightarrow[p \to \infty]{} 0 \qquad \operatorname{Re} p > \gamma \qquad (9\text{-}60)$$

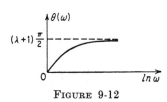

FIGURE 9-12

We shall then show that the inverse transform $f(t)$ of $F_\mathrm{I}(p)$ is zero for $t < a$

$$f(t) = 0 \qquad t < a \qquad (9\text{-}61)$$

Proof. Consider the closed curve C of Fig. 9-13; in its interior the function $F_\mathrm{I}(p)$ is analytic; hence

$$\int_\Gamma e^{pt} F_\mathrm{I}(p) \, dp + \int_L e^{pt} F_\mathrm{I}(p) \, dp = 0 \qquad (9\text{-}62)$$

But $$\int_\Gamma e^{pt} F_\mathrm{I}(p) \, dp = \int_\Gamma e^{(t-a)p} e^{ap} F_\mathrm{I}(p) \, dp$$

Hence from (9-60) and (II-54) it follows that, for $t < a$,

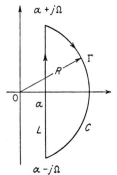

$$\int_\Gamma e^{pt} F_\mathrm{I}(p) \, dp \xrightarrow[R \to \infty]{} 0 \qquad (9\text{-}63)$$

and since the integral along L tends to the Bromwich path with Ω tending to infinity, we conclude from (9-62) and (9-63) that (9-61) is true.

9-6. The Bilateral Laplace Transform

Given a function $f(t)$, defined for every t, we form the integral†

FIGURE 9-13

$$F_\mathrm{II}(p) = \int_{-\infty}^{\infty} e^{-pt} f(t) \, dt \qquad (9\text{-}64)$$

† B. Van der Pol and H. Bremmer, "Operational Calculus Based on the Two sided Laplace Integral," Cambridge University Press, New York, 1950.

For the values of the complex variable p for which it exists, it defines a function $F_{II}(p)$ known as the *bilateral Laplace transform* of $f(t)$. To establish the region of convergence of (9-64), we write it in the form

$$F_{II}(p) = \int_0^\infty e^{-pt} f(t)\, dt + \int_{-\infty}^0 e^{-pt} f(t)\, dt$$

The first integral above is the unilateral Laplace transform of $f(t)U(t)$; therefore it converges in a region Re $p > \gamma_1$ and is an analytic function of p in that region. One can similarly show that the second

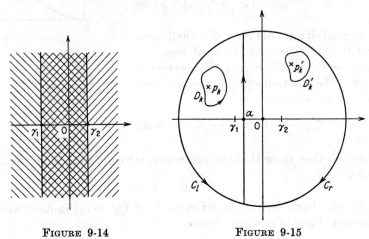

FIGURE 9-14   FIGURE 9-15

integral is analytic for Re $p < \gamma_2$; therefore the function $F_{II}(p)$ exists and is an analytic function of p in the vertical strip

$$\gamma_1 < \text{Re } p < \gamma_2 \qquad (9\text{-}65)$$

of Fig. 9-14.

With $p = \alpha + j\omega$, we have from (9-64)

$$F_{II}(\alpha + j\omega) = \int_{-\infty}^\infty e^{-\alpha t} f(t) e^{-j\omega t}\, dt \qquad (9\text{-}66)$$

Thus $F_{II}(\alpha + j\omega)$ is the Fourier transform of the function $e^{-\alpha t} f(t)$

$$e^{-\alpha t} f(t) \leftrightarrow F_{II}(\alpha + j\omega) \qquad (9\text{-}67)$$

Therefore we have from (2-3)

$$e^{-\alpha t} f(t) = \frac{1}{2\pi} \int_{-\infty}^\infty F_{II}(\alpha + j\omega) e^{j\omega t}\, d\omega \qquad (9\text{-}68)$$

Proceeding as in (9-18), we can write (9-68) in the form of a line integral

$$f(t) = \frac{1}{2\pi j} \int_{\alpha - j\infty}^{\alpha + j\infty} F_{II}(p) e^{pt}\, dp \qquad (9\text{-}69)$$

along the vertical line $\operatorname{Re} p = \alpha$ in the region (9-65) of existence of $F_{\mathrm{II}}(p)$. This is the inversion formula of the bilateral Laplace transform. The above integral can be evaluated by a contour integration, as in the case of (9-18). Suppose that the only singularities of $F_{\mathrm{II}}(p)$ are poles and that

$$F_{\mathrm{II}}(p) \to 0 \qquad \text{for } p \to \infty \qquad (9\text{-}70)$$

Then, reasoning as in (9-24), we conclude that, with C_l the closed path of Fig. 9-15 to the left of $\operatorname{Re} p = \alpha$, we have

$$f(t) = \frac{1}{2\pi j} \int_{C_l} e^{pt} F_{\mathrm{II}}(p) \, dp \qquad t > 0 \qquad (9\text{-}71)$$

and with C_r the path closing to the right of $\operatorname{Re} p = \alpha$

$$f(t) = \frac{1}{2\pi j} \int_{C_r} e^{pt} F_{\mathrm{II}}(p) \, dp \qquad t < 0 \qquad (9\text{-}72)$$

Denoting by $p_1, \ldots, p_n$ the poles of $F_{\mathrm{II}}(p)$ in the region $\operatorname{Re} p < \alpha$, and by $D_1, \ldots, D_n$ the closed curves of Fig. 9-15, each of which contains only one pole p_k, we obtain from (9-71) and Cauchy's theorem

$$f(t) = \frac{1}{2\pi j} \sum_{k=1}^{n} \int_{D_k} e^{pt} F_{\mathrm{II}}(p) \, dp \qquad t > 0 \qquad (9\text{-}73)$$

Similarly with $p_1', \ldots, p_m'$ the poles in the region $\operatorname{Re} p > \alpha$, we have

$$f(t) = \frac{-1}{2\pi j} \sum_{k=1}^{m} \int_{D_k'} e^{pt} F_{\mathrm{II}}(p) \, dp \qquad t < 0 \qquad (9\text{-}74)$$

The value of the integrals along D_k or D_k' is given by (9-27) or (9-31). The negative sign in (9-74) is due to the fact that the integrations along C_r and D_k' are in opposite directions.

Example 9-8

$$F_{\mathrm{II}}(p) = \frac{1}{(p+1)(p+2)(p-3)}$$

We should point out that $F_{\mathrm{II}}(p)$ is not specified by the above fraction; the region of its existence must also be given. Suppose that this region lies between the poles -1 and -2; then α in (9-65) must satisfy

$$-2 < \alpha < -1$$

$F_{\mathrm{II}}(p)$ has one pole $p_1 = -2$ for $\operatorname{Re} p < \alpha$, and since the residue of $e^{pt} F_{\mathrm{II}}(p)$ at $p = -2$ is given by

$$\left. \frac{e^{pt}}{(p+1)(p-3)} \right|_{p=-2} = \frac{e^{-2t}}{5}$$

we conclude from (9-73) that

$$f(t) = \frac{e^{-2t}}{5} \qquad \text{for } t > 0$$

For $\operatorname{Re} p > \alpha$ the poles of $e^{pt} F_{\mathrm{II}}(p)$ are $p_1' = -1$, $p_2' = 3$ with residues

$$\left. \frac{e^{pt}}{(p+2)(p-3)} \right|_{p=-1} = -\frac{e^{-t}}{4} \qquad \left. \frac{e^{pt}}{(p+1)(p+2)} \right|_{p=3} = \frac{e^{3t}}{20}$$

We thus have from (9-74)

$$f(t) = -\frac{e^{-t}}{4} - \frac{e^{3t}}{20} \qquad \text{for } t < 0$$

If the region of existence of $F_{\mathrm{II}}(p)$ is the strip

$$-1 < \alpha < 3$$

then its inverse transform is

$$f(t) = -\frac{e^{-t}}{4} + \frac{e^{-2t}}{5} \qquad t > 0$$

$$f(t) = -\frac{e^{3t}}{20} \qquad t < 0$$

as one can similarly see.

Example 9-9

$$F_{\mathrm{II}}(p) = \frac{1}{(k^2 - p^2)^2} \qquad -k < \operatorname{Re} p < k$$

Since the strip $-k < \operatorname{Re} p < k$ contains the imaginary axis, we see from (9-64) that the Fourier transform of $f(t)$ is given by

$$F(\omega) = F_{\mathrm{II}}(j\omega) = \frac{1}{(k^2 + \omega^2)^2}$$

$F_{\mathrm{II}}(p)$ has a double pole $p_1 = -k$ for $\operatorname{Re} p < \alpha$ and, since [see (9-31)]

$$\left. \frac{d}{dp} \frac{e^{pt}}{(k-p)^2} \right|_{p=-k} = \frac{e^{-kt}}{4k^3} (kt + 1)$$

we have

$$f(t) = \frac{e^{-kt}}{4k^3} (kt + 1) \qquad t > 0$$

Similarly

$$f(t) = \frac{e^{kt}}{4k^3} (1 - kt) \qquad t < 0$$

Even transforms. If the function $f(t)$ is even, then the corresponding transform $F_{\mathrm{II}}(p)$ is also even and the strip (9-65) of convergence of (9-64) has the imaginary axis as its axis of symmetry. This follows readily from

$$F_{\mathrm{II}}(-p) = \int_{-\infty}^{\infty} e^{pt} f(t) \, dt = \int_{-\infty}^{\infty} e^{-pt} f(-t) \, dt = F_{\mathrm{II}}(p)$$

because $f(-t) = f(t)$. Conversely, if $F_{\mathrm{II}}(p)$ is even and has no imaginary poles, and the strip of its convergence contains the imaginary axis, then its inverse $f(t)$ is even. Indeed, since (9-64) exists for $p = j\omega$, we have

$$F(\omega) = F_{\text{II}}(j\omega)$$

where $F(\omega)$ is the Fourier transform of $f(t)$. From the evenness of $F_{\text{II}}(p)$ it follows that $F(\omega)$ is real; hence $f(t)$ is even.

Utilizing the evenness of $F_{\text{II}}(p)$, one can, with the techniques of Sec. 10-2, simplify the evaluation of its inverse transform. Indeed, suppose that $F_{\text{II}}(p)$ is an even rational function of p:

$$F_{\text{II}}(p) = \frac{A(p^2)}{B(p^2)}$$

where $A(p^2)$ and $B(p^2)$ are polynomials in p^2. We can write $F_{\text{II}}(p)$ as a sum [see (10-21)]

$$F_{\text{II}}(p) = \frac{A(p^2)}{B(p^2)} = \frac{N(p)}{D(p)} + \frac{N(-p)}{D(-p)}$$

where $D(p)$ is a polynomial with all its roots in the Re $p < 0$ region. Denoting by $f^+(t)$ the inverse transform of

$$F_{\text{II}}^+(p) = \frac{N(p)}{D(p)}$$

we observe that since $F_{\text{II}}^+(p)$ is analytic in the interior of the region C_r of Fig. 9-15, we must have [see (9-72) and page 214]

$$f^+(t) = 0 \qquad \text{for } t < 0$$

Since $f(t)$ is even and $f^+(t)$ is causal, we conclude that

$$f(t) = f^+(t) + f^+(-t)$$

Thus the problem of determining $f(t)$ is reduced to the evaluation of $f^+(t)$, i.e., the inversion of the unilateral transform $N(p)/D(p)$.

Example 9-10

$$F_{\text{II}}(p) = \frac{1 - p^2}{1 + p^4} \qquad \text{existing for Re } p = 0$$

As in Example 10-2, we obtain

$$\frac{N(p)}{D(p)} = \frac{\sqrt{2}p + 1}{p^2 + \sqrt{2}p + 1}$$

The inverse $f^+(t)$ of $N(p)/D(p)$ can be found with elementary techniques (see Sec. 9-4):

$$f^+(t) = \sqrt{2}e^{-t/\sqrt{2}} \cos(t/\sqrt{2}) U(t)$$

Therefore
$$f(t) = \sqrt{2}e^{-|t|/\sqrt{2}} \cos(|t|/\sqrt{2})$$

Chapter 10. Integral Theorems

In Chap. 2 we showed that the transform $H(\omega) = R(\omega) + jX(\omega)$ of a causal function $h(t)$ is uniquely determined in terms of $R(\omega)$ or $X(\omega)$. In the following we establish this dependence explicitly by a set of equations known as Hilbert or Wiener-Lee transforms. We further show that $H(\omega)$ can be found if $R(\omega)$ and $X(\omega)$ are specified in different parts of the ω axis.† Similar results are obtained in terms of the attenuation and phase shift of $H(\omega)$ for the class of minimum-phase-shift functions. Finally, we examine the conditions on $H(\omega)$ for the causality of its inverse transform. The results are based on the analytic properties of $H_I(p) = H(p/j)$ in the right-hand plane.

10-1. Integral Theorems

For the purpose of introducing the techniques of this chapter, we shall first present two theorems of some interest. We shall also prove them by a simple argument not involving contour integration. It will be assumed that $h(t)$ is real, causal, and contains no impulses.

With $H(\omega) = R(\omega) + jX(\omega)$, the Fourier transform of $h(t)$, we shall show that

$$h(0^+) = \frac{1}{\pi} \int_{-\infty}^{\infty} R(\omega) \, d\omega \tag{10-1}$$

First Proof: Inversion Formula. From the inversion formula (2-3) we have, since $X(\omega)$ is odd,

$$\frac{h(0^+) + h(0^-)}{2} = \frac{1}{2\pi} \int_{-\infty}^{\infty} H(\omega) \, d\omega = \frac{1}{2\pi} \int_{-\infty}^{\infty} R(\omega) \, d\omega$$

and because $h(0^-) = 0$, Eq. (10-1) follows.

† H. W. Bode, "Network Analysis and Feedback Amplifier Design," D. Van Nostrand Company, Inc., Princeton, N.J., 1945.

The above theorem is also written in the form

$$\lim_{\omega \to \infty} [j\omega H(\omega)] = \frac{1}{\pi} \int_{-\infty}^{\infty} R(\omega)\, d\omega \qquad (10\text{-}2)$$

This follows from (10-1) and

$$h(0^+) = \lim_{\omega \to \infty} j\omega H(\omega)$$

[see (9-59)].

If $H(\omega)$ is the input impedance of a lumped parameter network, and C its input capacity, then

$$\lim j\omega H(\omega) = \frac{1}{C} \qquad \text{for } \omega \to \infty \qquad (10\text{-}3)$$

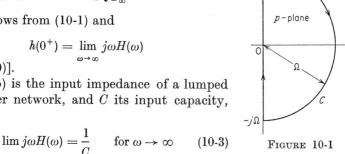

FIGURE 10-1

Hence

$$\int_{0}^{\infty} R(\omega)\, d\omega = \frac{\pi}{2C} \qquad (10\text{-}4)$$

This result is known as the *resistance integral theorem* (for an application see Prob. 34).

Second Proof: Contour Integration. The Laplace transform

$$H_I(p) = H(p/j)$$

of $h(t)$ is analytic in the right-hand plane: hence its integral along the closed path C of Fig. 10-1, consisting of the semicircle Γ and the portion $(-j\Omega, j\Omega)$ of the $j\omega$ axis, equals zero:

$$\int_C H_I(p)\, dp = \int_{-j\Omega}^{j\Omega} H_I(p)\, dp + \int_\Gamma H_I(p)\, dp = 0 \qquad (10\text{-}5)$$

From

$$\lim_{p \to \infty} p H_I(p) = h(0^+) \qquad p \text{ on } \Gamma \qquad (10\text{-}6)$$

we conclude that [see (II-21)]

$$\int_\Gamma H_I(p)\, dp \to -j\pi h(0^+) \qquad \Omega \to \infty \qquad (10\text{-}7)$$

But

$$\int_{-j\Omega}^{j\Omega} H_I(p)\, dp = j \int_{-\Omega}^{\Omega} H_I(j\omega)\, d\omega = j \int_{-\Omega}^{\Omega} H(\omega)\, d\omega$$

and with $\Omega \to \infty$ in (10-5), the desired result (10-1) follows.

The second theorem is the following identity:

$$\int_{-\infty}^{\infty} R^2(\omega)\, d\omega = \int_{-\infty}^{\infty} X^2(\omega)\, d\omega \qquad (10\text{-}8)$$

First Proof: Parseval's Formula. With

$$h(t) = h_e(t) + h_0(t)$$

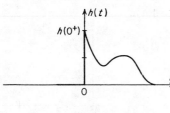

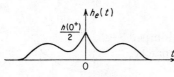

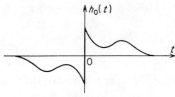

FIGURE 10-2

the decomposition of $h(t)$ into its even and odd parts (Fig. 10-2), we have from (2-26)

$$h_e(t) \to R(\omega) \qquad h_0(t) \to jX(\omega)$$

Therefore [see (2-75)]

$$\int_{-\infty}^{\infty} h_e{}^2(t)\, dt = \frac{1}{2\pi}\int_{-\infty}^{\infty} R^2(\omega)\, d\omega$$
$$\int_{-\infty}^{\infty} h_0{}^2(t)\, dt = \frac{1}{2\pi}\int_{-\infty}^{\infty} X^2(\omega)\, d\omega \tag{10-9}$$

From the causality of $h(t)$ it follows that

$$|h_e(t)| = |h_0(t)| \tag{10-10}$$

and (10-8) is proved.

Since $A^2(\omega) = R^2(\omega) + X^2(\omega)$, we conclude from (10-8) that for causal functions Parseval's formula (2-75) can be written in the form

$$\int_0^{\infty} h^2(t)\, dt = \frac{2}{\pi}\int_0^{\infty} R^2(\omega)\, d\omega \tag{10-11}$$

Second Proof: Contour Integration.   With C as in Fig. 10-1, we have

$$\int_C H_I{}^2(p)\, dp = \int_{-j\Omega}^{j\Omega} H_I{}^2(p)\, dp + \int_\Gamma H_I{}^2(p)\, dp = 0 \tag{10-12}$$

From (10-6) we conclude that

$$\lim pH_I{}^2(p) = 0 \qquad p \to \infty$$

Therefore [see (II-42)]

$$\lim \int_\Gamma H_I{}^2(p)\, dp = 0 \qquad \Omega \to \infty \tag{10-13}$$

But $H_I(j\omega) = R(\omega) + jX(\omega)$; hence

$$\int_{-j\Omega}^{j\Omega} H_I{}^2(p)\, dp = j\int_{-\Omega}^{\Omega} [R^2(\omega) - X^2(\omega)]\, d\omega - 2\int_{-\Omega}^{\Omega} R(\omega)X(\omega)\, d\omega \tag{10-14}$$

Making Ω tend to infinity in (10-12), we conclude, with (10-13) and (10-14), that (10-8) is true.

Example 10-1.   If $h(t) = e^{-\alpha t}U(t)$, then

$$H(\omega) = \frac{1}{\alpha + j\omega} = \frac{\alpha}{\alpha^2 + \omega^2} - j\frac{\omega}{\alpha^2 + \omega^2}$$

Therefore

$$\int_{-\infty}^{\infty} \frac{\alpha^2\, d\omega}{(\alpha^2 + \omega^2)^2} = \int_{-\infty}^{\infty} \frac{\omega^2\, d\omega}{(\alpha^2 + \omega^2)^2}$$

10-2. Relationship between $R(\omega)$ and $X(\omega)$

The discussion of the next two sections is based on the relationship, on the ω axis, between the real and imaginary parts of a function that is analytic for Re $p \geq 0$. Before developing the general case, we shall show that, if $H(\omega)$ is rational, then it can be simply determined from $R(\omega)$ or $X(\omega)$ by algebraic operations involving factorization of polynomials.

Rational $H(\omega)$. The system function $H_1(p)$ of the important class of networks with a finite number of elements is a rational function of p:

$$H_1(p) = \frac{N(p)}{D(p)} \tag{10-15}$$

From the analyticity of $H_1(p)$ in the right-hand plane it follows that $D(p)$ has no zeros for Re $p > 0$; such a polynomial is called *Hurwitz*. We shall first make the assumption that $D(p)$ has no imaginary zeros and that the degree of $N(p)$ does not exceed the degree of $D(p)$. Our problem is to determine the function $H_1(p)$ from the knowledge of its real part $R(\omega)$ on the $j\omega$ axis. We have

$$H_1(j\omega) = R(\omega) + jX(\omega) \qquad H_1(-j\omega) = R(\omega) - jX(\omega)$$

Hence
$$R(\omega) = \tfrac{1}{2}[H_1(j\omega) + H_1(-j\omega)]$$

and
$$R\left(\frac{p}{j}\right) = \tfrac{1}{2}[H_1(p) + H_1(-p)] \tag{10-16}$$

The rational function $R(\omega)$ contains only even powers of ω; therefore $R(p/j)$ is even, with real coefficients:

$$R\left(\frac{p}{j}\right) = \frac{A(p^2)}{B(p^2)} = \frac{1}{2}\left[\frac{N(p)}{D(p)} + \frac{N(-p)}{D(-p)}\right] \tag{10-17}$$

We expand $A(p^2)/B(p^2)$ into partial fractions in $p^2 = x$

$$\frac{A(x)}{B(x)} = k_0 + \frac{k_1}{x - x_1} + \cdots + \frac{k_n}{x - x_n} \tag{10-18}$$

where $x_1, \ldots, x_n$ are the roots of $B(x)$; hence $\pm\sqrt{x_1} \cdots \pm\sqrt{x_n}$ are the roots of $B(p^2)$. Since $H_1(p)$ has no purely imaginary poles, $B(p^2)$ has no imaginary zeros; therefore one of the square roots of x_i lies in the left-hand plane. We denote this root by p_i

$$p_i = \sqrt{x_i} \qquad \text{Re } p_i < 0 \tag{10-19}$$

But
$$\frac{k_i}{x - x_i} = \frac{k_i}{p^2 - p_i^2} = \frac{k_i/2p_i}{p - p_i} + \frac{k_i/2p_i}{-p - p_i} \tag{10-20}$$

Therefore (10-18) can be written in the form

$$\frac{A(p^2)}{B(p^2)} = \left(\frac{k_0}{2} + \frac{k_1/2p_1}{p - p_1} + \cdots + \frac{k_n/2p_n}{p - p_n}\right)$$
$$+ \left(\frac{k_0}{2} + \frac{k_1/2p_1}{-p - p_1} + \cdots + \frac{k_n/2p_n}{-p - p_n}\right) \quad (10\text{-}21)$$

The second parenthesis in (10-21) is obtained from the first by replacing p by $-p$; also, the denominator of the quantity in the first parenthesis is Hurwitz, as we see from (10-19). Comparing (10-16), (10-17), and (10-21), we conclude that the unknown system function is given by

$$H_I(p) = k_0 + \frac{k_1/p_1}{p - p_1} + \cdots + \frac{k_n/p_n}{p - p_n} \quad (10\text{-}22)$$

Example 10-2. Given $R(\omega) = (1 + \omega^2)/(1 + \omega^4)$, we want to find $H_I(p)$ and $X(\omega)$. We have

$$R\left(\frac{p}{j}\right) = \frac{1 - p^2}{1 + p^4} = \frac{1 - x}{1 + x^2} = \frac{1/\sqrt{2}\,e^{-j(3\pi/4)}}{x - j} + \frac{1/\sqrt{2}\,e^{j(3\pi/4)}}{x + j}$$

and since $p_1 = \sqrt{j} = e^{-j(3\pi/4)}$, $p_2 = \sqrt{-j} = e^{j(3\pi/4)}$, we obtain from (10-22)

$$H_I(p) = \frac{1/\sqrt{2}}{p - e^{-j(3\pi/4)}} + \frac{1/\sqrt{2}}{p - e^{j(3\pi/4)}} = \frac{\sqrt{2}p + 1}{p^2 + \sqrt{2}p + 1}$$

Hence $H(\omega) = \dfrac{1 + j\sqrt{2}\omega}{1 - \omega^2 + j\sqrt{2}\omega}$ and $X(\omega) = \dfrac{-\omega^3}{1 + \omega^4}\,\sqrt{2}$

The function $H_I(p)$ can be similarly determined if $X(\omega)$ is given. Since $2jX(\omega) = H(\omega) - H(-\omega)$, we obtain as in (10-16)

$$jX\left(\frac{p}{j}\right) = \tfrac{1}{2}[H_I(p) - H_I(-p)] \quad (10\text{-}23)$$

The function $jX(p/j)$ is odd with real coefficients; hence it can be written in the form

$$jX\left(\frac{p}{j}\right) = p\,\frac{A_1(p^2)}{B(p^2)} \quad (10\text{-}24)$$

where the degree of $A_1(p^2)$ is smaller than the degree of $B(p^2)$. With $p^2 = x$ we have

$$\frac{A_1(x)}{B(x)} = \frac{m_1}{x - x_1} + \cdots + \frac{m_n}{x - x_n} \quad (10\text{-}25)$$

and

$$p\,\frac{A_1(p^2)}{B(p^2)} = \left(\frac{pm_1/2p_1}{p - p_1} + \cdots + \frac{pm_n/2p_n}{p - p_n}\right)$$
$$- \left(\frac{-pm_1/2p_1}{-p - p_1} + \cdots + \frac{-pm_n/2p_n}{-p - p_n}\right) \quad (10\text{-}26)$$

and, since the second expression in parentheses can be obtained from the first by replacing p by $-p$, we conclude from (10-23), (10-24), and (10-26) that

$$H_I(p) = \frac{pm_1/2p_1}{p - p_1} + \cdots + \frac{pm_n/2p_n}{p - p_n} + C \qquad (10\text{-}27)$$

where C is an arbitrary constant. Thus, from $X(\omega)$, the function $R(\omega)$ is determined only within a constant.

Example 10-3. Given $X(\omega) = (\omega + \omega^3)/(1 + \omega^4)$, we want to find $R(\omega)$; we have

$$jX\left(\frac{p}{j}\right) = p\frac{1 - p^2}{1 + p^4}$$

and, with $(1 - x)/(1 + x^2)$ expanded as in Example 10-2, we obtain from (10-27)

$$H_I(p) = \frac{\sqrt{2}p^2 + p}{p^2 + \sqrt{2}p + 1} + C$$

Hence $H(\omega) = \dfrac{-\sqrt{2}\omega^2 + j\omega}{1 - \omega^2 + j\sqrt{2}\omega} + C$ $R(\omega) = \dfrac{\sqrt{2}\omega^4}{1 + \omega^4} + C$

The above results can be extended to functions with purely imaginary poles. If $H_I(p)$ has imaginary poles, then $R(\omega)$ or $X(\omega)$ will contain impulses and real poles. If these quantities are extracted, then the system function corresponding to the remainder can be found as above. It suffices therefore to examine the correspondence among $R(\omega)$, $H_I(p)$, and $X(\omega)$ when only such terms are present. From the relationship between the Fourier integral of causal functions and their Laplace transform developed in Sec. 9-2, we readily establish the following correspondence:

$$R(\omega): \quad \pi\delta(\omega - \omega_0) \qquad \frac{1}{\omega - \omega_0}$$

$$H_I(p): \quad \frac{1}{p - j\omega_0} \qquad \frac{j}{p - j\omega_0}$$

$$X(\omega): \quad \frac{1}{\omega_0 - \omega} \qquad \pi\delta(\omega - \omega_0)$$

From the above we see that if

$$R(\omega) = \frac{k}{\omega^2 - \omega_0^2} = \frac{k/2\omega_0}{\omega - \omega_0} - \frac{k/2\omega_0}{\omega + \omega_0}$$

then $$2\omega_0 X(\omega) = k\pi\delta(\omega - \omega_0) - k\pi\delta(\omega + \omega_0)$$

Example 10-4. $R(\omega) = 1/(\omega^2 + 1)(\omega^2 - 4)$. We have

$$R(\omega) = \frac{\frac{1}{5}}{\omega^2 - 4} - \frac{\frac{1}{5}}{\omega^2 + 1}$$

The system function corresponding to $\frac{1}{5}/(\omega^2 + 1)$ is given by $\frac{1}{5}/(p + 1)$, as we can find from (10-22); hence

$$X(\omega) = \frac{\pi}{20} [\delta(\omega - 2) - \delta(\omega + 2)] - \frac{\omega}{5(\omega^2 + 1)}$$

Hilbert transforms. In the following we shall give explicit equations relating the real and imaginary parts of a causal system function. If the causal function $h(t)$ contains no singularities at the origin, then with $H(\omega) = R(\omega) + jX(\omega)$ its Fourier integral, $R(\omega)$ and $X(\omega)$ satisfy the equations

$$X(\omega) = -\frac{1}{\pi} \int_{-\infty}^{\infty} \frac{R(y)}{\omega - y} \, dy \qquad (10\text{-}28)$$

$$R(\omega) = \frac{1}{\pi} \int_{-\infty}^{\infty} \frac{X(y)}{\omega - y} \, dy \qquad (10\text{-}29)$$

known as Hilbert transforms.

First Proof: Convolution Theorem. We denote by $h_e(t)$ and $h_0(t)$ the even and odd parts of $h(t)$ shown in Fig. 10-2; since $h_e(t) = h_0(t)$ for $t > 0$ and $h_e(t) = -h_0(t)$ for $t < 0$, we conclude that

$$h_0(t) = h_e(t) \, \text{sgn} \, t \qquad (10\text{-}30)$$

$$h_e(t) = h_0(t) \, \text{sgn} \, t \qquad (10\text{-}31)$$

where sgn t is the sign function of Fig. 2-8. The Fourier integrals of $h_e(t)$ and $h_0(t)$ are $R(\omega)$ and $jX(\omega)$ respectively, and the Fourier integral of sgn t equals $2/j\omega$:

$$h_e(t) \leftrightarrow R(\omega) \qquad h_0(t) \leftrightarrow jX(\omega) \qquad \text{sgn} \, t \leftrightarrow \frac{2}{j\omega} \qquad (10\text{-}32)$$

Since $h_0(t)$ is the product of $h_e(t)$ and sgn t, we conclude from the frequency convolution theorem (2-74) that

$$jX(\omega) = \frac{1}{2\pi} R(\omega) * \frac{2}{j\omega}$$

from which (10-28) follows. We similarly obtain (10-29) from (10-31).

Comment. The assumption that $h(t)$ contains no impulses at the origin imposes the restriction $H(\omega) \to 0$ for $\omega \to \infty$, as we see from (10-3). If $H(\omega) \to k_0$ for $\omega \to \infty$, then the even term $k_0\delta(t)$ is present in $h(t)$ and $h_e(t)$, but not in $h_0(t)$. Therefore (10-30) is still valid, but $k_0\delta(t)$ is lost in (10-31). We conclude that in this case $X(\omega)$ is given by (10-28), but from (10-29) only the quantity $R(\omega) - k_0$ is recovered. Thus $R(\omega)$ can be determined from $X(\omega)$ only within an arbitrary constant. Clearly, since k_0 is real, we have $R(\omega) \to k_0$ for $\omega \to \infty$, and (10-29) should be written in the form

$$R(\omega) = R(\infty) + \frac{1}{\pi} \int_{-\infty}^{\infty} \frac{X(y)}{\omega - y} \, dy \qquad (10\text{-}33)$$

This lack of symmetry between (10-29) and (10-33) is due to our assumption that $h(t)$ is real.

Second Proof: Contour Integration. We shall prove that if

$$H(\omega) \to k_0 = R(\infty) \qquad \text{for } \omega \to \infty \qquad (10\text{-}34)$$

then Eqs. (10-28) and (10-33) hold. We form the function

$$\frac{H_I(p)}{p - j\omega_0}$$

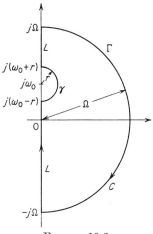

FIGURE 10-3

where ω_0 is an arbitrary real constant. This function is analytic in the interior of the closed curve C of Fig. 10-3, consisting of the semicircle Γ with center at the origin and radius Ω, the semicircle γ with center at $j\omega_0$ and radius r, and the portion L of the imaginary axis from $-j\Omega$ to $j(\omega_0 - r)$ and from $j(\omega_0 + r)$ to $j\Omega$. Therefore

$$\int_C \frac{H_I(p)}{p - j\omega_0}\, dp = 0$$

From (10-34) we conclude that $H_I(p) \to R(\infty)$ for $p \to \infty$; therefore

$$p\,\frac{H_I(p)}{p - j\omega_0} \to R(\infty) \qquad \text{for } p \to \infty$$

From (II-21) and the above we obtain

$$\int_\Gamma \frac{H_I(p)}{p - j\omega_0}\, dp \to -j\pi R(\infty) \qquad \text{for } \Omega \to \infty \qquad (10\text{-}35)$$

where the direction of integration is as in Fig. 10-3. As r tends to zero, the function $H_I(p)$ tends to $H_I(j\omega_0)$ for p on γ; therefore

$$\int_\gamma \frac{H_I(p)\,dp}{p - j\omega_0} \xrightarrow[r \to 0]{} H_I(j\omega_0) \int_\gamma \frac{dp}{p - j\omega_0} = j\pi H_I(j\omega_0) \qquad (10\text{-}36)$$

The line L tends to the entire imaginary axis for $r \to 0$ and $\Omega \to \infty$, and since $p = j\omega$ on L, we have

$$\int_L \frac{H_L(p)}{p - j\omega_0}\, dp \to j \int_{-\infty}^{\infty} \frac{H_L(j\omega)}{j\omega - j\omega_0}\, d\omega \qquad (10\text{-}37)$$

With $H_I(j\omega) = R(\omega) + jX(\omega)$, we conclude from (10-34) to (10-37) that

$$\int_{-\infty}^{\infty} \frac{R(\omega) + jX(\omega)}{\omega - \omega_0}\, d\omega + j\pi\,|R(\omega_0) + jX(\omega_0)| - j\pi R(\infty) = 0 \qquad (10\text{-}38)$$

Therefore

$$X(\omega_0) = \frac{-1}{\pi} \int_{-\infty}^{\infty} \frac{R(\omega)}{\omega_0 - \omega} \, d\omega \qquad R(\omega_0) = R(\infty) + \frac{1}{\pi} \int_{-\infty}^{\infty} \frac{X(\omega)}{\omega_0 - \omega} \, d\omega$$

and this completes the proof.

In the above proof it was assumed for simplicity that $H_I(p)$ has no poles on the imaginary axis; this assumption, however, is not necessary for the validity of the Hilbert transforms, as one can easily show with a suitable modification of the contour C.

Example 10-5. $R(\omega) = \pi\delta(\omega)$. The function $X(\omega)$ can be found from (10-28)

$$X(\omega) = -\frac{1}{\pi} \int_{-\infty}^{\infty} \frac{\pi\delta(y)}{\omega - y} \, dy = -\frac{1}{\omega}$$

and the pair of Fig. 10-4 results.

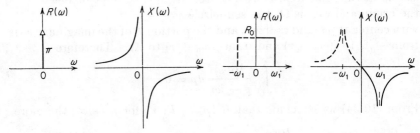

FIGURE 10-4   FIGURE 10-5

Example 10-6. $R(\omega) = R_0 p_{\omega_1}(\omega)$ as in Fig. 10-5. From (10-28) we have

$$X(\omega) = -\frac{1}{\pi} \int_{-\omega_1}^{\omega_1} \frac{R_0}{\omega - y} \, dy = \frac{R_0}{\pi} \ln \left| \frac{\omega - \omega_1}{\omega + \omega_1} \right|$$

shown in Fig. 10-5. From (2-31) we see that the corresponding time function is given by

$$f(t) = \frac{2R_0}{\pi} \frac{\sin \omega_1 t}{t} \, U(t)$$

The Hilbert transforms (10-28) and (10-29) can also be written in the form

$$X(\omega) = -\frac{2\omega}{\pi} \int_0^{\infty} \frac{R(y)}{\omega^2 - y^2} \, dy \qquad R(\omega) = \frac{2}{\pi} \int_0^{\infty} \frac{yX(y)}{\omega^2 - y^2} \, dy \quad (10\text{-}39)$$

if $H(\omega)$ contains no impulses at the origin. This follows easily from the fact that $R(\omega)$ is even and $X(\omega)$ odd.

Causality Resulting from Hilbert Transforms. If the real and imaginary parts of the Fourier integral of a function $h(t)$ satisfy either of Eqs. (10-28) and (10-29), then $h(t)$ is causal.

Indeed, suppose that (10-29) is true; this equation can be written in the form

$$R(\omega) = \frac{1}{2\pi} jX(\omega) * \frac{2}{j\omega}$$

From (10-32) and the convolution theorem (2-74) we conclude that the even and odd parts of $h(t) = h_e(t) + h_0(t)$ are related by

$$h_e(t) = h_0(t) \operatorname{sgn} t$$

Therefore $h_e(t) = -h_0(t)$ for $t < 0$

Hence $h(t)$ is causal.   A similar result can be established if (10-28) is true.

Wiener-Lee transforms.   The numerical evaluation of (10-28) or (10-29) is in general complicated.   In the following we shall show

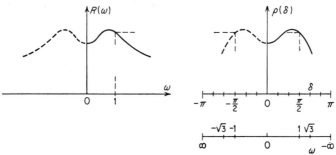

FIGURE 10-6

that, by a change in the independent variable, a simpler set of equations, known as Wiener-Lee transforms, can be derived.   This method will also permit the direct evaluation of the time function $h(t)$ in terms of $R(\omega)$ or $X(\omega)$.

We introduce the variable δ defined by

$$\omega = -\tan\frac{\delta}{2} \tag{10-40}$$

and express the system function $H(\omega) = R(\omega) + jX(\omega)$ in terms of δ.

$$H(\omega) = H\left(-\tan\frac{\delta}{2}\right) = R\left(-\tan\frac{\delta}{2}\right) + jX\left(-\tan\frac{\delta}{2}\right) \tag{10-41}$$

With $$\rho(\delta) = R\left(\tan\frac{\delta}{2}\right) \quad \chi(\delta) = X\left(\tan\frac{\delta}{2}\right) \tag{10-42}$$

we obtain $$H\left(-\tan\frac{\delta}{2}\right) = \rho(\delta) - j\chi(\delta) \tag{10-43}$$

since $R(\omega)$ is even and $X(\omega)$ odd.   Clearly, knowing $R(\omega)$ and $X(\omega)$, we can readily find $\rho(\delta)$ and $\chi(\delta)$.   In Fig. 10-6 we have shown $R(\omega)$

and the corresponding $\rho(\delta)$. We next expand the even function $\rho(\delta)$ and odd function $\chi(\delta)$ into a Fourier series in the $(-\pi,\pi)$ interval:

$$\rho(\delta) = a_0 + a_1 \cos \delta + \cdots + a_n \cos n\delta + \cdots \qquad (10\text{-}44)$$

$$\chi(\delta) = b_1 \sin \delta + \cdots + b_n \sin n\delta + \cdots \qquad (10\text{-}45)$$

where

$$a_n = \frac{1}{\pi} \int_{-\pi}^{\pi} \rho(\delta) \cos n\delta \, d\delta \qquad b_n = \frac{1}{\pi} \int_{-\pi}^{\pi} \chi(\delta) \sin n\delta \, d\delta \qquad n > 0 \quad (10\text{-}46)$$

We shall prove that if $h(t)$ is causal, then the coefficients in (10-44) and (10-45) are related by

$$b_n = -a_n \qquad (10\text{-}47)$$

Thus to determine $X(\omega)$ from $R(\omega)$, we form the function $\rho(\delta)$ and evaluate its Fourier coefficients a_n. The constants $b_n = -a_n$ are then determined, and $\chi(\delta)$ is given as a sine series (10-45). Finally $X(\omega)$ can be evaluated from $\chi(\delta)$, with the scale change (10-40). Conversely if $X(\omega)$ is known, $R(\omega)$ can be found; however, a_0 cannot be found from the series (10-45); $R(\omega)$ is determined only within the constant a_0.

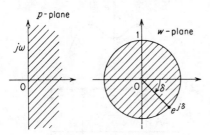

FIGURE 10-7

The transformation

$$w = \frac{1-p}{1+p} \qquad p = \frac{1-w}{1+w} \qquad (10\text{-}48)$$

maps the right-hand side of the p plane into the interior of the unit circle in the w plane (Fig. 10-7). For $p = j\omega$ we have

$$w = \frac{1 - j\omega}{1 + j\omega}$$

Hence $|w| = 1$ and with $w = e^{j\delta}$

$$\omega = -\tan \frac{\delta}{2}$$

The function $H_{\mathrm{I}}(p)$ is analytic for $\operatorname{Re} p \geq 0$; hence the function $H_{\mathrm{I}}[(1 - w)/(1 + w)]$ is analytic for $|w| \leq 1$. Expanding into a power series, we obtain

$$H_{\mathrm{I}}\left(\frac{1-w}{1+w}\right) = A_0 + A_1 w + \cdots + A_n w^n + \cdots \qquad (10\text{-}49)$$

and (10-49) converges for $|w| \leq 1$ because H_{I} is analytic in the interior of the unit circle. For $w = e^{j\delta}$ we have

$$H_{\mathrm{I}}\left(-j \tan \frac{\delta}{2}\right) = (A_0 + A_1 \cos \delta + \cdots + A_n \cos n\delta + \cdots)$$

$$+ j(A_1 \sin \delta + \cdots + A_n \sin n\delta + \cdots)$$

But
$$H_I\left(-j\tan\frac{\delta}{2}\right) = H\left(-\tan\frac{\delta}{2}\right) = \rho(\delta) - j\chi(\delta)$$

Hence
$$\rho(\delta) = A_0 + A_1\cos\delta + \cdots + A_n\cos n\delta + \cdots \tag{10-50}$$

$$\chi(\delta) = -A_1\sin\delta - \cdots - A_n\sin n\delta - \cdots \tag{10-51}$$

Comparing coefficients in (10-44) and (10-50) and in (10-45) and (10-51), we conclude that

$$a_n = A_n \qquad b_n = -A_n \tag{10-52}$$
and (10-47) follows.

Example 10-7. Suppose that $R(\omega)$ is such that the corresponding $\rho(\delta)$ is given by

$$\rho(\delta) = \cos n\delta$$

We then obtain from (10-45) and (10-47)

$$\chi(\delta) = -\sin n\delta$$

Therefore
$$H(\omega) = \cos n\delta + j\sin n\delta = e^{jn\delta} = \left(\frac{1-j\omega}{1+j\omega}\right)^n$$

With $h_n(t)$ the inverse transform of the above function

$$h_n(t) \leftrightarrow \left(\frac{1-j\omega}{1+j\omega}\right)^n \tag{10-53}$$

we can easily show that

$$(-1)^n h(t) = \delta(t) + U(t)e^t\frac{d^n}{dt^n}\left[\frac{t^{n-1}}{(n-1)!}e^{-2t}\right] \tag{10-54}$$

This function is related to the tabulated Laguerre polynomials.

Laguerre Expansion of a Causal Function. Using the previous results, we shall express the inverse transform of $H(\omega)$ in terms of $R(\omega)$ or $X(\omega)$. From (10-48) and (10-49) we have

$$H_I(p) = A_0 + \cdots + A_n\left(\frac{1-p}{1+p}\right)^n + \cdots \tag{10-55}$$
Hence [see (10-53)]

$$h(t) = \sum_{n=0}^{\infty} A_n h_n(t) \tag{10-56}$$

The constants A_n can be found from (10-46) and (10-52), if $R(\omega)$ or $X(\omega)$ is given.

In the next section we shall derive similar results between the attenuation $\alpha(\omega)$ and the phase shift $\theta(\omega)$ of minimum-phase-shift functions. We shall further show that $H(\omega)$ can be found if $\alpha(\omega)$ and $\theta(\omega)$ are specified in complementary parts of the ω axis. This can also be done for $R(\omega)$ and $X(\omega)$ by the same method.

10-3. Minimum-phase-shift Functions

Relationship between $\alpha(\omega)$ and $\theta(\omega)$. Consider the transform

$$H(\omega) = e^{-\alpha(\omega) - j\theta(\omega)} \qquad (10\text{-}57)$$

of the causal function $h(t)$. In general, $\theta(\omega)$ is not uniquely determined from $\alpha(\omega)$ if no additional assumptions are made about $H(\omega)$. Indeed, with

$$h_1(t) = e^{-2t} U(t) \qquad h_2(t) = (3e^{-2t} - 2e^{-t})U(t) \qquad (10\text{-}58)$$

the corresponding transforms are given by

$$H_1(\omega) = \frac{1}{2 + j\omega} \qquad H_2(\omega) = \frac{1}{2 + j\omega}\frac{j\omega - 1}{j\omega + 1} \qquad (10\text{-}59)$$

Clearly the amplitudes of these functions are equal:

$$|H_1(\omega)| = |H_2(\omega)| = \frac{1}{\sqrt{4 + \omega^2}}$$

but their phases are quite different from each other. The question arises whether it is possible, by imposing certain conditions on $H(\omega)$, to obtain a class of causal functions in which $\theta(\omega)$ is uniquely determined from $\alpha(\omega)$. In Sec. 10-2 it was shown that $X(\omega)$ could be uniquely determined from $R(\omega)$, because $R(\omega)$ and $X(\omega)$ were the real and imaginary parts of an analytic function in the right-hand plane. $\alpha(\omega)$ and $\theta(\omega)$ are also the real and imaginary parts of $\ln H_1(j\omega)$

$$\alpha(\omega) + j\theta(\omega) = -\ln H_{\mathrm{I}}(j\omega)$$

but $\ln H_{\mathrm{I}}(p)$ is not, in general, analytic for $\operatorname{Re} p \geq 0$. Indeed, $H_{\mathrm{I}}(p)$ might have zeros with positive real parts, and these zeros are singularity points of $\ln H_{\mathrm{I}}(p)$. However, if we assume that $H_{\mathrm{I}}(p)$ is not only analytic but has no zeros for $\operatorname{Re} p \geq 0$, then $\ln H_{\mathrm{I}}(p)$ will also be analytic in the right-hand plane, and $\theta(\omega)$ will be uniquely determined from $\alpha(\omega)$. The class of functions with this property is called *minimum-phase-shift*. From the two functions in (10-59), only $H_1(\omega)$ is minimum-phase-shift.

In the following we shall consider only minimum-phase-shift functions and shall develop equations, similar to the ones in Sec. 10-2, relating $\alpha(\omega)$ to $\theta(\omega)$.

We remark that, if $H_1(\omega)$ is minimum-phase-shift and

$$|H_2(\omega)| = |H_1(\omega)|$$

then $H_2(\omega)$ can be written in the form

$$H_2(\omega) = H_1(\omega)H_a(\omega) \qquad (10\text{-}60)$$

where

$$|H_a(\omega)| = 1 \qquad (10\text{-}61)$$

A causal system function $H_a(\omega)$ satisfying (10-61) is called *all-pass*. Conversely if a function $H_2(\omega)$ is not minimum-phase-shift, then it can be written as a product of minimum-phase-shift function $H_1(\omega)$ and an all-pass term $H_a(\omega)$ as in (10-60). This can be done by selecting, for $H_a(\omega)$, a function whose zeros z_i are the zeros of $H_2(\omega)$ in the right-hand part of the $p = j\omega$ plane, and whose poles are symmetrical to the z_i's with respect to the imaginary axis.

   Rational $H(\omega)$. Suppose that a function $H_1(p)$ is rational,

$$H_1(p) = \frac{N(p)}{D(p)}$$

where the polynomials $N(p)$ and $D(p)$ are Hurwitz. We shall show that if its amplitude $A(\omega) = |H_1(j\omega)|$ is known, then $H_1(p)$ can be determined with simple algebraic operations.

Since $H_1(-j\omega) = \overset{*}{H}_1(j\omega)$, we have

$$A^2(\omega) = H_1(j\omega)H_1(-j\omega)$$

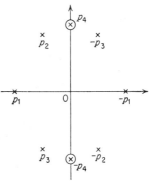

Therefore $A^2\left(\dfrac{p}{j}\right) = H_1(p)H_1(-p)$ (10-62)

From the evenness of $A^2(\omega)$ it follows that $A^2(p/j)$ is even in p, rational, with real coefficients:

$$A^2\left(\frac{p}{j}\right) = \frac{S(p^2)}{T(p^2)}$$ (10-63)

FIGURE 10-8

We shall express S/T as a product of the form (10-62). The roots of the even polynomial $S(p^2)$ are symmetrical about the imaginary axis, as in Fig. 10-8, because if p_i is a root, then $-p_i$ is also a root. Thus these roots can be grouped into $p_1, \ldots, p_n$ and $-p_1, \ldots, -p_n$, where the first group lies in the left-hand plane, and they are real or complex conjugate. If $S(p^2)$ has also purely imaginary roots, then, since $A^2(\omega) \geq 0$, these roots must be double, and half of them are assigned to the first group and the other half to the second; hence $S(p^2)$ can be written in the form

$$S(p^2) = k(p - p_1) \cdots (p - p_n)(-p - p_1) \cdots (-p - p_n)$$

and with

$$S_1(p) = \sqrt{k}(p - p_1) \cdots (p - p_n)$$
$$S_2(p) = \sqrt{k}(-p - p_1) \cdots (-p - p_n) = S_1(-p)$$

we have $S(p^2) = S_1(p)S_1(-p)$ (10-64)

Factoring $T(p^2)$ in the same way, we obtain

$$\frac{S(p^2)}{T(p^2)} = \frac{S_1(p)}{T_1(p)} \frac{S_1(-p)}{T_1(-p)}$$ (10-65)

where $S_1(p)$ and $T_1(p)$ are Hurwitz polynomials with real coefficients. From (10-62) and (10-65) we conclude that

$$H_I(p) = \frac{S_1(p)}{T_1(p)} \qquad (10\text{-}66)$$

Example 10-8

$$A(\omega) = \frac{1}{\sqrt{1 + \omega^6}}$$

We have

$$A^2\left(\frac{p}{j}\right) = \frac{1}{1 - p^6}$$

The denominator can be easily factored into the form (10-64)

$$1 - p^6 = (p^3 + 2p^2 + 2p + 1)(-p^3 + 2p^2 - 2p + 1)$$

Therefore

$$H_I(p) = \frac{1}{p^3 + 2p^2 + 2p + 1}$$

Comment. In order to factor a polynomial $S(p^2)$ as in (10-64), it is necessary to determine its roots. If these roots are complex, it is numerically simpler to evaluate the quadratic factors of $S(p^2)$; the corresponding quadratic factors of $S_1(p)$ can then be found directly. Indeed, suppose that $p^4 + ap^2 + b$ is a factor of $S(p^2)$; denoting by $p^2 + Ap + B$ the corresponding factor of $S_1(p)$, we can easily show that

$$B = \sqrt{b} \qquad A = \sqrt{2B - a}$$

If, for example, $S(p^2) = 1 + p^4$, then $a = 0$, $b = 1$; hence $B = 1$, $A = \sqrt{2}$, and the Hurwitz part of $S(p^2)$ is given by

$$S_1(p) = p^2 + \sqrt{2}p + 1.$$

Hilbert transforms. If the function $H(\omega)$ is minimum-phase-shift, then $\ln H_I(p)$ is analytic in the right-hand plane; in this case the attenuation and phase of $H(\omega) = e^{-\alpha(\omega)-j\theta(\omega)}$ are related by the following set of equations similar to (10-28) and (10-29):

$$\theta(\omega_0) = \frac{\omega_0}{\pi} \int_{-\infty}^{\infty} \frac{\alpha(\omega)}{\omega^2 - \omega_0^2} \, d\omega \qquad (10\text{-}67)$$

$$\alpha(\omega_0) = \alpha(0) - \frac{\omega_0^2}{\pi} \int_{-\infty}^{\infty} \frac{\theta(\omega)}{\omega(\omega^2 - \omega_0^2)} \, d\omega \qquad (10\text{-}68)$$

Thus $\theta(\omega)$ can be uniquely determined from $\alpha(\omega)$, and for the determination of $\alpha(\omega)$ one needs not only $\theta(\omega)$ but also the constant $\alpha(0)$. From the proof it will become clear that the above equations are not the only ones relating $\alpha(\omega)$ to $\theta(\omega)$; other sets of similar relationships can be derived (see H. W. Bode, op. cit.).

Proof. Contour Integration. From the asymptotic form (9-59) of $H_I(p)$ it follows that

$$\ln H_I(p) \to -k \ln p + \ln a \qquad \text{for } p \to \infty \qquad (10\text{-}69)$$

The above limit is infinite if $k \neq 0$; therefore, to account for the contribution of the semicircle Γ of Fig. 10-9 to the contour integration, it is not sufficient to divide $\ln H_I(p)$ by a linear term in p as in Sec. 10-4, but by a higher power of p. For this reason we form the function

$$F(p) = \frac{\ln H_I(p)}{p^2 + \omega_0{}^2} \qquad (10\text{-}70)$$

$F(p)$ is analytic in the right-hand plane, since $H_I(p)$ has neither zeros nor poles for Re $p > 0$. Assuming that $H_I(p)$ has no imaginary zeros, we conclude that the only singularities of $F(p)$ on the $j\omega$ axis are the points $\pm j\omega_0$; therefore $F(p)$ is analytic in the region bounded by the semicircle Γ, the semicircles γ_1 and γ_2 with centers $\pm j\omega_0$ and radius r, and the portion L of the imaginary axis as in Fig. 10-9. From Cauchy's theorem we obtain

$$\int_{L+\gamma_1+\gamma_2+\Gamma} F(p)\, dp = 0 \qquad (10\text{-}71)$$

But [see (10-69)]

$$pF(p) \to 0 \qquad \text{for } p \to \infty$$

Hence

$$\int_\Gamma F(p)\, dp \to 0 \qquad \text{for } \Omega \to \infty \qquad (10\text{-}72)$$

With a continuity argument we can easily show that

$$\int_{\gamma_1} \frac{\ln H_I(p)\, dp}{p^2 + \omega_0{}^2} \xrightarrow[r \to 0]{} \frac{\ln H_I(j\omega_0)}{j\omega_0 + j\omega_0} \int_{\gamma_1} \frac{dp}{p - j\omega_0} = \frac{\ln H_I(j\omega_0)}{2j\omega_0} j\pi \quad (10\text{-}73)$$

Similarly

$$\int_{\gamma_2} \frac{\ln H_I(p)\, dp}{p^2 + \omega_0{}^2} \xrightarrow[r \to 0]{} \frac{\ln H_I(-j\omega_0)}{-2j\omega_0} j\pi \qquad (10\text{-}74)$$

and since (see footnote on page 10)

$$\int_L \frac{\ln H_I(p)\, dp}{p^2 + \omega_0{}^2} \to \int_{-\infty}^{\infty} \frac{\ln H_I(j\omega)}{-\omega^2 + \omega_0{}^2} j\, d\omega \quad \text{for } r \to 0 \text{ and } \Omega \to \infty \quad (10\text{-}75)$$

we conclude from (10-71) to (10-75) and

$$\ln H_I(j\omega) = \ln H(\omega) = -\alpha(\omega) - j\theta(\omega)$$

that

$$j\int_{-\infty}^{\infty} \frac{\alpha(\omega) + j\theta(\omega)}{-\omega^2 + \omega_0{}^2}\, d\omega + \frac{\pi}{2\omega_0} [\alpha(\omega_0) + j\theta(\omega_0) - \alpha(-\omega_0) - j\theta(-\omega_0)] = 0$$

$$(10\text{-}76)$$

Equation (10-67) follows from (10-76) because

$$\alpha(-\omega) = \alpha(\omega) \qquad \theta(-\omega) = -\theta(\omega)$$

To prove (10-68) we integrate the function

$$F(p) = \frac{\ln H_{\mathrm{I}}(p)}{p(p^2 + \omega_0{}^2)}$$

along the closed curve C of Fig. 10-10, in the interior of which it is analytic. The contribution of Γ, γ_1, γ_2, and L to the integral can be

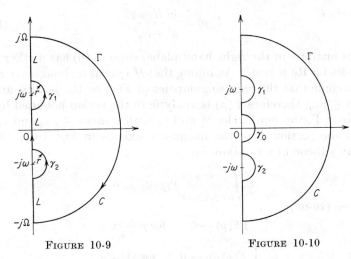

FIGURE 10-9   FIGURE 10-10

evaluated as above; an equation similar to (10-76) results, in which the integration along γ_0 must be added.   Since

$$\int_{\gamma_0} \frac{\ln H_{\mathrm{I}}(p)\, dp}{p(p^2 + \omega_0{}^2)} \xrightarrow[r \to 0]{} \frac{\ln H_{\mathrm{I}}(0)}{\omega_0{}^2} \int_{\gamma_0} \frac{dp}{p} = \frac{\alpha(0)}{\omega_0{}^2} j\pi$$

the desired result (10-68) readily follows.†

Equations (10-67) and (10-68) are valid even if $H(\omega)$ has imaginary zeros.   Indeed, if the curve of Fig. 10-10 is modified by introducing semicircles of radius r_i around the zeros of $H(\omega)$, then their contribution to the integral (10-71) tends to zero with $r_i \to 0$ because the integrand $F(p)$ tends to infinity logarithmically with r_i, whereas the length of the path of integration equals πr_i.

Example 10-9.   Consider a low-pass minimum-phase-shift filter with amplitude characteristic

$$A(\omega) = \begin{cases} 1 & |\omega| < \omega_c \\ A_s & |\omega| > \omega_c \end{cases}$$

† For the determination of $H_{\mathrm{I}}(p)$ in the region $\operatorname{Re} p > 0$ in terms of $\alpha(\omega)$ or $\theta(\omega)$, see Prob. 37.

as in Fig. 10-11a. Its phase shift $\theta(\omega)$ cannot be assigned arbitrarily but must be given by (10-67)

$$\theta(\omega) = \frac{-2\omega}{\pi} \int_{\omega_c}^{\infty} \frac{\ln A_s}{y^2 - \omega_c^2} \, dy = \frac{\ln A_s}{\pi} \ln \left| \frac{\omega - \omega_c}{\omega + \omega_c} \right|$$

as shown in Fig. 10-11b. The resulting group delay in the passband, equals

$$t_{gr}(\omega) = \theta'(\omega) = \frac{2 \ln A_s}{\pi} \frac{\omega_c}{\omega^2 - \omega_c^2} \qquad |\omega| < \omega_c$$

Thus the group delay is proportional to the attenuation in the stop band and it tends to infinity as A_s tends to zero. The quantity

$$t_{gr}(0) = \frac{2 \, |\ln A_s|}{\pi \omega_c}$$

equals the delay of the center of gravity of the input as it passes through the filter, and it is proportional to $|\ln A_s|$.

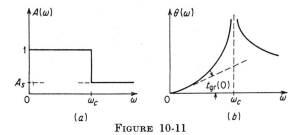

(a)   (b)

FIGURE 10-11

Example 10-10. As an application of (10-68), we shall evaluate the attenuation of a filter with a phase shift given by the curve

$$\theta(\omega) = \begin{cases} t_0\omega & |\omega| < \omega_c \\ t_0\omega_c & \omega > \omega_c \end{cases}$$

of Fig. 10-12a. A noncausal filter with the same phase was discussed in Sec. 6-4. Inserting the above characteristic into (10-68), we obtain

$$\alpha(\omega) = \alpha(0) - \frac{2\omega^2}{\pi} \int_0^{\omega_c} \frac{t_0 \, dy}{(y^2 - \omega^2)} - \frac{2\omega^2}{\pi} \int_{\omega_c}^{\infty} \frac{t_0\omega_c \, dy}{y(y^2 - \omega^2)}$$

$$= \alpha(0) + \frac{t_0\omega_c}{\pi} \left[\left(1 + \frac{\omega}{\omega_c} \right) \ln \left(1 + \frac{\omega}{\omega_c} \right) + \left(1 - \frac{\omega}{\omega_c} \right) \ln \left| 1 - \frac{\omega}{\omega_c} \right| \right]$$

The corresponding amplitude $A(\omega) = e^{-\alpha(\omega)}$ is shown in Fig. 10-12b for $t_0\omega_c = \pi$.

Wiener-Lee transforms. The attenuation $\alpha(\omega)$ can be related to the phase shift $\theta(\omega)$ by a Fourier series expansion similar to the method of Sec. 10-4.   With the change of variable

$$\omega = -\tan\frac{\delta}{2} \qquad (10\text{-}40)$$

the functions $\alpha(\omega)$ and $\theta(\omega)$ become $\bar{\alpha}(\delta)$ and $\bar{\phi}(\delta)$ defined by

$$\alpha(\omega) + j\theta(\omega) = \alpha\left(\tan\frac{\delta}{2}\right) - j\theta\left(\tan\frac{\delta}{2}\right) = \bar{\alpha}(\delta) - \bar{\theta}(\delta) \qquad (10\text{-}77)$$

Expanding $\bar{\alpha}(\delta)$ and $\bar{\theta}(\delta)$ into a cosine and a sine series respectively, we obtain

$$\bar{\alpha}(\delta) = d_0 + d_1 \cos\delta + \cdots + d_n \cos n\delta + \cdots \qquad (10\text{-}78)$$

$$\bar{\theta}(\delta) = e_1 \sin\delta + \cdots + e_n \sin n\delta + \cdots \qquad (10\text{-}79)$$

Proceeding as in Sec. 10-4, we can show that

$$d_n = -e_n \qquad (10\text{-}80)$$

Thus, knowing $\bar{\alpha}(\delta)$, we readily find $\bar{\theta}(\delta)$ from (10-79) and (10-80).   To determine $\bar{\alpha}(\delta)$ from $\bar{\theta}(\delta)$ we need also the constant d_0.

Specification of $\alpha(\omega)$ and $\theta(\omega)$ in different parts of the ω axis.   We shall now show that $H(\omega)$ is uniquely determined if $\alpha(\omega)$ and $\theta(\omega)$ are specified in complementary parts of the ω axis.   We shall first consider the case

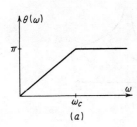

(a)

$$\alpha(\omega) \qquad \text{given for } |\omega| < \omega_c$$
$$\theta(\omega) \qquad \text{given for } |\omega| > \omega_c \qquad (10\text{-}81)$$

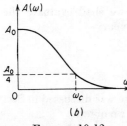

(b)

FIGURE 10-12

To find the unknown parts of $\alpha(\omega)$ and $\theta(\omega)$, we shall use a modified form of the Hilbert transforms.   We note first that Eqs. (10-67) and (10-68) establish the ω-axis relationship between the real and imaginary parts of a function that is analytic in the right-hand plane.   Our problem will therefore be solved if we can find a function whose real part can be determined from the available information (10-81).   For this purpose we shall multiply $\alpha(\omega) + j\theta(\omega)$ by a known quantity that is real for $|\omega| < \omega_c$ and imaginary for $|\omega| > \omega_c$.   An obvious choice is the radical

$$\sqrt{p^2 + \omega_c{}^2}$$

Indeed [see (II-15)], for $p = j\omega$,

if $|\omega| < \omega_c$,

then $$\sqrt{p^2 + \omega_c{}^2} = \sqrt{\omega_c{}^2 - \omega^2}$$

if $\omega > \omega_c$,

then
$$\sqrt{p^2 + \omega_c^2} = j\sqrt{\omega^2 - \omega_c^2}$$

if $\omega < -\omega_c$,

then
$$\sqrt{p^2 + \omega_c^2} = -j\sqrt{\omega^2 - \omega_c^2}$$

For the validity of (10-72) the factor $\sqrt{p^2 + \omega_c^2}$ must be in the denominator. We shall thus form the function

$$\frac{\ln H_{\mathrm{I}}(p)}{\sqrt{p^2 + \omega_c^2}}$$

On the $j\omega$ axis it is given by

$$\frac{\alpha(\omega) + j\theta(\omega)}{\sqrt{\omega_c^2 - \omega^2}} = \frac{\alpha(\omega)}{\sqrt{\omega_c^2 - \omega^2}} + j\,\frac{\theta(\omega)}{\sqrt{\omega_c^2 - \omega^2}} \qquad |\omega| < \omega_c$$

$$= \frac{\theta(\omega)}{\sqrt{\omega^2 - \omega_c^2}} - j\,\frac{\alpha(\omega)}{\sqrt{\omega^2 - \omega_c^2}} \qquad \omega > \omega_c$$

$$= -\frac{\theta(\omega)}{\sqrt{\omega^2 - \omega_c^2}} + j\,\frac{\alpha(\omega)}{\sqrt{\omega^2 - \omega_c^2}} \qquad \omega < -\omega_c$$

Hence its real part is known [see (10-81)], and its imaginary part can be found† from (10-67). Using the fact that $\alpha(\omega)$ is even and $\theta(\omega)$ odd, we obtain

$$\frac{2\omega}{\pi} \int_0^{\omega_c} \frac{\alpha(y)\,dy}{\sqrt{\omega_c^2 - y^2}(y^2 - \omega^2)} + \frac{2\omega}{\pi} \int_{\omega_c}^{\infty} \frac{\theta(y)\,dy}{\sqrt{y^2 - \omega_c^2}(y^2 - \omega^2)}$$

$$= \begin{cases} \theta(\omega)/\sqrt{\omega_c^2 - \omega^2} & 0 < \omega < \omega_c \\ -\alpha(\omega)/\sqrt{\omega^2 - \omega_c^2} & \omega > \omega_c \end{cases} \qquad (10\text{-}82)$$

and $H(\omega)$ is determined.

Specification of $\alpha(\omega)$ and $\theta(\omega)$ in other frequency intervals can be treated similarly. As an example consider the case

$$\theta(\omega) \qquad \text{given for } |\omega| < \omega_c$$
$$\alpha(\omega) \qquad \text{given for } |\omega| > \omega_c \qquad (10\text{-}83)$$

Applying (10-67) to the real and imaginary parts of the function

$$\left. \frac{p \ln H_{\mathrm{I}}(p)}{\sqrt{p^2 + \omega_c^2}} \right|_{p = j\omega}$$

† The radical introduces the imaginary axis singularities $p = \pm j\omega_c$; therefore, the path of Fig. 10-9 must be indented around $\pm j\omega_c$. However, the integrations along the indentations tend to zero with their radius.

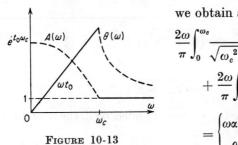

FIGURE 10-13

we obtain as in (10-82)

$$\frac{2\omega}{\pi} \int_0^{\omega_c} \frac{-y\theta(y)\,dy}{\sqrt{\omega_c^2 - y^2}(y^2 - \omega^2)}$$

$$+ \frac{2\omega}{\pi} \int_{\omega_c}^{\infty} \frac{y\alpha(y)\,dy}{\sqrt{y^2 - \omega_c^2}(y^2 - \omega^2)} \tag{10-84}$$

$$= \begin{cases} \omega\alpha(\omega)/\sqrt{\omega_c^2 - \omega^2} & 0 < \omega < \omega_c \\ \omega\theta(\omega)/\sqrt{\omega^2 - \omega_c^2} & \omega > \omega_c \end{cases}$$

Similar equations can be established for $R(\omega)$ and $X(\omega)$.

Example 10-11. We shall use the above result (10-84) to determine a causal, low-pass filter, with linear phase shift $\theta(\omega) = \omega t_0$ in the bandpass $|\omega| < \omega_c$, and a constant amplitude in the stopband $|\omega| > \omega_c$ as in Fig. 10-13. We shall assume, without loss of generality, that $A(\omega) = 1$ for $|\omega| > \omega_c$. With this assumption, the second integral in (10-84) vanishes, and the unknown parts of $\alpha(\omega)$ and $\theta(\omega)$ are given by

$$\frac{-2t_0\sqrt{\omega_c^2 - \omega^2}}{\pi} \int_0^{\omega_c} \frac{y^2\,dy}{\sqrt{\omega_c^2 - y^2}(y^2 - \omega^2)} = \begin{cases} \alpha(\omega) & 0 < \omega < \omega_c \\ \theta(\omega) & \omega > \omega_c \end{cases}$$

The above integral can be easily evaluated; the result is

$$\alpha(\omega) = -t_0\sqrt{\omega_c^2 - \omega^2} \qquad 0 < \omega < \omega_c$$

$$\theta(\omega) = t_0\omega - t_0\sqrt{\omega^2 - \omega_c^2} \qquad \omega > \omega_c$$

The computed parts of the functions $\theta(\omega)$ and

$$A(\omega) = e^{t_0\sqrt{\omega_c^2 - \omega^2}}$$

are shown by the dotted lines in Fig. 10-13.

10-4. Energy of a Signal

One is often interested in evaluating the energy

$$E = \int_0^{\infty} f^2(t)\,dt$$

of a causal function in terms of its transform $F(\omega)$. This can be simply done as follows: using the techniques of Sec. 10-5, we form the function $Z(p)$ such that

$$\operatorname{Re} Z(j\omega) = |F(\omega)|^2 \tag{10-85}$$

The unknown energy E is then given by

$$E = \tfrac{1}{2} \lim_{p \to \infty} pZ(p) \tag{10-86}$$

Proof. From (10-2) and (10-85) we have

$$\lim_{p \to \infty} pZ(p) = \frac{1}{\pi} \int_{-\infty}^{\infty} \text{Re } Z(j\omega) \, d\omega = \frac{1}{\pi} \int_{-\infty}^{\infty} |F(\omega)|^2 \, d\omega$$

and since [see (2-75)]

$$E = \frac{1}{2\pi} \int_{-\infty}^{\infty} |F(\omega)|^2 \, d\omega$$

Eq. (10-86) follows.

The above method is related to Darlington's synthesis; $Z(p)$ is the input impedance of a network and $F(\omega)$ its transfer function.

Example 10-12. We shall evaluate the energy of a Butterworth response. Given

$$|F(\omega)|^2 = \frac{1}{1 + \omega^6}$$

we have, with $\omega^2 = -p^2 = -x$ [see (10-18)],

$$\frac{1}{1 - x^3} = -\frac{1}{3}\left(\frac{1}{x - 1} + \frac{e^{j2\pi/3}}{x - e^{j2\pi/3}} + \frac{e^{-j2\pi/3}}{x - e^{-j2\pi/3}} \right)$$

Therefore from (10-21) we obtain

$$Z(p) = -\frac{1}{3}\left[\frac{1/(-1)}{p + 1} + \frac{e^{j2\pi/3}/e^{-j2\pi/3}}{p - e^{-j2\pi/3}} + \frac{e^{-j2\pi/3}/e^{j2\pi/3}}{p - e^{j2\pi/3}} \right]$$

$$= \frac{1}{3}\frac{2p^2 + 4p + 3}{p^3 + 2p^2 + 2p + 1}$$

and since

$$\lim_{p \to \infty} pZ(p) = \tfrac{2}{3}$$

the unknown energy equals $\tfrac{1}{3}$.

10-5. Causality Conditions

An important problem in the theory of the Fourier integral and in the study of linear systems is the determination of the conditions that a given function $H(\omega) = R(\omega) + jX(\omega) = A(\omega)e^{-j\theta(\omega)}$ must satisfy in order to be the Fourier integral of a causal function $h(t)$. There is no completely general solution to this problem; however, if certain assumptions are made about $H(\omega)$, then one can establish necessary and sufficient conditions for its inverse transform to be zero for negative t. A not too restrictive assumption, leading to simple results, is the requirement that $A(\omega)$ be square-integrable:

$$\int_{-\infty}^{\infty} A^2(\omega) \, d\omega < \infty \tag{10-87}$$

i.e., that the corresponding $h(t)$ has finite energy. This excludes from our investigation the singularity functions discussed in Chap. 3; if such functions are present, however, they can be easily recognized and subtracted from $H(\omega)$; the causality criteria can then be applied to the reduced $H(\omega)$. Suppose, for example, that $H(\omega)$ contains the term $k\delta(\omega - \omega_0)$; the corresponding causal term in $h(t)$ is $(k/\pi)e^{j\omega_0 t}U(t)$, and its Fourier integral $k\delta(\omega - \omega_0) + k/j\pi(\omega - \omega_0)$ must be subtracted from $H(\omega)$. If $H(\omega)$ tends to B for $\omega \to \infty$, then (10-87) is not satisfied. In this case the analysis can be applied to $H(\omega) - B$, whose inverse is $h(t) - B\delta(t)$.

The following basic theorem relates the analyticity of a square-integrable function $H(\omega)$ to the causality of its inverse Fourier transform $h(t)$.

Theorem. If $H(p/j)$ is analytic for Re $p \geq 0$ and

$$H\left(\frac{p}{j}\right) \to 0 \qquad \text{for } p \to \infty \tag{10-88}$$

then $H(\omega)$ is the Fourier transform of a causal function.

Indeed, the function

$$h(t) = \frac{1}{2\pi} \int_{-\infty}^{\infty} H(\omega)e^{j\omega t}\, d\omega \tag{10-89}$$

exists for every t because of (10-87). To show that it equals zero for $t < 0$, we form the contour integral of $H(p/j)$ along the curve C of Fig. 10-1. From the analyticity of $H(p/j)$ in its interior, it follows that

$$j\int_{-\Omega}^{\Omega} H(\omega)e^{j\omega t}\, d\omega + \int_{\Gamma} e^{pt}H\left(\frac{p}{j}\right) dp = 0$$

From (10-88) and Jordan's lemma we conclude that, for $t < 0$, the integral along Γ tends to zero as $\Omega \to \infty$; therefore the first integral above tends also to zero, and the causality of $h(t)$, as given by (10-89), is established.

If $H(p/j)$ is analytic in the right-hand plane but not at infinity, i.e., if (10-88) is not true, then $h(t)$ is not causal. However, if

$$e^{-pt_0}H\left(\frac{p}{j}\right) \to 0 \qquad \text{for } p \to \infty \text{ on } \Gamma$$

then $$h(t) = 0 \qquad \text{for } t < -t_0$$

The proof is similar to the above [see also (9-61)].

Often $H(\omega)$ is only partially given, and the question arises whether it is possible to complete its specification so as to obtain a causal transform. We shall consider the two most common cases: (1) given $R(\omega)$ and (2) given $A(\omega)$. If $R(\omega)$ is given, then one can always associate to it a suitable $X(\omega)$ so that the function $R(\omega) + jX(\omega)$ has a causal inverse. Indeed, the function

$$h(t) = \frac{2U(t)}{\pi} \int_0^\infty R(\omega) \cos \omega t \, d\omega$$

exists for every t, as we see from (10-87), is causal, and the real part of its Fourier integral is $R(\omega)$ [see (2-31)]. The problem is not trivial, however, if $A(\omega)$ is specified; in this case (10-87) is not sufficient and $A(\omega)$ must satisfy the Paley-Wiener condition.

Paley-Wiener condition.† A necessary and sufficient condition for a square-integrable function $A(\omega) \geq 0$ to be the Fourier spectrum of a causal function is the convergence of the integral

$$\int_{-\infty}^\infty \frac{|\ln A(\omega)|}{1 + \omega^2} \, d\omega < \infty \tag{10-90}$$

We remark that if the amplitude of a function $H(\omega)$ satisfies (10-90), it does not follow that $H(\omega)$ has a causal inverse. The above says that to $|H(\omega)| = A(\omega)$ a suitable phase can be associated, so that the resulting function has a causal inverse. We further note that if $A(\omega)$ is not square-integrable, then (10-90) is neither necessary nor sufficient. We shall attempt to give a simple justification of (10-90).

Sufficiency. Suppose that $A(\omega)$ is square-integrable and satisfies (10-90); we shall show that a $\theta(\omega)$ can be found so that

$$H(\omega) = A(\omega)e^{-j\theta(\omega)}$$

is analytic in the right-hand plane. It will then follow from the theorem of this section that $H(\omega)$ has a causal inverse. With the change of variable

$$\omega = -\tan \delta/2$$

the function $\alpha(\omega) = \ln A(\omega)$ becomes $\alpha(-\tan \delta/2) = \bar{\alpha}(\delta)$ as in (10-77). Since

$$d\omega = -\frac{d\delta}{2 \cos^2 \delta/2} = -\frac{(1 + \omega^2)}{2} \, d\delta$$

we obtain from (10-90)

$$\int_{-\pi}^\pi |\bar{\alpha}(\delta)| \, d\delta < \infty \tag{10-91}$$

$\bar{\alpha}(\delta)$ is therefore absolutely integrable; hence it can be expanded into a convergent Fourier series

$$\bar{\alpha}(\delta) = d_0 + d_1 \cos \delta + \cdots + d_n \cos n\delta + \cdots \tag{10-92}$$

† Raymond E. A. C. Paley and Norbert Wiener, "Fourier Transforms in the Complex Domain," American Mathematical Society Colloquium Publication 19, New York, 1934.

We form the series

$$F(w) = d_0 + d_1 w + \cdots + d_n w^n + \cdots \qquad (10\text{-}93)$$

We have $F(1) = \bar{\alpha}(0)$; hence $F(w)$ converges for every $|w| < 1$ and defines a function $F(w)$ analytic in the interior of the unit circle. Therefore, with

$$w = \frac{1-p}{1+p}$$

the function

$$F\left(\frac{1-p}{1+p}\right) = Z(p) \qquad (10\text{-}94)$$

is analytic in the right-hand p plane, since the transformation maps the interior of the unit circle $|w| = 1$ in the right-hand part of the p plane. We maintain that

$$H(\omega) = e^{Z(j\omega)} \qquad (10\text{-}95)$$

is the unknown function; indeed,

$$H\left(\frac{p}{j}\right) = e^{Z(p)}$$

is analytic in the right-hand plane, and since [see (10-93)]

$$Z(j\omega) = F\left(\frac{1-j\omega}{1+j\omega}\right) = F(e^{-j\delta}) = d_0 + \cdots + d_n e^{-jn\delta} + \cdots \qquad (10\text{-}96)$$

we have

$$\ln|H(\omega)| = \operatorname{Re} Z(j\omega) = d_0 + \cdots + d_n \cos n\delta + \cdots = \bar{\alpha}(\delta) = \alpha(\omega)$$

Thus $|H(\omega)| = A(\omega)$ and the sufficiency of the condition is established.

Necessity. The function $H(\omega)$ is the Fourier integral of a causal function, and its amplitude $|H(\omega)| = A(\omega)$ satisfies (10-87). We shall show that it satisfies (10-90). With $\alpha(\omega) = \ln A(\omega)$ and $\alpha^+(\omega)$ defined by

$$\alpha^+(\omega) = \begin{cases} \alpha(\omega) & \text{if } \alpha(\omega) > 0 \\ 0 & \text{if } \alpha(\omega) < 0 \end{cases}$$

we have for $A(\omega) \geq 1$, $A(\omega) = e^{\alpha^+(\omega)} \geq \alpha^+(\omega)$; hence

$$\infty > \int_{-\infty}^{\infty} A^2(\omega)\, d\omega > \int_{-\infty}^{\infty} 2\alpha^+(\omega)\, d\omega > 2 \int_{-\infty}^{\infty} \frac{\alpha^+(\omega)}{1+\omega^2}\, d\omega \qquad (10\text{-}97)$$

Therefore, to prove (10-90), it suffices to show that

$$\int_{-\infty}^{\infty} \frac{\ln A(\omega)}{1+\omega^2}\, d\omega < \infty \qquad (10\text{-}98)$$

since the positive part of the integrand in (10-90) converges because of (10-97).   We shall now assume that $H(p/j)$ has no zeros in the right-hand plane; if this is not the case, then we write it as a product $H(\omega) = H_1(\omega)H_a(\omega)$ of a minimum-phase-shift function $H_1(\omega)$ and an all-pass term $H_a(\omega)$ [see (10-60)].   And since $|H(\omega)| = |H_1(\omega)|$, it suffices to prove (10-98) for $H_1(\omega)$.   We now form the function

$$\frac{\ln H_{\mathrm{I}}(p)}{1 - p^2}$$

where $H_{\mathrm{I}}(p) = H_1(p/j)$.   This function is analytic everywhere in the right-hand plane except at the point $p = 1$, where it has a simple pole with residue

$$- \frac{\ln H_{\mathrm{I}}(1)}{2}$$

Therefore its integral along the closed curve C of Fig. 10-1 equals $j\pi \ln H_{\mathrm{I}}(1)$.   The contribution of the semicircle Γ to this integral tends to zero with $\Omega \to \infty$; hence

$$\int_{-\infty}^{\infty} \frac{\ln H_{\mathrm{I}}(j\omega)}{1 + \omega^2}\, d\omega = \pi \ln H_{\mathrm{I}}(1) < \infty$$

and (10-98) follows.

Problems

29. Show that
$$\Gamma(\tfrac{1}{2}) = \sqrt{\pi}$$
where $\Gamma(\alpha)$ is the gamma function
$$\Gamma(\alpha + 1) = \int_0^\infty e^{-x} x^\alpha \, dx$$

30. Prove that
$$\int_{-\infty}^\infty e^{jy^2} \, dy = \sqrt{\pi j} \tag{i}$$

$$\int_{-\infty}^\infty \cos \frac{\pi}{2} x^2 \, dx = \int_{-\infty}^\infty \sin \frac{\pi}{2} x^2 \, dx = 1 \tag{ii}$$

Hence $\qquad C(\infty) = \tfrac{1}{2} \qquad S(\infty) = \tfrac{1}{2}$

where $C(\tau)$ and $S(\tau)$ are the Fresnel integrals (see Prob. 25).

31. A signal $h(t)$ is causal and its Laplace transform $H_I(p)$ contains no singularities on the $j\omega$ axis. Show that with $H_I(j\omega) = R(\omega) + jX(\omega)$
$$\int_{-\infty}^\infty \frac{[R(\omega) - R(0)]^2}{\omega^2} \, d\omega = \int_{-\infty}^\infty \frac{X^2(\omega)}{\omega^2} \, d\omega \tag{i}$$

32. With $h(t)$ and $H_I(p)$ as in Prob. 31, show that
$$\frac{1}{\pi} \int_{-\infty}^\infty \frac{R(\omega) - R(0)}{\omega^2} \, d\omega = X'(0) = -\int_0^\infty t \, h(t) \, dt \tag{i}$$

33. $R(\omega)$ is the real part of the Fourier transform of a real causal function, and $F_I(p)$ its Laplace transform; show that [see (9-16)]
$$F_I(p) = \frac{2p}{\pi} \int_0^\infty \frac{R(y) \, dy}{p^2 + y^2} \tag{i}$$

using the identity $\qquad R(\omega) = \int_{-\infty}^\infty R(y) \, \delta(\omega - y) \, dy \tag{ii}$

34. *Gain-bandwidth theorem.* (a) With $i(t)$ the input current, $v(t)$ the output voltage, and $A(\omega)$ the amplitude of the transfer impedance $V(j\omega)/I(j\omega)$ of the passive network of Fig. P-34a, show that
$$\int_0^\infty A^2(\omega) \, d\omega \le \frac{\pi R}{2C} \tag{i}$$

218

If the network is lossless and its total input capacity equals C, then (i) holds with the equality sign.

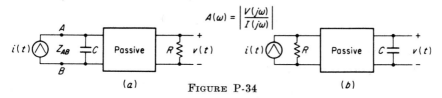

$$A(\omega) = \left| \frac{V(j\omega)}{I(j\omega)} \right|$$

(a)   FIGURE P-34   (b)

(b) Show that (i) is true also for the network of Fig. P-34b.

35. The amplitude characteristic of a causal signal $h(t)$ is a Butterworth function

$$A(\omega) = \frac{1}{\sqrt{1 + (\omega/\omega_c)^{2n}}}$$

Show that the energy of the signal is given by

$$E = \int_0^\infty h^2(t)\, dt = \frac{\omega_c}{2n \sin(\pi/2n)} \tag{i}$$

36. The phase lag of a minimum-phase-shift filter is given by

$$\theta(\omega) = \begin{cases} 0 & |\omega| < \omega_c \\ \theta_0 & \omega > \omega_c \end{cases}$$

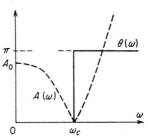

as in Fig. P-36. Determine its amplitude $A(\omega)$.
Answer.

$$A(\omega) = A_0 \left| 1 - \frac{\omega^2}{\omega_c^2} \right|^{\theta_0/\pi} \tag{i}$$

FIGURE P-36

(dotted line in Fig. P-36 plotted for $\theta_0 = \pi$).

37. The system function $H_I(p)$ exists for Re $p \geq 0$ and is minimum-phase-shift. With $H_I(j\omega) = e^{-\alpha(\omega) - j\theta(\omega)}$ prove that

$$\ln H_I(p) = -\frac{p}{\pi} \int_{-\infty}^\infty \frac{\alpha(\omega)}{\omega^2 + p^2}\, d\omega \qquad \text{Re } p > 0 \tag{i}$$

$$\ln H_I(p) = \ln H_I(0) - \frac{p^2}{\pi} \int_{-\infty}^\infty \frac{\theta(\omega)\, d\omega}{\omega(\omega^2 + p^2)} \qquad \text{Re } p > 0 \tag{ii}$$

38. The phase lag of a minimum-phase-shift function $H_I(p)$ is given by

$$\theta(\omega) = \begin{cases} k\omega^2/\omega_c^2 & 0 < \omega < \omega_c \\ k & \omega > \omega_c \end{cases}$$

Using the result of Prob. 37, show that for Re $p > 0$

$$H_I(p) = H_I(0) \left(\frac{p^2}{p^2 + \omega_c^2} \right)^{kp^2/\pi\omega_c^2} \left(\frac{\omega_c^2}{p^2 + \omega_c^2} \right)^{k/\pi} \tag{i}$$

39. (a) Show that if a signal $f(t)$ with finite energy is causal, then its Fourier transform cannot be zero in any interval

$$F(\omega) = 0 \text{ for } \omega_1 < \omega < \omega_2 \qquad \omega_1 \neq \omega_2$$

(b) Prove that a signal $f(t)$ cannot be time- and frequency-limited; i.e., it is impossible to have

$$f(t) = 0 \quad \text{for} \quad |t| > T$$

and

$$F(\omega) = 0 \quad \text{for} \quad |\omega| > \Omega$$

Solutions

29. From the definition it follows that

$$\Gamma(\tfrac{1}{2}) = \int_0^\infty \frac{e^{-x}}{\sqrt{x}}\, dx$$

and with $x = t^2$

$$\Gamma(\tfrac{1}{2}) = 2\int_0^\infty e^{-t^2}\, dt = \sqrt{\pi}$$

30. From (9-5) we conclude that (see comment on page 185 and Prob. 29)

$$\frac{U(t)}{\sqrt{t}} \leftrightarrow \frac{\Gamma(\tfrac{1}{2})}{\sqrt{j\omega}} = \sqrt{\frac{\pi}{j\omega}}$$

Hence

$$\sqrt{\frac{\pi}{j\omega}} = \int_0^\infty \frac{e^{-j\omega t}}{\sqrt{t}}\, dt$$

and with $\omega = -1$, $t = y^2$,

$$\frac{\sqrt{\pi}}{\sqrt{-j}} = 2\int_0^\infty e^{jy^2}\, dy = \int_{-\infty}^\infty e^{jy^2}\, dy$$

and (i) is proved; since $\sqrt{\pi j} = \sqrt{\pi/2} + j\sqrt{\pi/2}$, Eq. (ii) follows from (i) with $y^2 = \pi x^2/2$.

31. The function

$$F(p) = \frac{[H_I(p) - H_I(0)]^2}{p^2}$$

is analytic for $\operatorname{Re} p \geq 0$; hence its integral along the closed pass C of Fig. 10.1 equals zero. Since $pF(p) \to 0$ for $p \to \infty$ on Γ, the integration along Γ tends to zero with $\Omega \to \infty$; hence

$$j\int_{-\infty}^\infty \frac{[H_I(j\omega) - H_I(0)]^2}{-\omega^2}\, d\omega = 0$$

and (i) follows from the imaginary part of the above.

32. The function $F(p) = [H_I(p) - H_I(0)]/p^2$ is analytic for $\operatorname{Re} p \geq 0$ except at $p = 0$. Integrating along the curve C of Fig. 10-3 where $\omega_0 = 0$, we have since $\lim p\,F(p) = 0$ for $p \to \infty$ and

$$\int_\gamma \frac{H_I(p) - H_I(0)}{p^2}\, dp \simeq H_I'(0)\int_\gamma \frac{dp}{p} = j\pi H_I'(0)$$

$$j\int_\infty^\infty \frac{H_I(j\omega) - H_I(0)}{-\omega^2}\, d\omega + j\pi H_I'(0) = 0$$

But $H_I'(0) = -jH'(0) = X'(0) = \int_0^\infty (-t)\, h(t)\, dt$

and (i) follows.

33. If $R(\omega) = \pi\delta(\omega - y)$ then [see (9-8)] $F_I(p) = \dfrac{1}{p - jy}$.   Therefore, for an arbitrary $R(\omega)$ as in (ii), the corresponding $F_I(p)$ is given by

$$F_I(p) = \frac{1}{\pi} \int_{-\infty}^{\infty} \frac{R(y)\, dy}{p - jy}$$

and (i) follows from the above and from the evenness of $R(\omega)$.

34. (a) We denote by $Z_{AB}(j\omega)$ the input impedance of the network. If $i(t)$ is a sine current, the power delivered to the system equals $\frac{1}{2}|I(j\omega)|^2 \operatorname{Re} Z_{AB}(j\omega)$ and the power dissipated in the output resistance equals $|V(j\omega)|^2/2R$; hence

$$\operatorname{Re} Z_{AB}(j\omega) \geq \left|\frac{V(j\omega)}{I(j\omega)}\right|^2 \frac{1}{R} = \frac{A^2(\omega)}{R} \quad .$$

Since $\lim\limits_{p\to\infty} pZ_{AB}(p) \leq \dfrac{1}{C}$, we conclude from the resistance integral theorem (10-2) that

$$\frac{1}{C} \geq \lim_{p\to\infty} pZ_{AB}(p) = \frac{2}{\pi}\int_0^{\infty} \operatorname{Re} Z_{AB}(j\omega)\, d\omega \geq \frac{2}{\pi R}\int_0^{\infty} A^2(\omega)\, d\omega$$

and (i) is proved.

(b). Follows from (a) and the reciprocity theorem.

35. From (10-86) or with a contour integration, we see that E equals the sum of the residues A_i of the left-hand plane poles of $A^2(p/j) = 1/[1 + (-p^2/\omega_c{}^2)^n]$. Since

$$A_i = \frac{1}{-2np_i^{2n-1}/\omega_c{}^{2n}} = \frac{p_i}{2n}$$

we have $2nE = \Sigma p_i$ where the summation extends over all left-hand plane poles. It is easy to see that this sum equals $\omega_c/\sin (\pi/2n)$ and (i) follows.

36. From (10-68) we have

$$\ln A(\omega) - \ln A_0 = -\frac{\omega^2}{\pi}\int_{\omega_c}^{\infty} \frac{\theta_0\, dy}{y(y^2 - \omega^2)} = \frac{\theta_0}{\pi}\ln\left|\frac{\omega_c{}^2 - \omega^2}{\omega_c{}^2}\right|$$

and (i) follows.

37. (i) For $\operatorname{Re} p > 0$ the function $\ln H_I(s)/(s^2 - p^2)$ is analytic everywhere in the interior of the curve C_r of Fig. P-37a except at the point $s = p$.   The integration along the arc Γ_r tends to zero with $R \to \infty$ and the residue at $s = p$ equals $\ln H_I(p)/2p$; hence

$$\int_{-\infty}^{\infty} \frac{\ln H_I(j\omega)\, j\, d\omega}{-\omega^2 - p^2} + \frac{2\pi j}{2p}\ln H_I(p) = 0$$

Inserting $\ln H_I(j\omega) = -\alpha(\omega) - j\theta(\omega)$ into the above, we obtain because of the oddness of $\theta(\omega)/(\omega^2 + p^2)$ the desired result.

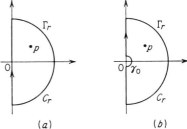

(ii) For $\operatorname{Re} p > 0$ the function $\ln H_I(s)/$ $s(s^2 - p^2)$ is analytic in the interior of the curve C_r of Fig. P-37b.   The contribution of the semicircle γ_0 to the integral tends to $-j\pi \ln H_I(0)/p^2$.   The residue at p equals $\ln H_I(p)/2p^2$; hence

$$\int_{-\infty}^{\infty} \frac{\ln H_I(j\omega)j}{j\omega(-\omega^2 - p^2)}\, d\omega - \frac{j\pi \ln H_I(0)}{p^2} + \frac{2\pi j \ln H_I(p)}{2p^2} = 0$$

and (ii) follows.   The above results are not true if $\operatorname{Re} p = 0$.

FIGURE P-37

38. We have

$$\int_0^\infty \frac{\theta(\omega)}{\omega(p^2 + \omega^2)}\, d\omega = \frac{k}{\omega_c{}^2}\int_0^{\omega_c} \frac{\omega\, d\omega}{p^2 + \omega^2} + k\int_{\omega_c}^\infty \frac{d\omega}{\omega(p^2 + \omega^2)}$$

$$= \frac{k}{2\omega_c{}^2}\ln\frac{p^2 + \omega_c{}^2}{p^2} + \frac{k}{2p^2}\ln\frac{p^2 + \omega_c{}^2}{\omega_c{}^2}$$

and (i) follows from Prob. 37 and the oddness of $\theta(\omega)$.

39. (a) If $F(\omega) = 0$ for $\omega_1 < \omega < \omega_2$, then

$$\int_{-\infty}^\infty \frac{\ln|A(\omega)|}{1 + \omega^2}\, d\omega = \infty$$

Therefore the Paley-Wiener condition is violated.

(b) If $f(t) = 0$ for $|t| > T$, then $f(t - T)$ is causal; hence its Fourier transform $e^{j\omega t}F(\omega)$ cannot equal zero for $|\omega| > \Omega$.

PART FOUR

Chapter 11. Positive Functions and Limit Theorems

In this chapter we shall consider only positive functions of time; these functions are extensively used in probability theory, but they also frequently appear in the study of linear systems. We shall discuss briefly some of the properties of their Fourier transform; however, the main part of the chapter will be devoted to the central-limit theorem and the approximate evaluation of the convolution of n time functions.

11-1. The Density Function

We shall assume in the entire chapter that all time functions $f(t)$ are real and positive,

$$f(t) \geq 0 \tag{11-1}$$

and that they might contain impulses but no other singularities. Without loss of generality we shall also impose the condition

$$\int_{-\infty}^{\infty} f(t)\, dt = 1 \tag{11-2}$$

Borrowing from the terminology of probability theory, we shall call a function satisfying (11-1) and (11-2) *density*. Denoting by $F(\omega)$ the Fourier transform of $f(t)$, we readily obtain from (11-1) and (2-1)

$$F(0) = 1 \tag{11-3}$$

Furthermore we have

$$|F(\omega)| \leq 1 \tag{11-4}$$

because [see (11-1)]

$$|F(\omega)| = \left| \int_{-\infty}^{\infty} f(t)e^{-j\omega t}\, dt \right| \leq \int_{-\infty}^{\infty} f(t)\, dt = 1$$

It is of interest to show that the equality sign in (11-4) cannot be satisfied for $\omega \neq 0$ unless $f(t)$ is a sequence of equidistant pulses.† Indeed, suppose that for $\omega_1 \neq 0$ we have $|F(\omega_1)| = 1$. We can then assume that $F(\omega_1) = 1$ because the possible phase angle in $F(\omega_1)$ can be made equal to zero by a proper shifting of $f(t)$ [see (2-36)]. We thus have

$$F(\omega_1) = \int_{-\infty}^{\infty} f(t)e^{-j\omega_1 t}\, dt = 1 = \int_{-\infty}^{\infty} f(t)\cos \omega_1 t\, dt$$

and, because of (11-2),

$$\int_{-\infty}^{\infty} (1 - \cos \omega_1 t)f(t)\, dt = 0$$

The integrand above is nonnegative; hence

$$(1 - \cos \omega_1 t)f(t) \equiv 0$$

This is possible only if $f(t)$ consists of impulses at the zeros of $1 - \cos \omega_1 t$, as we see from the property (I-23)

$$g(t)\delta(t - t_n) = g(t_n)\delta(t - t_n)$$

of the delta function. And since $1 - \cos \omega_1 t = 0$ for $t = 2\pi n/\omega_1$, we conclude, with (11-2), that if $F(\omega_1) = 0$ for $\omega_1 \neq 0$, then

$$f(t) = \sum_{n=-\infty}^{\infty} a_n \delta\left(t - \frac{2\pi n}{\omega_1}\right) \qquad \sum_{n=-\infty}^{\infty} a_n = 1 \qquad (11\text{-}5)$$

A density consisting of a sequence of pulses as in (11-5) is called *lattice* (Fig. 11-1). As we know from Sec. 3-4, its Fourier transform is a periodic function of ω.

We give below a number of common densities and their Fourier transforms, known also as *characteristic functions*. The transform pairs are taken from the examples in Chap. 2.

$$\text{Rectangular:} \quad \frac{p_T(t)}{2T} \leftrightarrow \frac{\sin \omega T}{\omega T}$$

$$\text{Triangular:} \quad \frac{q_T(t)}{T} \leftrightarrow \frac{4 \sin^2 (\omega T/2)}{\omega^2 T^2}$$

$$\text{Normal:} \quad \sqrt{\alpha/\pi}\, e^{-\alpha t^2} \leftrightarrow e^{-\omega^2/4\alpha}$$

$$\text{Laplace:} \quad \frac{\alpha}{2} e^{-\alpha|t|} \leftrightarrow \frac{\alpha^2}{\alpha^2 + \omega^2}$$

$$\text{Cauchy:} \quad \frac{\alpha/\pi}{\alpha^2 + t^2} \leftrightarrow e^{-\alpha|\omega|}$$

† E. Lukacs, "Characteristic Functions," Hafner Publishing Company, New York, 1960.

The following two examples of lattice densities were discussed in Sec. 3-4.

Poisson: $e^{-\lambda} \sum_{n=0}^{\infty} \frac{\lambda^n}{n!} \delta(t - n) \leftrightarrow e^{\lambda(e^{-j\omega} - 1)}$

Binomial: $\sum_{k=0}^{n} \binom{n}{k} p^k q^{n-k} \delta(t - k) \leftrightarrow (p + qe^{j\omega})^n$

where $p + q = 1$.

Conditions (11-3) and (11-4) are necessary but by no means sufficient for a function $F(\omega)$ to be the Fourier transform of a density. Indeed, $p_1(\omega)$ satisfies (11-3) and (11-4), but its inverse $\sin t/\pi t$ is not positive. The problem of determining the positiveness of $f(t)$ from the properties of $F(\omega)$ is important but has no simple solution. In Sec. 5-3 we gave a sufficient condition [Eq. (5-43)] for $f(t)$ to be positive,

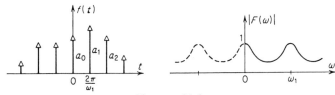

FIGURE 11-1

but this condition is too special. The following is a necessary and sufficient condition for $F(\omega)$ to be the Fourier integral of a positive function.

For any choice of the complex constants $a_1, a_2, \ldots, a_n$, the real constants $\omega_1, \omega_2, \ldots, \omega_n$, and any value of n, we must have

$$\sum_{m=1}^{n} \sum_{k=1}^{n} F(\omega_m - \omega_k) a_m \overset{*}{a}_k \geq 0 \tag{11-6}$$

A function $F(\omega)$ satisfying (11-6) is called *positive definite*. The proof of the necessity of (11-6) is simple. Indeed, if $F(\omega)$ is the Fourier transform of a positive function $f(t)$, then

$$0 \leq \int_{-\infty}^{\infty} \left| \sum_{m=1}^{n} a_m e^{j\omega_m t} \right|^2 f(t) \, dt$$

$$= \int_{-\infty}^{\infty} \left(\sum_{m=1}^{n} a_m e^{-j\omega_m t} \right) \left(\sum_{k=1}^{n} \overset{*}{a}_k e^{j\omega_k t} \right) f(t) \, dt$$

$$= \sum_{m=1}^{n} \sum_{k=1}^{n} a_m \overset{*}{a}_k F(\omega_m - \omega_k)$$

and (11-6) follows. The proof of the sufficiency of the above condition, known as Bochner's theorem, is more difficult and will be omitted. The above condition is mainly of theoretical value; the testing of (11-6) is in general not easy.

11-2. Repeated Convolution

We now turn to the problem of evaluating the convolution

$$f(t) = f_1(t) * f_2(t) * \cdots * f_n(t) \tag{11-7}$$

of n densities $f_1(t), \ldots, f_n(t)$. In probability theory, $f(t)$ is the density of the sum of n independent random variables of respective densities $f_i(t)$. In network theory, $f(t)$ is the impulse response of an n-stage amplifier if $f_i(t)$ is the response of each stage. Denoting by

$$F(\omega) = A(\omega)e^{j\phi(\omega)} \qquad F_i(\omega) = A_i(\omega)e^{j\phi_i(\omega)} \qquad i = 1, \ldots, n \tag{11-8}$$

the Fourier transforms of the above functions, we obtain from (11-7) and (2-71)

$$F(\omega) = F_1(\omega) \cdots F_n(\omega) \tag{11-9}$$

Hence

$$A(\omega) = A_1(\omega) \cdots A_n(\omega) \tag{11-10}$$

$$\phi(\omega) = \phi_1(\omega) + \cdots + \phi_n(\omega) \tag{11-11}$$

It is clear from (11-7) that $f(t) \geq 0$, and since $F(0) = 1$ [see (11-9)], we conclude that $f(t)$ satisfies (11-1) and (11-2); therefore it is a density. In the next section we shall need the moments

$$m_k = \int_{-\infty}^{\infty} t^k f(t)\, dt \tag{11-12}$$

of the function $f(t)$. These moments can be determined from the moments of the given functions $f_i(t)$. This is done by expanding $F(\omega)$ and $F_i(\omega)$ in (11-9) into power series and equating equal powers of ω [see (2-48)]. We shall carry out the details for the mean

$$\eta = \int_{-\infty}^{\infty} tf(t)\, dt = m_1$$

and dispersion

$$\sigma^2 = \int_{-\infty}^{\infty} (t - \eta)^2 f(t)\, dt = m_2 - m_1^2$$

With η_i and σ_i the corresponding quantities of $f_i(t)$, we have

$$\begin{aligned} \eta &= \eta_1 + \eta_2 + \cdots + \eta_n \\ \sigma^2 &= \sigma_1^2 + \sigma_2^2 + \cdots + \sigma_n^2 \end{aligned} \tag{11-13}$$

This follows easily from

$$A(\omega) = 1 - \frac{\sigma^2 \omega^2}{2} + \cdots \qquad A_i(\omega) = 1 - \frac{\sigma_i^2 \omega^2}{2} + \cdots \tag{11-14}$$

$$\phi(\omega) = -\eta\omega + \cdots \qquad \phi_i(\omega) = -\eta_i\, \omega + \cdots \tag{11-15}$$

[see (2-51)] and Eqs. (11-10) and (11-11).

11-3. The Central-limit Theorem†

This fundamental theorem is extensively treated in the mathematical literature; its formulation is based, however, on probabilistic concepts, and the proofs are rather involved. Furthermore the continuous and the lattice cases are presented as two different results. In the following we shall give a simple discussion of the central-limit theorem and will show how it can be used for an approximate evaluation of the convolution of n functions. The theorem states that if the densities $f_1(t), \ldots, f_n(t)$ are not lattices, then, under certain general conditions, the product $F(\omega)$ of their Fourier transforms

$$F_1(\omega), \ldots, F_n(\omega)$$

tends to a Gaussian function as $n \to \infty$

$$F(\omega) \sim e^{-\sigma^2\omega^2/2 - j\eta\omega} \qquad \text{for } n \to \infty \qquad (11\text{-}16)$$

where σ and η are given by (11-13).

From (11-16) and a limit argument, known as Helly's theorem, one concludes that the inverse transform of $F(\omega)$, given by (11-7), tends also to a Gaussian function [see (2-68)]

$$f(t) \sim \frac{1}{\sigma\sqrt{2\pi}} e^{-(t-\eta)^2/2\sigma^2} \qquad \text{for } n \to \infty \qquad (11\text{-}17)$$

i.e., to the inverse of the limit of $F(\omega)$.

The following two conditions ensure the validity of (11-16); though they are not the most general, they cover a wide range of applications.

1. The third moment of the functions $f_i(t)$ is finite, and for every i it is smaller than an arbitrary constant C.

2. The sum of the dispersions of $f_i(t)$ tends to infinity with n

$$\sigma^2 = \sigma_1^2 + \sigma_2^2 + \cdots + \sigma_n^2 \underset{n \to \infty}{\to} \infty \qquad (11\text{-}18)$$

The first condition is satisfied, for example, if the functions $f_i(t)$ are zero above $|t| = T$. The second condition is true if all dispersions are larger than a positive constant; this is certainly the case if the functions $f_i(t)$ are all equal.

Clearly, for $\omega \neq 0$, the function $F(\omega)$ in (11-16) tends to zero with $n \to \infty$; for this reason, in the usual formulation of the theorem, the change of variable $y = \sigma\omega$ is introduced and (11-16) takes the form

$$F\left(\frac{y}{\sigma}\right) \sim e^{-y^2/2 - j\eta y/\sigma} \qquad \text{for } n \to \infty$$

---

† B. V. Gnedenko and A. Kolmogorov, "Limit Distributions for Sums of Independent Random Variables," Addison-Wesley Publishing Company, Reading, Mass., 1954.; J. V. Uspensky, "Introduction to Mathematical Probability," McGraw-Hill Book Company, Inc., New York, 1937.

Example 11-1. Before proving the theorem, we shall demonstrate, in a simple example, the rapidity with which $f(t)$ tends to its asymptotic value (11-17). We assume that the functions $f_i(t)$ are all equal to the rectangular pulse of Fig. 11-2a

$$f_1(t) = \cdots = f_n(t) = U(t) - U(t - 1) \tag{11-19}$$

It is easy to show that their convolution $f(t)$ is given by

$$f(t) = \sum_{k=0}^{n} \binom{n}{k}(-1)^k U_{-n}(t - k) \tag{11-20}$$

where

$$U_{-n}(t) = U(t) * \cdots * U(t) = \frac{t^{n-1}}{(n-1)!}\, U(t) \tag{11-21}$$

Indeed, with $V(\omega)$ the Fourier transform of the step function $U(t)$, we have from (11-19) and (2-36)

$$F_i(\omega) = V(\omega)(1 - e^{-j\omega})$$

$$F(\omega) = V^n(\omega)(1 - e^{-j\omega})^n$$

Expanding $(1 - e^{-j\omega})^n$ and using the fact that $V^n(\omega)$ is the Fourier transform of $U_{-n}(t)$, we obtain (11-20). The function $f(t)$ is given by the triangle of

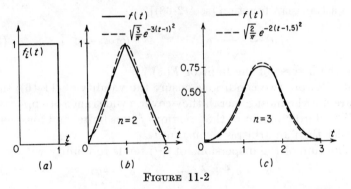

FIGURE 11-2

Fig. 11-2b for $n = 2$, and for $n = 3$ it consists of three parabolic pieces, as in Fig. 11-2c.

The first and second moments of $f_i(t)$ are $\tfrac{1}{2}$ and $\tfrac{1}{3}$ respectively; hence [see (2-55)]

$$\eta_i = \tfrac{1}{2} \qquad \sigma_i^2 = \tfrac{1}{3} - \tfrac{1}{4} = \tfrac{1}{12}$$

Therefore for

$$n = 2: \quad \eta = \eta_1 + \eta_2 = 1 \qquad \sigma^2 = \sigma_1^2 + \sigma_2^2 = \tfrac{1}{6}$$

$$n = 3: \quad \eta = \tfrac{3}{2} \qquad\qquad\quad \sigma^2 = \tfrac{1}{4}$$

and the corresponding Gaussian curves in (11-17) are given by

$$n = 2: \quad \sqrt{3/\pi}\, e^{-3(t-1)^2} \qquad\qquad n = 3: \quad \sqrt{2/\pi}\, e^{-2(t-1.5)^2}$$

as shown in Fig. 11-2. Thus, even for such small values of n, the asymptotic form (11-17) is a good estimate of $f(t)$.

We return to the justification of (11-16).   Since $f_i(t)$ is not a lattice, its Fourier spectrum $A_i(\omega)$ is smaller than one for every $\omega \neq 0$

$$A_i(\omega) < 1 \qquad \omega \neq 0 \tag{11-22}$$

Therefore (see Fig. 11-3) a sufficiently small constant ω_0 can be found, such that

$$A_i(\omega) < A_i(\omega_0) \qquad \text{for } |\omega| > \omega_0 \tag{11-23}$$

We now select ω_0 so small that $A_i(\omega)$ and $e^{-\sigma_i^2\omega^2/2}$ can be approximated by a parabola in the $|\omega| < \omega_0$ interval [see (11-14)]:

$$\left. \begin{aligned} A_i(\omega) &\simeq 1 - \frac{\sigma_i^2\omega^2}{2} \\[2mm] e^{-\sigma_i^2/2} &\simeq 1 - \frac{\sigma_i^2\omega^2}{2} \end{aligned} \right\} \quad \text{for } |\omega| < \omega_0 \tag{11-24}$$

It can be seen from Conditions 1 and 2, above that it is possible to satisfy (11-23) and (11-24) for every i with the same constant ω_0.   For $\omega > \omega_0$ we have [see (11-23) and (11-24)]

$$\begin{aligned} A(\omega) &= A_1(\omega) \cdots A_n(\omega) < A_1(\omega_0) \cdots A_n(\omega_0) \\ &\simeq e^{-\sigma_1^2\omega_0^2/2} \cdots e^{-\sigma_n^2\omega_0^2/2} = e^{-\sigma^2\omega_0^2/2} \end{aligned}$$

But $\sigma \to \infty$ with $n \to \infty$ [see (11-18)]; hence

$$A(\omega) \xrightarrow[n \to \infty]{} 0 \qquad \text{for } |\omega| > \omega_0$$

In the $|\omega| < \omega_0$ interval

$$\begin{aligned} A(\omega) &= A_1(\omega) \cdots A_n(\omega) \simeq e^{-\sigma_1^2\omega^2/2} \cdots e^{-\sigma_n^2\omega^2/2} \\ &= e^{-\sigma^2\omega^2/2} \end{aligned}$$

and since $e^{-\sigma^2\omega^2/2}$ tends also to zero for $|\omega| > \omega_0$ with $n \to \infty$, we conclude that

$$A(\omega) \sim e^{-\sigma^2\omega^2/2} \tag{11-25}$$

The phase part of (11-16) can be justified by noticing that the interval in which $F(\omega)$ has significant values tends to zero for $n \to \infty$, and in a small interval near the origin

$$\phi(\omega) \simeq -\eta\omega$$

In the above reasoning we attempted to bring out the main factors for the validity of the central-limit theorem; following the same line of thought, one can give a complete proof.

Sequence of equidistant pulses.   We shall now investigate the asymptotic form of $f(t)$ when the functions $f_i(t)$ are sequences of equidistant pulses with total area equal to one (lattices)

$$f_i(t) = \sum_{k=-\infty}^{\infty} a_{ik}\,\delta(t - kT) \qquad \sum_{k=-\infty}^{\infty} a_{ik} = 1 \tag{11-26}$$

It is easy to see that their mean η_i and variance σ_i^2 are given by

$$\eta_i = \sum_{k=-\infty}^{\infty} kT a_{ik} \qquad \sigma_i^2 = \sum_{k=-\infty}^{\infty} (kT - \eta_i)^2 a_{ik} \qquad (11\text{-}27)$$

From the discussion of Sec. 3-4 we know that the transforms $F_i(\omega)$ of $f_i(t)$ are periodic, with period

$$\Omega = 2\pi/T$$

Therefore their product $F(\omega) = F_1(\omega) \cdots F_n(\omega)$ is also periodic with the same period. Forming the convolution $f(t)$ of $f_i(t)$, as in (11-7), we conclude that since $\delta(t - t_1) * \delta(t - t_2) = \delta[t - (t_1 + t_2)]$ [see (I-29)], $f(t)$ is also lattice

$$f(t) = \sum_{k=-\infty}^{\infty} a_k\, \delta(t - kT) \qquad (11\text{-}28)$$

with mean $\eta = \eta_1 + \cdots + \eta_n$ and variance $\sigma^2 = \sigma_1^2 + \cdots + \sigma_n^2$.

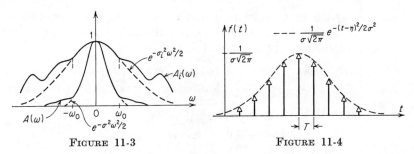

FIGURE 11-3   FIGURE 11-4

From (11-17) and Sec. 3-4 one can derive the following version of the central-limit theorem: the asymptotic form of the lattice $f(t)$ is given by

$$f(t) \sim \frac{1}{\sigma\sqrt{2\pi}} e^{-(t-\eta)^2/2\sigma^2} s_T(t) \qquad \text{for } n \to \infty \qquad (11\text{-}29)$$

where

$$s_T(t) = \sum_{n=-\infty}^{\infty} \delta(t - nT)$$

is a sequence of equidistant pulses as in Fig. 3-12. Thus $f(t)$ tends to a lattice whose envelope is a Gaussian curve (Fig. 11-4). This result follows from (11-16) and the discussion of Sec. 3-4. Indeed, if $F_i(\omega)$ is truncated above $|\omega| = \pi/T$, the function $F_{i0}(\omega)$ of Fig. 11-5b results, whose inverse transform $f_{i0}(t)$ is continuous; in fact, it is the envelope of $f_i(t)$. According to (11-16), the product

$$F_0(\omega) = F_{10}(\omega) \cdots F_{n0}(\omega)$$

tends to a Gaussian curve (Fig. 11-5c); therefore, its periodic repetition

$$F(\omega) = \sum_{-\infty}^{\infty} F_0\left(\omega + \frac{2n\pi}{T}\right)$$

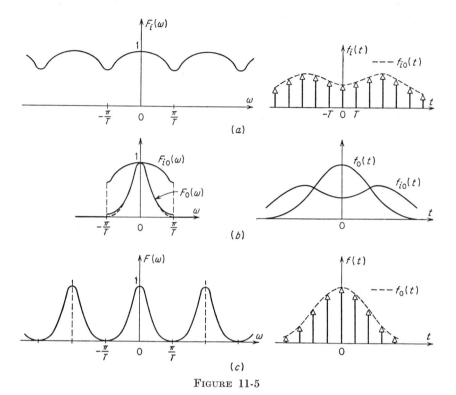

Figure 11-5

Figure 11-6

is the transform of a sequence of pulses of the form (11-29). By taking the moments of both sides in (11-29), one can also show that η and σ have the proper values (11-13).

Example 11-2. *Bernoulli's Theorem.* Suppose that the functions $f_i(t)$ are all equal and that each consists of two impulses

$$f_1(t) = \cdots = f_n(t) = p\delta(t) + q\delta(t - T) \qquad p + q = 1 \qquad (11\text{-}30)$$

as in Fig. 11-6a. Bernoulli's theorem states that their convolution tends to a lattice with Gaussian envelope

$$f(t) \sim \frac{1}{T\sqrt{2\pi npq}}\, e^{-(t-nqT)^2/2nT^2pq} s_T(t) \qquad (11\text{-}31)$$

for $n \rightarrow \infty$ (Fig. 11-6b). This follows readily from (11-29), because, as we see from (11-27),

$$\eta_i = qT \qquad \sigma_i{}^2 = T^2pq$$

Hence

$$\eta = nqT \qquad \sigma^2 = nT^2pq$$

From the proof of the central-limit theorem, one sees that (11-31) gives a good estimate to $f(t)$ if not only n, but also nq is large.

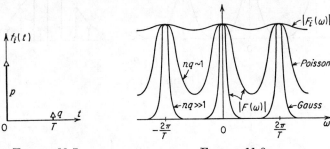

FIGURE 11-7   FIGURE 11-8

A case of particular interest is the following. Suppose that

$$q \ll 1$$

as in Fig. 11-7. If n is so large that $nq \gg 1$, then the estimate (11-31) holds. We shall now examine $f(t)$ for n of the order of $1/q$. We have from (11-30)

$$F_i(\omega) = p + qe^{-j\omega T} \qquad (11\text{-}32)$$

Since $q \ll 1$, $F_i(\omega)$ can be approximated by

$$F_i(\omega) \simeq e^{q(e^{-j\omega T}-1)} \qquad (11\text{-}33)$$

Indeed $$e^{q(e^{-j\omega T}-1)} \simeq 1 + q(e^{-j\omega T} - 1) = p + qe^{-j\omega T}$$

Hence $$F(\omega) \simeq e^{nq(e^{-j\omega T}-1)} \qquad (11\text{-}34)$$

Thus $f(t)$ is given by the *Poisson* density [see (3-70)], with $\lambda = nq$. We repeat, if n is so large that $nq \gg 1$, then (11-34) is equivalent to (11-31). In Fig. 11-8 we have shown $|F_i(\omega)|$, $|F(\omega)|$ for n of the order of $1/q$, and $|F(\omega)|$ for $n \gg 1/q$.

11-4. Error Correction

A numerical computation of the convolution $f(t)$ of n functions as in (11-7) is in general complicated. However, since $f(t)$ tends to a known function (11-17) as $n \to \infty$, it is natural to utilize the asymptotic form of $f(t)$ for its evaluation. If $f(t)$ is approximated by the Gaussian curve in (11-17), an error

$$\epsilon(t) = f(t) - \frac{1}{\sigma} G\left(\frac{t - \eta}{\sigma}\right) \tag{11-35}$$

is introduced, where $\qquad G(t) = \frac{1}{\sqrt{2\pi}} e^{-t^2/2} \tag{11-36}$

For large n this error is small, but for moderate values of n it cannot be neglected. In the following we shall give a method for evaluating $\epsilon(t)$ in terms of the moments of $f(t)$.

Hermite polynomials. We shall assume $\eta = 0$; this can be accomplished by a shift of the time origin. The error $\epsilon(t)$ will be written as a sum

$$\epsilon(t) = \frac{1}{\sigma} G\left(\frac{t}{\sigma}\right) \sum_{k=3}^{\infty} C_k H_k\left(\frac{t}{\sigma}\right) \tag{11-37}$$

where the functions $H_k(t)$ are the well-known Hermite polynomials

$$H_k(t) = (-1)^k e^{t^2/2} \frac{d^k}{dt^k} e^{-t^2/2}$$

$$= t^k - \binom{k}{2} t^{k-2} + 1 \cdot 3 \binom{k}{4} t^{k-4} + \cdots \tag{11-38}$$

forming a complete orthogonal system

$$\int_{-\infty}^{\infty} e^{-t^2/2} H_n(t) H_m(t)\, dt = \begin{cases} n!\, \sqrt{2\pi} & \text{for } n = m \\ 0 & \text{for } n \neq m \end{cases} \tag{11-39}$$

To evaluate the constants C_k, we shall take the moments of both sides of (11-37). The moments A_k of the Gaussian function $G(t)$ are given by [see (2-66)]

$$A_{2n} = \frac{1}{\sqrt{2\pi}} \int_{-\infty}^{\infty} t^{2n} e^{-t^2/2}\, dt = \frac{(2n)!}{2^n \cdot n!} \qquad A_{2n+1} = 0 \tag{11-40}$$

and the nth moment a_{nk}

$$a_{nk} = \frac{1}{\sqrt{2\pi}} \int_{-\infty}^{\infty} t^n e^{-t^2/2} H_k(t)\, dt \tag{11-41}$$

of the function $G(t) H_k(t)$ can be found from (11-38) to (11-40).

The following equations easily result:

$$a_{nk} = A_{n+k} - \binom{k}{2} A_{n+k-2} + 1 \cdot 3 \binom{k}{4} A_{n+k-4} + \cdots \qquad (11\text{-}42)$$

$$a_{nk} = 0 \qquad \text{if } n + k \text{ odd or } k > n \qquad (11\text{-}43)$$

$$a_{nn} = n! \qquad (11\text{-}44)$$

With m_n the nth moment of $f(t)$ [see (11-12)], the corresponding moment of $\epsilon(t)$ is given by $m_n - \sigma^n A_n$; therefore, equating moments in (11-37), we obtain

$$3! \, \sigma^3 C_3 = m_3$$

$$4! \, \sigma^4 C_4 = m_4 - \sigma^4 A_4 \qquad (11\text{-}45)$$

$$n! \, \sigma^n C_n + a_{n(n-2)} \sigma^n C_{n-2} + \cdots = m_n - \sigma^n A_n$$

The coefficients C_n can thus be determined successively. If $f(t)$ is even, then $C_{2n+1} = 0$; retaining only the first term in (11-37), we thus have the following approximation to an even function:

$$f(t) \simeq \bar{f}(t) = \frac{1}{\sigma} G\left(\frac{t}{\sigma}\right) \left\{ 1 + \frac{1}{4!} \left(\frac{m_4}{\sigma^4} - 3 \right) \left[\left(\frac{t}{\sigma} \right)^4 - 6 \left(\frac{t}{\sigma} \right)^2 + 3 \right] \right\} \qquad (11\text{-}46)$$

because $\qquad A_4 = 3 \qquad H_4(t) = t^4 - 6t^2 + 3$

Example 11-3. Suppose that the functions $f_i(t)$ are all equal to the rectangular pulse of Fig. 11-9a, and $n = 3$. Their convolution is obtained by properly shifting the corresponding function (11-20) of Example 11-1 (solid line in Fig. 11-9a). Since $\sigma = \frac{1}{2}$, the Gaussian approximation equals

$$\sqrt{\frac{2}{\pi}} \, e^{-2t^2}$$

(dotted line in Fig. 11-9a). Clearly $f(t)$ is even, and its fourth moment is given by $m_4 = \frac{13}{80}$; from (11-46) we obtain its first-order correction

$$f(t) \simeq \sqrt{\frac{2}{\pi}} \, e^{-2t^2} \left(1 - \frac{4}{15} t^4 + \frac{2}{5} t^2 - \frac{1}{20} \right) = \bar{f}(t)$$

shown in Fig. 11-9a. The error $\epsilon(t)$ and the first-order correction error $f(t) - \bar{f}(t)$ are shown in Fig. 11-9b.

Causal form of the central-limit theorem. Often the functions $f_i(t)$ are causal and, obviously, the same is true for their convolution; however, the Gaussian estimate (11-17) of $f(t)$ is not zero for $t < 0$, no matter how large η is. In this case it would be desirable to modify the central-limit theorem so that $f(t)$ tends to a causal function as

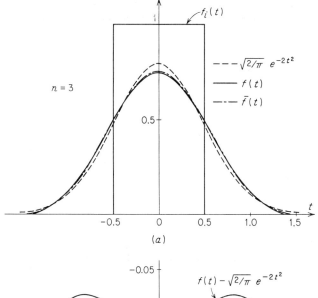

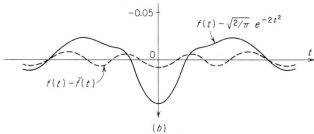

Figure 11-9

$n \to \infty$.   This can be done as follows.   Consider the function

$$s(t) = \frac{t^\alpha e^{-t/\beta}}{\beta^{\alpha+1}\Gamma(\alpha + 1)} U(t) \tag{11-47}$$

known as χ^2 *density*.   Its transform is given by

$$S(\omega) = \frac{1}{(1 + j\beta\omega)^{\alpha+1}} \tag{11-48}$$

as we can easily see from (9-5).

Clearly $S(\omega)$ can be written as the product of n characteristic functions, where n is the nearest integer to α.   Therefore, according to (11-16) and (11-17), $s(t)$ is a causal approximation to a Gaussian function, provided α is large.   We thus have the following version of the central-limit theorem for causal functions:

$$f(t) \simeq \frac{t^\alpha e^{-t/\beta}}{\beta^{\alpha+1}\Gamma(\alpha + 1)} \qquad \text{for } n \to \infty \tag{11-49}$$

where the constants α and β are such that the mean and variance of the two sides of (11-49) are equal. To determine them, we first find the kth moment B_k of $s(t)$:

$$
\begin{aligned}
B_k &= \frac{1}{\beta^{\alpha+1}\Gamma(\alpha+1)} \int_0^\infty t^k t^\alpha e^{-t/\beta} \, dt \\
&= (\alpha+1)(\alpha+2) \cdots (\alpha+k)\beta^k
\end{aligned} \tag{11-50}
$$

[see (9-5)]. Thus, with η and σ as in (11-13),

$$
(\alpha+1)\beta = \eta \qquad (\alpha+1)(\alpha+2)\beta^2 = \sigma^2 + \eta^2
$$

Therefore α and β are given by

$$
(\alpha+1) = \frac{\eta^2}{\sigma^2} \qquad \beta = \frac{\sigma^2}{\eta} \tag{11-51}
$$

and (11-49) gives a causal estimate to $f(t)$.

From the above results we easily obtain the following interesting approximation:

$$
\frac{1}{(\beta_1 p + 1)(\beta_2 p + 1) \cdots (\beta_n p + 1)} \simeq \frac{1}{(\beta p + 1)^{\alpha+1}} \tag{11-52}
$$

where

$$
\alpha + 1 = \frac{\left(\sum\limits_{i=1}^n \beta_i\right)^2}{\sum\limits_{i=1}^n \beta_i^2} \qquad \beta = \frac{\sum\limits_{i=1}^n \beta_i^2}{\sum\limits_{i=1}^n \beta_i}
$$

subject to the condition $\alpha \gg 1$.

If we replace α in (11-49) by its nearest integer, we introduce a small error, since $\alpha \gg 1$; we thus obtain an estimate of $f(t)$ with rational transform.

Error Correction with Laguerre Polynomials. Proceeding as in (11-37), we shall express the error

$$
\epsilon(t) = f(t) - s(t) \tag{11-53}
$$

as a series

$$
\epsilon(t) = s(t) \sum_{k=3}^\infty D_k L_k^{(\alpha)}\left(\frac{t}{\beta}\right) \tag{11-54}
$$

in terms of the generalized Laguerre polynomials

$$
\begin{aligned}
L_k^{(\alpha)}(t) &= \frac{t^{-\alpha}e^{-t}}{k!} \frac{d^k}{dt^k} (t^{\alpha+k}e^{-t}) \\
&= \frac{(-1)}{k!}\left[t^k - \binom{k}{1}(k+\alpha)t^{k-1} + \cdots \pm (k+\alpha)(k-1+\alpha)\cdots(1+\alpha) \right]
\end{aligned} \tag{11-55}
$$

To determine the constants D_k, we shall need the moments b_{nk} of the function $s(t)L_k^{(\alpha)}(t/\beta)$

$$b_{nk} = \int_0^\infty t^n s(t) L_k^{(\alpha)}\left(\frac{t}{\beta}\right) dt \tag{11-56}$$

From (11-55) and (11-50) we easily find

$$b_{nk} = \frac{(-1)^k}{k!}\left[B_{n+k} - \binom{k}{1}(k+\alpha)B_{n+k-1} + \cdots\right] \tag{11-57}$$

and since the polynomials $L_k^{(\alpha)}(t)$ are orthogonal

$$\int_0^\infty e^{-t}t^\alpha L_n^{(\alpha)}(t)L_m^{(\alpha)}(t)\ dt = \begin{cases} \dfrac{\Gamma(n+\alpha+1)}{n!} & \text{for } n = m \\ 0 & \text{for } n \neq m \end{cases} \tag{11-58}$$

we conclude that

$$b_{nn} = (-1)^n B_n \qquad b_{nk} = 0 \qquad \text{for } n < k \tag{11-59}$$

Taking the moments of both sides of (11-53), we finally obtain

$$(-1)^3 B_3 D_3 = m_3 - B_3$$

$$\cdots\cdots\cdots\cdots\cdots$$

$$(-1)^n B_n D_n + b_{n(n-1)}D_{n-1} + \cdots \pm b_{n3}D_3 = m_n - B_n \tag{11-60}$$

$$\cdots\cdots\cdots\cdots\cdots\cdots\cdots\cdots\cdots\cdots\cdots\cdots\cdots\cdots$$

where m_n are the moments of $f(t)$. If only the first term in (11-54) is retained, then the following approximation to $f(t)$ results:

$$f(t) \simeq s(t)\left[1 + \left(1 - \frac{m_3}{B_3}\right)L_3^{(\alpha)}\left(\frac{t}{\beta}\right)\right]U(t) \tag{11-61}$$

Example 11-4. We shall now give a causal estimate to the function $f(t)$ of Fig. 11-2c discussed in Example 11-1. Since

$$\eta = \tfrac{3}{2} \qquad \sigma^2 = \tfrac{1}{4}$$

we have from (11-51) $\alpha = 8 \qquad \beta = \tfrac{1}{6}$
Therefore [see (11-49)]

$$f(t) \simeq \frac{6^9}{8!} t^8 e^{-6t} U(t)$$

The first-order error correction (11-61) is given by

$$f(t) \simeq \frac{6^9}{8!} t^8 e^{-6t}[1 - \tfrac{3}{55}(2t^3 - 66t^2 + 110t - 55)]U(t)$$

since $m_3 = \tfrac{9}{2}$ and $B_3 = \tfrac{55}{12}$ [see (11-50)].

Time-limited functions. If the functions $f_i(t)$ are zero outside a finite interval (a_i, b_i), then their convolution $f(t)$ is zero outside the interval (a, b), where

$$a = \sum_{i=1}^{n} a_i \qquad b = \sum_{i=1}^{n} b_i$$

In this case one would like to formulate the central-limit theorem so that $f(t)$ tends to a function that equals zero outside the (a, b) interval. This can be done as follows.

By a linear change in the variable t, we can assume that

$$a = 0 \qquad b = 1 \tag{11-62}$$

The function

$$p(t) = \begin{cases} Mt^{\alpha}(1 - t)^{\beta} & \text{for } 0 < t < 1 \\ 0 & \text{elsewhere} \end{cases} \tag{11-63}$$

known as *beta density*, will be used to approximate $f(t)$ for large n

$$f(t) \sim p(t) \qquad n \to \infty \tag{11-64}$$

The justification of (11-64) is similar to (11-49). The constant M is determined by the requirement that the area of $p(t)$ must equal one; we easily find

$$M = \frac{\Gamma(\alpha + \beta + 2)}{\Gamma(\alpha + 1)\Gamma(\beta + 1)} \tag{11-65}$$

The exponents α and β are chosen so that the mean and variance of $p(t)$ equal η and σ^2. The nth moment F_n of $p(t)$ is easily found to be

$$F_n = \frac{(2\alpha + 2)(2\alpha + 4) \cdots (2\alpha + 2n)}{\nu(\nu + 2) \cdots (\nu + 2n - 2)} \qquad \nu = 2\alpha + 2\beta + 4 \tag{11-66}$$

Therefore α and β must be given by

$$\alpha + 1 = \frac{\eta(\eta - \eta^2 - \sigma^2)}{\sigma^2} \qquad \beta + 1 = \frac{(1 - \eta)(\alpha + 1)}{\eta} \tag{11-67}$$

as one can see from (11-66).

The error $f(t) - p(t)$ can be expressed as a series, similar to (11-54), in terms of the *Jacobi* polynomials

$$J_n(\alpha + 1, \alpha + \beta + 1; t)$$

because the function $p(t)$ is the orthogonalizing weight of these polynomials in the $(0,1)$ interval. An expansion similar to (11-54) results, but the details will be omitted.

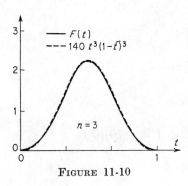

FIGURE 11-10

We shall apply (11-64) to the function $f(t)$ of Example 11-4. Since $f(t)$ is zero outside the (0,3) interval, we must change t to $3t$ in order to satisfy (11-62); the density

$$F(t) = 3f(3t)$$

results, whose mean and variance are given by

$$\eta = \tfrac{1}{2} \qquad \sigma^2 = \tfrac{1}{36}$$

From (11-65) and (11-67) we obtain

$$M = 140 \qquad \alpha + 1 = 4 \qquad \beta + 1 = 4$$

Hence
$$F(t) \simeq 140t^3(1 - t)^3 \qquad \text{for } 0 < t < 1$$

The two functions above are shown in Fig. 11-10.

Chapter 12. Generalized Harmonic Analysis, Correlation, and Power Spectra

We shall now investigate the important class of functions that have no Fourier integral but whose average power is finite. The concepts of correlation and power spectrum will be discussed, and their relationship to the *generalized* Fourier transform will be established.† ‡ These quantities are usually defined for *random variables* in books on probability theory§; however, they form a part of Fourier analysis and can be treated with deterministic considerations.

12-1. Introduction

The *average power* of a signal $f(t)$ is defined by the limit

$$\overline{f^2}(t) = \lim_{T \to \infty} \frac{1}{2T} \int_{-T}^{T} |f(t)|^2 \, dt \qquad (12\text{-}1)$$

For the purposes of this chapter, we shall classify all functions into the following three classes, depending on the value of their average power:

$$\text{I. } \overline{f^2}(t) = 0$$

$$\text{II. } 0 < \overline{f^2}(t) < \infty$$

$$\text{III. } \overline{f^2}(t) = \infty$$

Since the square of distributions is undefined, the concept of power is meaningless for signals containing singularities; we shall, nevertheless, classify functions of the form

$$f(t) = \sum_{n=-\infty}^{\infty} \alpha_n \delta(t - t_n)$$

---

† N. Wiener, "The Fourier Integral and Certain of Its Applications," Cambridge University Press, New York, 1933.

‡ Y. W. Lee, "Statistical Theory of Communications," John Wiley & Sons, Inc. New York, 1960.

§ W. B. Davenport, Jr., and W. L. Root, "Introduction to Random Signals and Noise," McGraw-Hill Book Company, Inc., New York, 1958.

as above, according to the value of the limit

$$\lim_{T \to \infty} \frac{1}{2T} \sum |\alpha_n|^2 \qquad (12\text{-}2)$$

$$n : |t_n| < T$$

where the summation extends over all n such that t_n is in the $(-T,T)$ interval.

Class I obviously contains all functions with finite energy

$$\int_{-\infty}^{\infty} |f(t)|^2 \, dt < \infty \qquad (12\text{-}3)$$

Class II contains all periodic functions

$$f(t) = \sum_{n=-\infty}^{\infty} \alpha_n e^{jn\omega_0 t} \qquad \sum_{n=-\infty}^{\infty} |\alpha_n|^2 < \infty \qquad (12\text{-}4)$$

and $\overline{f^2}(t)$ is given by their average power in one period

$$\overline{f^2}(t) = \frac{\omega_0}{2\pi} \int_{-\pi/\omega_0}^{\pi/\omega_0} f^2(t) \, dt = \sum_{-\infty}^{\infty} |\alpha_n|^2 \qquad (12\text{-}5)$$

Nonperiodic functions are also included in this class; however, their characterization is not always easy, and for this reason they are usually specified only by certain averages. One such average is their *autocorrelation* $R_f(t)$ defined by

$$R_f(t) = \lim_{T \to \infty} \frac{1}{2T} \int_{-T}^{T} f(\tau) f(t + \tau) \, d\tau \qquad (12\text{-}6)$$

In Sec. 12-3 we shall study the properties of $R_f(t)$ and of its Fourier transform. As a preparation, we shall also define the autocorrelation $\rho_f(t)$

$$\rho_f(t) = \int_{-\infty}^{\infty} f(t + \tau) f(\tau) \, d\tau \qquad (12\text{-}7)$$

for functions with finite energy. Functions of Class III will not be further considered.

12-2. Finite Energy Signals

In this section we consider signals with finite energy. Suppose that $f(t)$ is real and its Fourier integral exists and is given by

$$F(\omega) = A(\omega)e^{j\phi(\omega)}.$$

The inverse transform of their energy spectrum

$$E(\omega) = A^2(\omega),$$

known as autocorrelation, will be denoted by $\rho_f(t)$

$$\rho_f(t) = \frac{1}{2\pi}\int_{-\infty}^{\infty} A^2(\omega)e^{j\omega t}\,d\omega = \frac{1}{2\pi}\int_{-\infty}^{\infty} A^2(\omega)\cos\omega t\,d\omega \qquad (12\text{-}8)$$

We shall presently see that $\rho_f(t)$ can be written in the form (12-7). Indeed, from

$$\rho_f(t)\leftrightarrow A^2(\omega) = F(\omega)\overset{*}{F}(\omega) \qquad (12\text{-}9)$$

and from the convolution theorem (2-71) it follows that, since

$$f(-t)\leftrightarrow \overset{*}{F}(\omega) = F(-\omega),$$

$$\rho_f(t) = f(t) * f(-t) \qquad (12\text{-}10)$$

Thus

$$\rho_f(t) = \int_{-\infty}^{\infty} f(t-\tau)f(-\tau)\,d\tau$$

Changing $-\tau$ into τ, we obtain (12-7).

From the realness of $A^2(\omega)$ it follows that

$$\rho_f(-t) = \rho_f(t) \qquad (12\text{-}11)$$

and since $A^2(\omega) \geq 0$, we conclude that [see (12-8)]

$$|\rho_f(t)| \leq \rho_f(0) = \int_{-\infty}^{\infty} |f(t)|^2\,dt \qquad (12\text{-}12)$$

For a fixed $t = t_1$, the energy of the signal

$$f(t) + f(t+t_1)$$

can be expressed in terms of $\rho_f(t)$. Indeed,

$$\int_{-\infty}^{\infty} [f(t) \pm f(t+t_1)]^2\,dt = 2[\rho_f(0) \pm \rho_f(t_1)] \qquad (12\text{-}13)$$

For complex signals the autocorrelation is defined by (12-8); however, (12-7) is no longer valid. In this case $\rho_f(t)$ is given by

$$\rho_f(t) = \int_{-\infty}^{\infty} f(t+\tau)\overset{*}{f}(\tau)\,d\tau \qquad (12\text{-}14)$$

as we can easily see. Inequality (12-12) is still true, but (12-11) is replaced by

$$\rho_f(-t) = \overset{*}{\rho}_f(t)$$

Suppose now that $f(t)$ is the input to the linear system of Fig. 12-1 and $h(t)$ its impulse response. As we know, $g(t)$ is given by

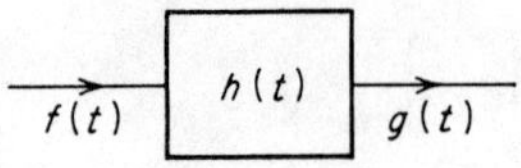

FIGURE 12-1

$$g(t) = \int_{-\infty}^{\infty} f(\tau)h(t-\tau)\,d\tau = f(t) * h(t) \qquad (12\text{-}15)$$

We assume that the functions $f(t)$ and $h(t)$ are of Class I; with $\rho_f(t)$ and $\rho_g(t)$ the autocorrelations of $f(t)$ and $g(t)$, we shall show that

$$\rho_g(t) = \rho_f(t) * h(t) * h(-t) \quad (12\text{-}16)$$

Indeed, with $F(\omega)$ and $H(\omega)$ the Fourier transform of $f(t)$ and $h(t)$ respectively, the Fourier transform of $g(t)$ is given by $F(\omega)\,H(\omega)$; hence

$$\rho_g(t) \leftrightarrow |F(\omega)|^2\,|H(\omega)|^2$$
$$= |F(\omega)|^2\,H(\omega)\overset{*}{H}(\omega)$$

and since

$$\rho_f(t) \leftrightarrow |F(\omega)|^2 \qquad h(t) \leftrightarrow H(\omega)$$
$$h(-t) \leftrightarrow H(-\omega) = \overset{*}{H}(\omega)$$

(12-16) follows by a repeated application of (2-71). In the above discussion the presence of the linear system is incidental; (12-16) relates the autocorrelations of two functions satisfying (12-15).

(a)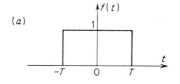

(b)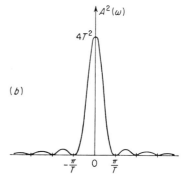

(c)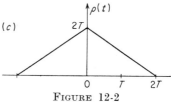

FIGURE 12-2

Example 12-1. We shall determine the autocorrelation and energy spectrum of the rectangular pulse

$$f(t) = p_T(t)$$

of Fig. 12-2a. Its Fourier transform is given by (2-56)

$$F(\omega) = \frac{2 \sin \omega T}{\omega}$$

and its energy spectrum $\qquad A^2(\omega) = \dfrac{4 \sin^2 \omega T}{\omega^2}$

is shown in Fig. 12-2b. $\rho_f(t)$ is the triangular pulse

$$\rho_f(t) = 2T q_{2T}(t)$$

of Fig. 12-2c, as we see from (12-9) and (2-58).

Example 12-2. $f(t) = e^{-\alpha|t|} \qquad \alpha > 0$
We have from (2-62)

$$F(\omega) = \frac{2\alpha}{\alpha^2 + \omega^2}$$

Hence $\qquad A^2(\omega) = \dfrac{4\alpha^2}{(\alpha^2 + \omega^2)^2}$

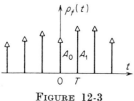

FIGURE 12-3

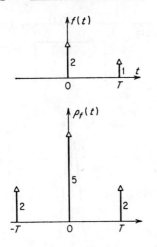

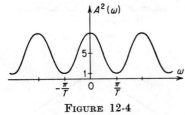

FIGURE 12-4

and (see Example 9-9)

$$\rho_f(t) = e^{-\alpha|t|}\left(|t| + \frac{1}{\alpha}\right)$$

Example 12-3. $f(t)$ equals the sum of real equidistant pulses (Fig. 12-3)

$$f(t) = \sum_{n=-\infty}^{\infty} \alpha_n \delta(t - nT)$$

$$\sum_{n=-\infty}^{\infty} |\alpha_n|^2 < \infty$$

and belongs to Class I. We shall determine its autocorrelation $\rho_f(t)$ directly from (12-7). Using the convolution between two impulses

$$\delta(t - t_1) * \delta(t - t_2) = \delta[t - (t_1 + t_2)]$$

[see (I-29)], we easily obtain

$$\rho_f(t) = \sum_{-\infty}^{\infty} A_n \delta(t - nT)$$

where $A_n = \sum_{m=-\infty}^{\infty} \alpha_m \alpha_{m+n}$

Thus if $f(t) = 2\delta(t) + \delta(t - T)$, then $\rho_f(t) = 2\delta(t - T) + 5\delta(t) + 2\delta(t + T)$, and $A^2(\omega) = 5 + 4 \cos T$. These functions are shown in Fig. 12-4.

Cross-correlation. Consider the real functions $f_1(t)$ and $f_2(t)$, their Fourier integrals $F_1(\omega)$ and $F_2(\omega)$, and their cross-energy spectrum $E_{12}(\omega) = \overset{*}{F}_1(\omega)F_2(\omega)$ [see (2-80)]. The inverse transform of $E_{12}(\omega)$ we denote by $\rho_{12}(t)$ and will call the *cross-correlation* between $f_1(t)$ and $f_2(t)$. We thus have

$$\rho_{12}(t) = \frac{1}{2\pi} \int_{-\infty}^{\infty} \overset{*}{F}_1(\omega)F_2(\omega)e^{j\omega t} \, d\omega \qquad (12\text{-}17)$$

Since $E_{12}(-\omega) = \overset{*}{E}_{12}(\omega)$, we conclude that $\rho_{12}(t)$ is real but in general not even; however,

$$\rho_{12}(-t) = \rho_{21}(t) \qquad (12\text{-}18)$$

as we can readily see. From

$$f_1(-t) \leftrightarrow F_1(-\omega) = \overset{*}{F}_1(\omega) \qquad f_2(t) \leftrightarrow F_2(\omega)$$

and (2-71) we obtain $\rho_{12}(t) = f_1(-t) * f_2(t) \qquad (12\text{-}19)$

which can also be written in the form

$$\rho_{12}(t) = \int_{-\infty}^{\infty} f_1(\tau)f_2(t + \tau) \, d\tau \qquad (12\text{-}20)$$

For a fixed $t = t_1$, the energy of the signal $f_1(t) \pm f_2(t + t_1)$ can be expressed in terms of $\rho_{f1}(0)$, $\rho_{f2}(0)$, and $\rho_{12}(t_1)$. Indeed,

$$\int_{-\infty}^{\infty} [f_1(t) \pm f_2(t + t_1)]^2 \, dt = \rho_{f1}(0) + \rho_{f2}(0) \pm 2\rho_{12}(t_1) \quad (12\text{-}21)$$

Since the left-hand side of (12-21) is obviously nonnegative, we must have

$$2 \, |\rho_{12}(t)| \leq \rho_{f1}(0) + \rho_{f2}(0) \quad (12\text{-}22)$$

From Schwarz's inequality (4-47) we also have

$$\int_{-\infty}^{\infty} f_1^2(\tau) \, d\tau \int_{-\infty}^{\infty} f_2^2(t + \tau) \, d\tau \geq |\int_{-\infty}^{\infty} f_1(\tau) f_2(t + \tau) \, d\tau|^2$$

Hence
$$|\rho_{12}(t)|^2 \leq \rho_{f1}(0)\rho_{f2}(0) \quad (12\text{-}23)$$

If the functions $f_1(t)$ and $f_2(t)$ are not real, then $\rho_{12}(t)$ is still defined as the inverse of $E_{12}(\omega)$ and is given by

$$\rho_{12}(t) = \int_{-\infty}^{\infty} \overset{*}{f_1}(\tau) f_2(t + \tau) \, d\tau \quad (12\text{-}24)$$

as we can easily see.

12-3. Finite Power Signals

We now turn to the main task of this chapter, the consideration of functions with finite nonzero power

$$\overline{f^2}(t) = \lim_{T \to \infty} \frac{1}{2T} \int_{-T}^{T} |f(t)|^2 \, dt < \infty \quad (12\text{-}1)$$

The average power $\overline{f^2}(t)$ can also be written in the form

$$\overline{f^2}(t) = \lim_{T \to \infty} \frac{1}{2T} \int_{-T-A}^{T-A} |f(t)|^2 \, dt = \lim_{T \to \infty} \frac{1}{2T} \int_{-T}^{T} |f(t + A)|^2 \, dt \quad (12\text{-}25)$$

where A is an arbitrary constant (see Prob. 43).

We shall first assume that $f(t)$ is real; its autocorrelation $R(t)$ is defined by

$$R(t) = \lim_{T \to \infty} \frac{1}{2T} \int_{-T}^{T} f(\tau) f(t + \tau) \, d\tau \quad (12\text{-}6)$$

[the subscript in $R(t)$ will be dropped whenever the corresponding function $f(t)$ is clearly understood]. This function exists for every t; indeed, from Schwarz's inequality (4-47) we have

$$\frac{1}{2T} \int_{-T}^{T} f^2(\tau) \, d\tau \frac{1}{2T} \int_{-T}^{T} f^2(t + \tau) \, d\tau \geq \left| \frac{1}{2T} \int_{-T}^{T} f(\tau) f(t + \tau) d\tau \right|^2 \quad (12\text{-}26)$$

Because of (12-1) and (12-25), each integral at the left of $\geq$ tends to the finite limit

$$R(0) = \lim_{T \to \infty} \frac{1}{2T} \int_{-T}^{T} |f(t)|^2 \, dt \qquad (12\text{-}27)$$

as T tends to infinity. Hence the limit $|R(t)|^2$ of the right-hand side is finite; in fact, from (12-26) we also obtain

$$|R(t)| \leq R(0) \qquad (12\text{-}28)$$

We shall now show that $R(t)$ is even

$$R(-t) = R(t) \qquad (12\text{-}29)$$

From (12-6) we have

$$R(-t) = \lim_{T \to \infty} \frac{1}{2T} \int_{-T}^{T} f(\tau) f(-t+\tau) \, d\tau = \lim_{T \to \infty} \frac{1}{2T} \int_{-T-t}^{T-t} f(t+\tau) f(\tau) \, d\tau$$

and (12-29) follows, because the last limit equals $R(t)$ (see Prob. 43). It is obvious from (12-6) that the autocorrelation of $af(t)$ equals $a^2 R(t)$:

$$R_{af}(t) = a^2 R_f(t) \qquad (12\text{-}30)$$

Power spectrum. The power spectrum $S(\omega)$ of a function $f(t)$ is defined as the Fourier transform of its autocorrelation $R(t)$

$$R(t) \leftrightarrow S(\omega) \qquad (12\text{-}31)$$

Since $R(t)$ is real and even, we conclude that $S(\omega)$ is also real and even and

$$S(\omega) = \int_{-\infty}^{\infty} R(t) \, \cos \omega t \, dt \qquad R(t) = \frac{1}{2\pi} \int_{-\infty}^{\infty} S(\omega) \cos \omega t \, d\omega \qquad (12\text{-}32)$$

We shall now show that the power spectrum $S(\omega)$ can be expressed directly in terms of $f(t)$:

$$S(\omega) = \lim_{T \to \infty} \frac{1}{2T} \left| \int_{-T}^{T} f(t) e^{-j\omega t} \, dt \right|^2 \geq 0 \qquad (12\text{-}33)$$

To prove (12-33), we form the function

$$f_T(t) = f(t) p_T(t) \qquad (12\text{-}34)$$

obtained by truncating $f(t)$ above $|t| = T$. The Fourier transform $F_T(\omega)$ of $f_T(t)$ is given by

$$F_T(\omega) = \int_{-T}^{T} f(t) e^{-j\omega t} \, dt \qquad (12\text{-}35)$$

and its average power in the $(-T, T)$ interval, by

$$S_T(\omega) = \frac{1}{2T} |F_T(\omega)|^2 \qquad (12\text{-}36)$$

With $R_T(t)$ the inverse transform of $S_T(\omega)$

$$R_T(t) \leftrightarrow S_T(\omega) \tag{12-37}$$

we easily see from (12-36) that

$$R_T(t) = \frac{1}{2T} f_T(t) * f_T(-t) \tag{12-38}$$

For $t > 0$, (12-38) can be written in the form

$$R_T(t) = \frac{1}{2T} \int_{-T}^{T-t} f(\tau)f(t + \tau)\, d\tau \tag{12-39}$$

From (12-39) we conclude that the limit of $R_T(t)$ as $T \to \infty$ equals $R(t)$

$$R_T(t) \xrightarrow[T \to \infty]{} R(t) \tag{12-40}$$

Therefore, the same is true for the Fourier transforms $S_T(\omega)$ and $S(\omega)$ (Helly's theorem)

$$S_T(\omega) \xrightarrow[T \to \infty]{} S(\omega) \tag{12-41}$$

and (12-33) is proved.

Comment. Since $S(\omega)$ is positive, we conclude as in (11-6) that $R(t)$ is positive-definite. It is shown on page 255 that this condition is also sufficient; i.e., if $R(t)$ is positive-definite, then it is an autocorrelation.

For a fixed $t = t_1$, the average power of $f(t) \pm f(t + t_1)$ can be written in terms of $R(t)$:

$$\overline{|f(t) \pm f(t + t_1)|^2} = 2[R(0) \pm R(t_1)] \tag{12-42}$$

This follows readily from (12-6).

Suppose now that $f(t)$ is of Class II and $h(t)$ of Class I; we form the function

$$g(t) = \int_{-\infty}^{\infty} f(t - \tau)h(\tau)\, d\tau = f(t) * h(t)$$

as in (12-15). We shall show that $g(t)$ is of Class II and its autocorrelation $R_g(t)$ and power spectrum $S_g(\omega)$ are given by

$$R_g(t) = R_f(t) * h(t) * h(-t) \tag{12-43}$$

$$S_g(\omega) = S_f(\omega)\, |H(\omega)|^2 \tag{12-44}$$

where $R_f(t)$, $S_f(\omega)$ are the corresponding quantities of $f(t)$, and $H(\omega)$ the Fourier transform of $h(t)$. To prove (12-43), we observe that

$$R_g(t) = \lim_{T \to \infty} \frac{1}{2T} \int_{-T}^{T} g(\tau)g(t + \tau)\, d\tau$$

$$= \lim_{T \to \infty} \frac{1}{2T} \int_{-T}^{T} d\tau \int_{-\infty}^{\infty} f(\tau - x)h(x)\, dx \int_{-\infty}^{\infty} f(t + \tau - y)h(y)\, dy$$

Changing the order of integration and using the identity

$$R_f(t + x - y) = \lim_{T \to \infty} \frac{1}{2T} \int_{-T}^{T} f(\tau - x)f(t + \tau - y)\, d\tau$$

we obtain (12-43). Equation (12-44) follows from (12-43) and the convolution integral; indeed, since

$$R_f(t) \leftrightarrow S_f(\omega) \qquad h(t) \leftrightarrow H(\omega) \qquad h(-t) \leftrightarrow \overset{*}{H}(\omega)$$

we have from (12-43)

$$S_g(\omega) = S_f(\omega)\overset{*}{H}(\omega)H(\omega)$$

and (12-44) is proved.

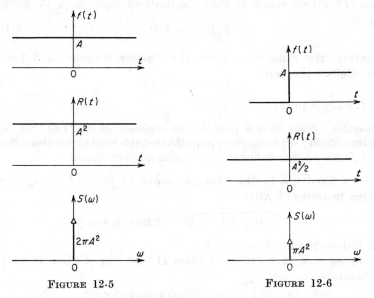

FIGURE 12-5   FIGURE 12-6

We note that if $R_f(t) = \delta(t)$, then $S_f(\omega) = 1$ and

$$S_g(\omega) = |H(\omega)|^2 \tag{12-45}$$

i.e., the power spectrum of the output $g(t)$ equals the energy spectrum of $h(t)$.

Example 12-4. If $f(t)$ is constant, as in Fig. 12-5,

$$f(t) = A$$

then $R(t)$ is also constant:

$$R(t) = \lim_{T \to \infty} \frac{1}{2T} \int_{-T}^{T} A^2\, d\tau = A^2$$

and $S(\omega)$ is given by the impulse

$$S(\omega) = 2\pi A^2 \delta(\omega)$$

of Fig. 12-5 [see (3-6)].

Example 12-5. If $f(t) = AU(t)$, then

$$R(t) = \lim_{T \to \infty} \frac{A^2}{2T} \int_{-T}^{T} U^2(t) \, dt = \lim \frac{A^2}{2T} \int_{0}^{T} dt = \frac{A^2}{2}$$

and $S(\omega) = \pi A^2 \delta(\omega)$

as in Fig. 12-6. We remark that, in general, $R(t)$ is not given by the limit

$$\lim_{T \to \infty} \frac{1}{T} \int_{0}^{T} f(\tau) f(t + \tau) \, d\tau$$

as we see from Example 12-5. However, for certain functions (see Example 12-4) the above integral equals $R(t)$.

Example 12-6. Suppose that $f(t)$ is a sequence of pulses

$$f(t) = \sum_{k=-\infty}^{\infty} \alpha_k \, \delta(t - kc) \tag{12-46}$$

of Class II [see (12-2)], with real coefficients α_n. We can show as in Example 12-3 that its autocorrelation $R(t)$ is also a sequence of pulses

$$R(t) = \sum_{k=-\infty}^{\infty} A_k \, \delta(t - kc) \tag{12-47}$$

where $A_k = \lim_{n \to \infty} \frac{1}{2nc} \sum_{m=-n}^{n} \alpha_m \alpha_{k+m}$ $\tag{12-48}$

From (12-47) and (12-32) it follows that the power spectrum of $f(t)$ is a periodic function of ω with period $\omega_0 = 2\pi/c$:

$$S(\omega) = \sum_{k=-\infty}^{\infty} A_k \cos kc\omega = A_0 + 2\sum_{k=1}^{\infty} A_k \cos kc\omega \tag{12-49}$$

since [see (12-48)]

$$A_{-k} = A_k$$

The following special case of (12-46) is of particular interest. If

$$f(t) = \sum_{k=-\infty}^{\infty} \delta(t - nc) = s_c(t) \tag{12-50}$$

(Fig. 12-7), then $A_k = 1/c$; hence

$$R(t) = \frac{1}{c} s_c(t) \tag{12-51}$$

and $S(\omega) = \frac{1}{c} \sum_{k=-\infty}^{\infty} e^{jkc\omega}$

$$= \frac{2\pi}{c^2} \sum_{k=-\infty}^{\infty} \delta(\omega - n\omega_0)$$

$$= \frac{2\pi}{c^2} s_{\omega 0}(\omega) \tag{12-52}$$

[see (3-42)].

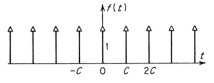
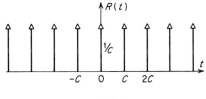
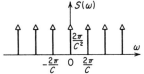

FIGURE 12-7

Example 12-7. In this example we shall show that the power spectrum $S(\omega)$ of a periodic function (Fig. 12-8a)

$$f(t) = \sum_{n=-\infty}^{\infty} \alpha_n e^{jn\omega_0 t} \qquad \omega_0 = \frac{2\pi}{c} \qquad (12\text{-}53)$$

of period c, is a sequence of equidistant pulses

$$S(\omega) = 2\pi \sum_{n=-\infty}^{\infty} |\alpha_n|^2 \, \delta(\omega - n\omega_0) \qquad (12\text{-}54)$$

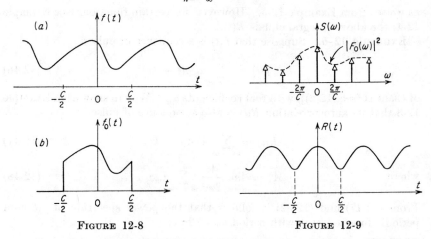

FIGURE 12-8   FIGURE 12-9

as in Fig. 12-9. It will then follow from (12-32) that the autocorrelation $R(t)$ is also a periodic function given by

$$R(t) = \sum_{n=-\infty}^{\infty} |\alpha_n|^2 \cos n\omega_0 t \qquad (12\text{-}55)$$

If $f(t)$ is real, then, since $|\alpha_{-n}| = |\alpha_n|$, (12-55) takes the form

$$R(t) = |\alpha_0|^2 + 2\sum_{n=1}^{\infty} |\alpha_n|^2 \cos n\omega_0 t \qquad (12\text{-}56)$$

To prove (12-54), we form the function

$$f_0(t) = f(t)p_{c/2}(t)$$

as in Fig. 12-8b. With $s_c(t)$ the train of pulses (12-50), $f(t)$ is given by

$$f(t) = \sum_{n=-\infty}^{\infty} f_0(t + nc) = f_0(t) * s_c(t) \qquad (12\text{-}57)$$

The function $f_0(t)$ is obviously of Class I; denoting by $F_0(\omega)$ its Fourier transform, we obtain from (12-44) and (12-52)

$$S(\omega) = \frac{2\pi}{c^2} |F_0(\omega)|^2 \sum_{n=-\infty}^{\infty} \delta(\omega - n\omega_0) \qquad (12\text{-}58)$$

As we know from (3-53), the coefficients α_n of (12-53) are related to $F_0(\omega)$ by

$$\alpha_n = \frac{F_0(n\omega_0)}{c}$$

and since [see (I-21)]

$$|F_0(\omega)|^2\, \delta(\omega - n\omega_0) = |F_0(n\omega_0)|^2\, \delta(\omega - n\omega_0)$$

Eq. (12-54) is proved.
For the special case

$$f(t) = a \cos(\omega_0 t + \phi)$$

we have $$\alpha_1 = \alpha_{-1} = \frac{a}{2}\, e^{j\phi}$$

Hence $$S(\omega) = \frac{\pi a^2}{2}\, [\delta(\omega - \omega_0) + \delta(\omega + \omega_0)] \qquad R(t) = \frac{a^2}{2}\cos\omega_0 t$$

Example 12-8. The impulse response of a linear system (Fig. 12-1) is given by

$$h(t) = e^{-\alpha t} U(t) \qquad \alpha > 0$$

and the autocorrelation of its input equals $R_f(t) = \delta(t)$. We want to find the

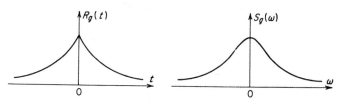

FIGURE 12-10

autocorrelation $R_g(t)$ and the power spectrum $S_g(\omega)$ of the output $g(t)$.
Since $H(\omega) = 1/(\alpha + j\omega)$, we obtain from (12-45)

$$S_g(\omega) = \frac{1}{\alpha^2 + \omega^2}$$

Hence [see (2-62)]

$$R_g(t) = \frac{1}{2\alpha}\, e^{-\alpha|t|}$$

as in Fig. 12-10.
We observe that the input to the above system was specified not in terms of $f(t)$, but only by its autocorrelation. It is actually not easy at all to construct a function with $R_f(t) = \delta(t)$. In Sec. 12-4 a rather elaborate example of a function with *flat* power spectrum will be given.

Correlation of $df(t)/dt$. We shall prove that if $R(t)$ is the autocorrelation of a differentiable function $f(t)$, and $S(\omega)$ its power spectrum, then the autocorrelation $R_{f'}(t)$ and power spectrum $S_{f'}(\omega)$ of its derivative $df(t)/dt$ are given by

$$R_{f'}(t) = -\frac{d^2R(t)}{dt^2} \qquad (12\text{-}59)$$

$$S_{f'}(\omega) = \omega^2 S(\omega) \qquad (12\text{-}60)$$

To prove (12-60), we write df/dt as a convolution [see (I-24)]

$$\frac{df(t)}{dt} = -\int_{-\infty}^{\infty} f(t-\tau)\frac{d\delta(\tau)}{d\tau}\,d\tau = -f(t)*\frac{d\delta(t)}{dt}$$

and apply (12-44). Since [see (3-32)]

$$\delta'(t) \leftrightarrow j\omega$$

and the power spectrum of $-f(t)$ equals $S(\omega)$, we obtain

$$S_{f'}(\omega) = S(\omega)\,|j\omega|^2 = \omega^2 S(\omega)$$

and (12-60) is proved; (12-59) follows from (12-60) and (2-41). By repeated application of (12-59) we conclude that the autocorrelation of $d^n f(t)/dt^n$ is given by $(-1)^n d^{2n}R(t)/dt^{2n}$.

Cross-correlation. Given two real functions $f_1(t)$ and $f_2(t)$ of Class II, we define their *cross-correlation* $R_{12}(t)$ by

$$R_{12}(t) = \lim \frac{1}{2T}\int_{-T}^{T} f_1(\tau)f_2(t+\tau)\,d\tau \qquad (12\text{-}61)$$

$R_{12}(t)$ is not, in general, even but

$$R_{12}(-t) = R_{21}(t) \qquad (12\text{-}62)$$

With $R_1(t)$ and $R_2(t)$ the autocorrelations of $f_1(t)$ and $f_2(t)$, the average power of

$$f_1(t) \pm f_2(t+t_1)$$

is given by

$$\overline{|f_1(t) \pm f_2(t+t_1)|^2} = R_1(0) + R_2(0) \pm 2R_{12}(t) \qquad (12\text{-}63)$$

as we can easily see from (12-61) and (12-27). From Schwarz's inequality we obtain as in (12-23)

$$|R_{12}(t)| \leq \sqrt{R_1(0)R_2(0)} \leq \frac{R_1(0)+R_2(0)}{2} \qquad (12\text{-}64)$$

The Fourier transform $S_{12}(\omega)$ of $R_{12}(t)$

$$R_{12}(t) \leftrightarrow S_{12}(\omega) \qquad (12\text{-}65)$$

is known as the *cross-power spectrum* of $f_1(t)$ and $f_2(t)$. From (12-61) and (12-65) we have

$$R_{12}(0) = \frac{1}{2\pi}\int_{-\infty}^{\infty} S_{12}(\omega)\,d\omega = \overline{f_1(t)f_2(t)}$$

Thus, if $f_1(t)$ is the current through a 1-ohm resistor and $f_2(t)$ the voltage across its terminals, then the average dissipated power equals the area of $S_{12}(\omega)/2\pi$.

The cross-power spectrum can be written in terms of the function $f(t)$:

$$S_{12}(\omega) = \lim_{T \to \infty} \frac{1}{2T} \int_{-T}^{T} f_1(t)e^{j\omega t}\, dt \int_{-T}^{T} f_2(t)e^{-j\omega t}\, dt \qquad (12\text{-}66)$$

The proof of (12-66) is similar to the proof of (12-33).

Consider now complex time signals. The autocorrelation of $f(t)$ is defined by

$$R(t) = \lim_{T \to \infty} \frac{1}{2T} \int_{-T}^{T} f(t + \tau)\overset{*}{f}(\tau)\, d\tau \qquad (12\text{-}67)$$

and the cross-correlation of $f_1(t)$ and $f_2(t)$ by

$$R_{12}(t) = \lim_{T \to \infty} \frac{1}{2T} \int_{-T}^{T} \overset{*}{f_1}(\tau) f_2(t + \tau)\, d\tau \qquad (12\text{-}68)$$

The corresponding power spectra are the Fourier transforms of $R(t)$ and $R_{12}(t)$ as in the real case

$$R(t) \leftrightarrow S(\omega) \qquad R_{12}(t) \leftrightarrow S_{12}(t) \qquad (12\text{-}69)$$

and are given by

$$S(\omega) = \lim_{T \to \infty} \frac{1}{2T} \left| \int_{-T}^{T} f(\tau)e^{-j\omega \tau}\, d\tau \right|^2 \qquad (12\text{-}70)$$

and

$$S_{12}(\omega) = \lim_{T \to \infty} \frac{1}{2T} \int_{-T}^{T} \overset{*}{f_1}(t)e^{j\omega t}\, dt \int_{-T}^{T} f_2(t)e^{-j\omega t}\, dt \qquad (12\text{-}71)$$

From (12-67) it readily follows that if $R(t)$ is the autocorrelation of $f(t)$ then the autocorrelation of $f(t)e^{-j\omega_0 t}$ is given by $R(t)e^{-j\omega_0 t}$.

Example 12-9. We shall find the cross-correlation of the functions

$$f_1(t) = a_0 e^{j\omega_0 t} \qquad f_2(t) = a_1 e^{j\omega_1 t}$$

Since

$$\frac{1}{2T} \int_{-T}^{T} \overset{*}{f_1}(\tau) f_2(t + \tau)\, d\tau = \frac{1}{2T} \int_{-T}^{T} \overset{*}{a_0} a_1 e^{j\omega_1 t} e^{j(\omega_1 - \omega_0)\tau}\, d\tau$$

we conclude from (12-68) that

$$R_{12}(t) = \lim_{T \to \infty} \frac{\overset{*}{a_0} a_1}{T} e^{j\omega_1 t} \frac{\sin (\omega_1 - \omega_0)T}{j(\omega_1 - \omega_0)}$$

$$\qquad (12\text{-}72)$$

$$= \begin{cases} 0 & \text{for } \omega_1 \neq \omega_0 \\ \overset{*}{a_0} a_1 e^{j\omega_0 t} & \text{for } \omega_1 = \omega_0 \end{cases}$$

Example 12-10. Consider the periodic functions

$$f_1(t) = \sum_{n=-\infty}^{\infty} \alpha_n e^{jn\omega_0 t} \qquad f_2(t) = \sum_{n=-\infty}^{\infty} \beta_n e^{jn\omega_1 t}$$

of periods $T_0 = 2\pi/\omega_0$ and $T_1 = 2\pi/\omega_1$. It can be easily seen from (12-72) that if ω_1/ω_0 is irrational, then

$$R_{12}(t) = 0$$

if $\omega_1 = \omega_0$, then

$$R_{12}(t) = \sum_{n=-\infty}^{\infty} \overset{*}{\alpha}_n \beta_n e^{jn\omega_0 t}$$

12-4. Functions with Arbitrary Power Spectra

In the preceding section we considered functions of Class II, but the only examples given had power spectra consisting of impulses. One might wonder whether it is possible to find time signals having other kinds of power spectra. Such signals appear in probability theory, but they are specified only statistically, and not by a definite rule. In this section we shall show that it is indeed possible, although by no means simple, to construct a signal $f(t)$, having as power spectrum a positive function $S(\omega)$. We shall first determine a function with a *flat* power spectrum

$$S(\omega) = 1 \qquad (12\text{-}73)$$

This can be done as follows. Suppose that a sequence of real numbers α_m can be found satisfying

$$\lim_{n\to\infty} \frac{1}{2n} \sum_{m=-n}^{n} \alpha_m{}^2 = 1$$

$$ \qquad (12\text{-}74)$$

$$\lim_{n\to\infty} \frac{1}{2n} \sum_{m=-n}^{n} \alpha_m \alpha_{k+m} = 0 \qquad \text{for } k \neq 0$$

These numbers might be called *uncorrelated*. With α_n as in (12-74), we form the function

$$w(t) = \sum_{k=-\infty}^{\infty} \alpha_k \delta(t-k) \qquad (12\text{-}75)$$

From (12-47) and (12-48) it follows that the autocorrelation $R_w(t)$ of the above sequence of uncorrelated impulses is given by

$$R_w(t) = \delta(t) \qquad (12\text{-}76)$$

Therefore its power spectrum $S_w(\omega)$ is flat

$$S_w(\omega) = 1 \qquad (12\text{-}77)$$

At the end of the section we shall construct a sequence of numbers satisfying (12-74); the corresponding function $w(t)$ is shown in Fig. 12-11.

We turn now to the general case. Given an arbitrary positive function $S(\omega)$, we can find, with the techniques of Sec. 10-3, a function $H(\omega)$ such that

$$|H(\omega)|^2 = S(\omega) \qquad (12\text{-}78)$$

We denote by $h(t)$ the inverse Fourier transform of $H(\omega)$

$$h(t) \leftrightarrow H(\omega) \qquad (12\text{-}79)$$

and form the function

$$g(t) = w(t) * h(t) = \sum_{k=-\infty}^{\infty} \alpha_k h(t - k) \qquad (12\text{-}80)$$

$g(t)$ is the response of the system $H(\omega)$ to the input $w(t)$. From (12-44)

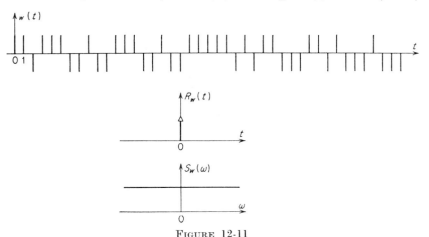

Figure 12-11

it follows that the power spectrum $S_g(\omega)$ of $g(t)$ is given by [see (12-78) and (12-77)]

$$S_g(\omega) = S_w(\omega)\, |H(\omega)|^2 = S(\omega) \qquad (12\text{-}81)$$

and our problem is solved.

We observe that $H(\omega)$ is not determined uniquely from (12-78), but is given by

$$H(\omega) = \sqrt{S(\omega)}\, e^{j\phi(\omega)}$$

where $\phi(\omega)$ is arbitrary. If $S(\omega)$ satisfies the Paley-Wiener condition, then $H(\omega)$ can be so chosen as to have a causal inverse.

The determination of a function having as an autocorrelation a positive-definite function $R(t)$ can be similarly achieved; we first determine the Fourier transform $S(\omega)$ of $R(t)$ and then proceed as above.

Example 12-11. We shall determine a function $g(t)$ whose autocorrelation equals the triangular pulse

$$R(t) = q_T(t)$$

of Fig. 12-12. The corresponding power spectrum $S(\omega)$ is given by [see (2-58)]

$$S(\omega) = \frac{4 \sin^2 (\omega T/2)}{\omega^2 T}$$

We first find $H(\omega)$ from

$$|H(\omega)| = \sqrt{S(\omega)} = \left| \frac{2 \sin (\omega T/2)}{\omega \sqrt{T}} \right|$$

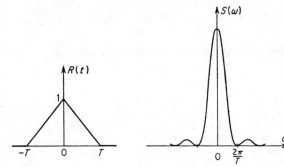

FIGURE 12-12

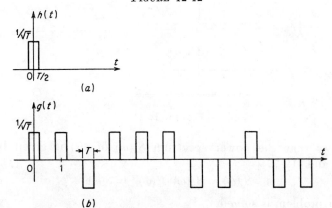

(a)

(b)

FIGURE 12-13

A possible choice is the function

$$H(\omega) = \frac{2 \sin (\omega T/2)}{\omega \sqrt{T}}$$

whose inverse $h(t)$ is the rectangular pulse

$$h(t) = \frac{1}{\sqrt{T}} p_{T/2}(t)$$

of Fig. 12-13a.   The unknown function $g(t)$ is obtained from (12-80)

$$g(t) = \sum_{k=-\infty}^{\infty} \alpha_k h(t - k)$$

with $h(t)$ as above.   It consists of a sequence of rectangular pulses shown in Fig. 12-13b for $T < 1$.

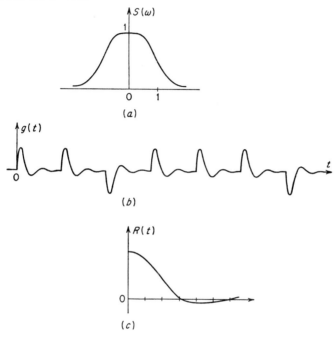

(a)

(b)

(c)

FIGURE 12-14

Example 12-12.   Suppose now that $S(\omega)$ is given by the rational function

$$S(\omega) = \frac{1}{1 + \omega^4}$$

of Fig. 12-14a.   To determine the corresponding $g(t)$, we shall first find a function $H(\omega)$ satisfying (12-78)

$$|H(\omega)| = \frac{1}{\sqrt{1 + \omega^4}}$$

and having a causal inverse.   As in Example 10-8, we obtain

$$H_1(p) = \frac{1}{p^2 + \sqrt{2p} + 1}$$

Hence

$$h(t) = \sqrt{2} e^{-t/\sqrt{2}} U(t) \sin \frac{t}{\sqrt{2}}$$

Inserting $h(t)$ into (12-80), we obtain the unknown function $g(t)$ shown in Fig. 12-14b. Its autocorrelation $R(t)$ is given by (Fig. 12-14c)

$$R(t) = \tfrac{1}{2}e^{-|t|/\sqrt{2}}\cos\left(\frac{|t|}{\sqrt{2}} - \frac{\pi}{4}\right)$$

as one can easily see with the techniques of Sec. 9-6.

The Wiener numbers.[†]   We shall now give a sequence of uncorrelated numbers developed by Wiener; these are not the only numbers satisfying (12-74), and we challenge the reader to construct a different sequence. We shall first put down the beginning of the sequence α_n and then describe the general law of their formation.

$$
\begin{array}{ccccc}
\overset{1}{\underline{1,\,1}} & \underline{1,\,-1} & & \underline{-1,\,1} & \overset{-1}{\underline{-1,\,-1}} \\
\end{array}
$$
repeated twice

$$\underline{1,\,1,\,1,} \;\; \underline{1,\,1,\,-1,} \;\; \underline{1,\,-1,\,1,} \;\; \underline{1,\,-1,\,-1;} \quad \underline{-1,\,1,\,1,} \;\; \underline{-1,\,1,\,-1,} \;\; \underline{-1,\,-1,\,1,} \;\; \underline{-1,\,-1,\,-1}$$
repeated four times

$$\underline{1,\,1,\,1,\,1,} \;\; \underline{1,\,1,\,1,\,-1,}\,\text{etc.}$$
repeated eight times
etc.

The general rule will become clear if we explain how the numbers of the third row were obtained. The second row contains four underlined groups of numbers; in front of each group we put 1, then -1; thus the two halves of the third row are formed. We repeat these 24 numbers four times. Putting 1 and -1 in front of each underlined group, we form the 72 numbers of the fourth row, and so we proceed. The sequence of numbers so constructed is denoted by α_n for $n = 1, 2, 3, \ldots$. To complete the definition of α_n for negative n, we assume

$$\alpha_{-n} = \alpha_n \qquad n = 1, 2, \ldots$$

and $\alpha_0 = 1$. It can be shown that the above numbers α_n are uncorrelated; i.e., they satisfy (12-74).

Note. It should be pointed out that, unlike the deterministic case, the construction of a random process with an arbitrary power spectrum $S(\omega)$ is simple. Indeed, consider the process

$$x(t) = ae^{j\omega t}$$

where a is a constant and ω a random variable with density $p(\omega)$. Since the

[†] N. Wiener, The Spectrum of an Array and Its Application to the Study of the Translation Properties of a Simple Class of Arithmetical Functions, *J. Math. and Phys.*, vol. 6, 1926–1927.

autocorrelation of $x(t)$ is given by

$$R(t) = E\{x(t + \tau)\overset{*}{x}(\tau)\} = |a|^2\, E\{e^{j\omega t}\} = |a|^2 \int_{-\infty}^{\infty} e^{j\omega t} p(\omega)\, d\omega$$

we conclude that if a and $p(\omega)$ are such that

$$|a|^2 = \frac{1}{2\pi} \int_{-\infty}^{\infty} S(\omega)\, d\omega, \qquad p(\omega) = \frac{S(\omega)}{2\pi\,|a|^2}$$

then the Fourier transform of $R(t)$, i.e., the power spectrum of $x(t)$, equals the given function $S(\omega)$. The same spectrum $S(\omega)$ can also be obtained with the real process $x(t) = a \cos(\omega t + \phi)$, where ϕ is a random variable uniformly distributed in the $(0,2\pi)$ interval, ω a random variable with density $p(\omega) = S(\omega)/\pi a^2$, and a a constant given by $a = \dfrac{1}{\pi} \int_{-\infty}^{\infty} S(\omega)\, d\omega$. The proof is similar.

From the above we draw the interesting conclusion that in all applications involving only autocorrelations (as, for example, in linear least-squares prediction) an arbitrary random process can be replaced by $ae^{j\omega t}$ or $a \cos(\omega t + \phi)$, provided a, ω, and ϕ are chosen as above.

12-5. Generalized Harmonic Analysis

From the preceding discussion we see that a function $f(t)$ whose power spectrum $S(\omega)$ is not identically zero cannot, in general, have a Fourier integral. In the following we shall show that these functions have an *integrated* transform which is related to their autocorrelation and power spectrum.

Consider first a function $f(t)$ having a Fourier transform $F(\omega)$. We define the *generalized* Fourier transform $G_\epsilon(\omega)$ of $f(t)$ by the integral

$$G_\epsilon(\omega) = \int_{\omega - \epsilon}^{\omega + \epsilon} F(y)\, dy \qquad (12\text{-}82)$$

Example 12-13. The function $f(t)$ is periodic, as in Fig. 12-15a,

$$f(t) = \sum_{n=-\infty}^{\infty} \alpha_n e^{jn\omega_0 t}$$

Its Fourier transform is given by the sequence of impulses

$$F(\omega) = \sum_{n=-\infty}^{\infty} \alpha_n \delta(\omega - n\omega_0)$$

of Fig. 12-15b. From (12-82) we easily find

$$G_\epsilon(\omega) = \sum_{n=-\infty}^{\infty} \alpha_n p_\epsilon(\omega)$$

as shown in Fig. 12-15c, for $\epsilon < \omega_0/2$ and α_n real.

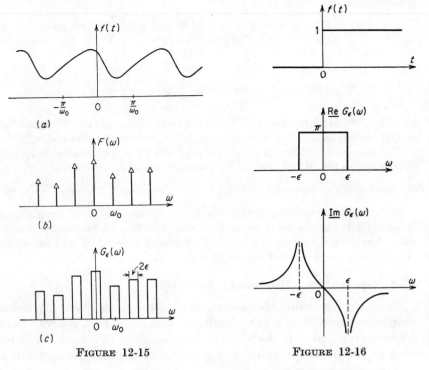

FIGURE 12-15   FIGURE 12-16

Example 12-14.   The Fourier integral of the unit step function

$$f(t) = U(t)$$

is given by

$$F(\omega) = \pi \, \delta(\omega) + \frac{1}{j\omega}$$

and its generalized transform by

$$G_\epsilon(\omega) = \pi p_\epsilon(\omega) + \frac{j}{\epsilon} \ln \left| \frac{\omega - \epsilon}{\omega + \epsilon} \right|$$

as we can easily see from (12-82).   The real and imaginary parts of $G_\epsilon(\omega)$ are shown in Fig. 12-16.

The function G_ϵ can be written directly in terms of $f(t)$.   Indeed, from (12-82) it follows that

$$G_\epsilon(\omega) = F(\omega) * p_\epsilon(\omega) \tag{12-83}$$

where $p_\epsilon(\omega)$ is a rectangular pulse, and since the inverse transform of $p_\epsilon(\omega)$ is the Fourier kernel $(\sin \epsilon t)/\pi t$ [see (2-57)]

$$\frac{\sin \epsilon t}{\pi t} \leftrightarrow p_\epsilon(\omega) \tag{12-84}$$

we conclude from (12-83) and (2-74) that $G_\epsilon(\omega)$ is the Fourier transform of $f(t)(2 \sin \epsilon t)/t$

$$\frac{2 \sin \epsilon t}{t} f(t) \leftrightarrow G_\epsilon(\omega) \qquad (12\text{-}85)$$

We turn now to the general case. If a function $f(t)$ is of Class II, then, for every $\epsilon \neq 0$, the function $f(t)(2 \sin \epsilon t)/t$ has finite energy; this follows easily from (12-1). Therefore the integral

$$G_\epsilon(\omega) = \int_{-\infty}^{\infty} \frac{2 \sin \epsilon t}{t} f(t) e^{-j\omega t}\, dt \qquad (12\text{-}86)$$

exists and defines the generalized Fourier transform of $f(t)$, as in (12-85).

Note. In the spectral analysis of random processes, the integrated transform $G(\omega)$ of a function $f(t)$ is defined by

$$G(\omega_2) - G(\omega_1) = \int_{-\infty}^{\infty} \frac{e^{-j\omega_2 t} - e^{-j\omega_1 t}}{-jt} f(t)\, dt$$

The above stochastic integral determines a random process $G(\omega)$ within a constant, and from (12-86) we see that $G_\epsilon(\omega) = G(\omega + \epsilon) - G(\omega - \epsilon)$. Consider a bandpass system with system function $H(\omega) = 2\pi[U(\omega_1 - \omega) - U(\omega_2 - \omega)]$. Its impulse response is given by

$$h(t) = \frac{e^{j\omega_2 t} - e^{j\omega_1 t}}{jt}$$

Therefore, with $f(t)$ as input, the output $g(t)$, evaluated as $t = 0$, satisfies

$$g(0) = \int_{-\infty}^{\infty} \frac{e^{-j\omega_2 t} - e^{-j\omega_1 t}}{-jt} f(t)\, dt$$

as we can easily see. Thus the output $g(0)$ equals the increment $G(\omega_2) - G(\omega_1)$ of the integrated transform $G(\omega)$ of the input $f(t)$.

Relationship between $G_\epsilon(\omega)$ and $S(\omega)$. If the real function $f(t)$ has a Fourier transform $F(\omega)$, then

$$\lim_{\epsilon \to 0} \frac{G_\epsilon(\omega)}{2\epsilon} = F(\omega) \qquad (12\text{-}87)$$

as we can easily see from (12-82); however, for the function of Class II, the limit in (12-87) is infinite. We shall show that $G_\epsilon(\omega)$ is related to the power spectrum $S(\omega)$ of $f(t)$ by the following equation:

$$\lim_{\epsilon \to 0} \frac{|G_\epsilon(\omega)|^2}{4\pi\epsilon} = S(\omega) \qquad (12\text{-}88)$$

or its equivalent [see (12-32)]

$$R(t) = \lim_{\epsilon \to 0} \frac{1}{2\pi} \int_{-\infty}^{\infty} \frac{|G_\epsilon(\omega)|^2}{4\pi\epsilon} \cos \omega t \, d\omega \qquad (12\text{-}89)$$

The proof of (12-89) is based on the equation

$$\lim_{T \to \infty} \frac{1}{2T} \int_{-T}^{T} f^2(t) \, dt = \lim_{\epsilon \to 0} \frac{1}{\pi\epsilon} \int_{-\infty}^{\infty} f^2(t) \frac{\sin^2 \epsilon t}{t^2} \, dt \qquad (12\text{-}90)$$

due to Wiener; the validity of this equation will be established at the end of the section.

From (12-85) and Parseval's formula (2-75) we have

$$\int_{-\infty}^{\infty} f^2(t) \frac{\sin^2 \epsilon t}{t^2} \, dt = \frac{1}{2\pi} \int_{-\infty}^{\infty} \frac{|G_\epsilon(\omega)|^2}{4} \, d\omega \qquad (12\text{-}91)$$

and from (12-27)

$$R(0) = \lim_{T \to \infty} \frac{1}{2T} \int_{-T}^{T} f^2(t) \, dt \qquad (12\text{-}92)$$

Therefore

$$R(0) = \lim_{\epsilon \to 0} \frac{1}{2\pi} \int_{-\infty}^{\infty} \frac{|G_\epsilon(\omega)|^2}{4\pi\epsilon} \, d\omega \qquad (12\text{-}93)$$

as we see from Eqs. (12-90) to (12-92); thus (12-89) is proved for $t = 0$.

To prove (12-89) for an arbitrary $t = t_1$, we use the following relatively simple argument. From (12-85) and (2-36) we have

$$\frac{2f(t + t_1) \sin \epsilon(t + t_1)}{t + t_1} \leftrightarrow G_\epsilon(\omega)e^{j\omega t_1} \qquad (12\text{-}94)$$

But the Fourier integral of

$$f(t)\left[\frac{\sin \epsilon(t + t_1)}{t + t_1} - \frac{\sin \epsilon t}{t}\right]$$

tends to zero with $\epsilon \to 0$, as we can easily see; therefore, for sufficiently small ϵ, (12-94) can be written in the form

$$\frac{2f(t + t_1) \sin \epsilon t}{t} \leftrightarrow \simeq G_\epsilon(\omega)e^{j\omega t_1} \qquad (12\text{-}95)$$

Similarly

$$\frac{2f(t - t_1) \sin \epsilon t}{t} \leftrightarrow \simeq G_\epsilon(\omega)e^{-j\omega t_1} \qquad (12\text{-}96)$$

Adding (12-95) and (12-96), we conclude that, for small enough ϵ, the generalized Fourier transform of

$$\frac{f(t + t_1) + f(t - t_1)}{2}$$

equals $G_\epsilon(\omega) \cos \omega t_1$. But the autocorrelation of

$$[f(t + t_1) + f(t - t_1)]/2$$

equals

$$\frac{2R(t) + R(t + 2t_1) + R(t - 2t_1)}{4}$$

(see Prob. 47). Applying (12-93) to this function, we obtain

$$\frac{2R(0) + R(2t_1) + R(-2t_1)}{4} = \lim_{\epsilon \to 0} \frac{1}{2\pi} \int_{-\infty}^{\infty} \frac{|G_\epsilon(\omega)|^2 \cos^2 \omega t_1}{4\pi\epsilon} \, d\omega \quad (12\text{-}97)$$

Because of $R(-2t_1) = R(2t_1)$ and

$$\cos^2 \omega t_1 = \frac{1 + \cos 2\omega t_1}{2}$$

we finally obtain, with the help of (12-93),

$$R(2t_1) = \lim_{\epsilon \to 0} \frac{1}{2\pi} \int_{-\infty}^{\infty} \frac{|G_\epsilon(\omega)|^2 \cos 2\omega t_1}{4\pi\epsilon} \, d\omega \quad (12\text{-}98)$$

and since t_1 is arbitrary, (12-89) is proved.

We come lastly to the proof, Wiener's equation (12-90). If we succeed to prove that

$$\lim_{T \to \infty} \frac{1}{T} \int_0^T f^2(t) \, dt = \lim_{\epsilon \to 0} \frac{2}{\pi\epsilon} \int_0^\infty f^2(t) \frac{\sin^2 \epsilon t}{t^2} \, dt$$

$$= \lim_{\epsilon \to 0} \frac{2}{\pi} \int_0^\infty f^2\left(\frac{x}{\epsilon}\right) \frac{\sin^2 x}{x^2} \, dx \quad (12\text{-}99)$$

then (12-90) will follow, because the same equation with upper limits $-T$ and $-\infty$ can be similarly proved. It is known that (see Example 2-7)

$$\frac{2}{\pi} \int_0^\infty \frac{\sin^2 x}{x^2} \, dx = 1$$

Hence, with A given by

$$\lim_{T \to \infty} \frac{1}{T} \int_0^T f^2(t) \, dt = A \quad (12\text{-}100)$$

(12-99) is reduced to

$$\lim_{\epsilon \to 0} \int_0^\infty \left[f^2\left(\frac{x}{\epsilon}\right) - A \right] \frac{\sin^2 x}{x^2} \, dx = 0 \quad (12\text{-}101)$$

It is easy to see from (12-100) that if T_1 is sufficiently large, then

$$\int_T^{T + T_1} f^2(t) \, dt \simeq A T_1 \quad (12\text{-}102)$$

We now select $x_2 - x_1$ so small that the function $(\sin^2 x)/x^2$ is approximately constant in the interval (x_1,x_2), so that

$$\int_{x_1}^{x_2}\left[f^2\left(\frac{x}{\epsilon}\right) - A\right]\frac{\sin^2 x}{x^2}\, dx \simeq \frac{\epsilon \sin^2 x_1}{x_1^2}\int_{x_1/\epsilon}^{x_2/\epsilon}[f^2(t) - A]\, dt$$

With x_1 and x_2 fixed, we make the interval $(x_2 - x_1)/\epsilon$ large, by properly choosing ϵ; it then follows from (12-102) that the last integral above tends to zero with ϵ, and since (12-101) can be written as a sum of such integrals, the validity of (12-90) is somehow established!

Problems

40. Show that if $f(t) \geq 0$, then the real part $R(\omega)$ of its Fourier integral satisfies

$$R(0) - R(\omega) \geq \tfrac{1}{4}[R(0) - R(2\omega)] \tag{i}$$

$$R(0) - R(\omega) \geq \frac{1}{4^n}[R(0) - R(2^n\omega)] \tag{ii}$$

41. Show that if $f(t) \geq 0$, then its Fourier spectrum $A(\omega)$ satisfies

$$A^2(0) - A^2(\omega) \geq \frac{1}{4^n}[A^2(0) - A^2(2^n\omega)] \tag{i}$$

Application. If the impulse response $f(t)$ of a low-pass filter is positive and the variation of its energy spectrum $A^2(\omega)$ in the passband equals ϵ, then the attenuation in the next octave cannot exceed 4ϵ:
If

$$A^2(0) - A^2(\omega) \leq \epsilon \qquad \text{for } |\omega| < \omega_c$$

then

$$A^2(0) - A^2(\omega) \leq 4\epsilon \qquad \text{for } |\omega| < 2\omega_c$$

42. An amplifier contains n identical stages (Fig. P-42). The input is a rectangular pulse $f(t) = E[U(t) - U(t - T)]$, and the impulse response of each

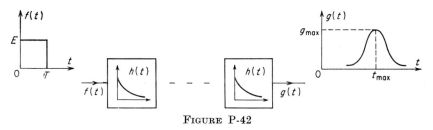

FIGURE P-42

stage is given by $h(t) = Ae^{-\alpha t}U(t)$. Using the central-limit theorem, find the maximum value of the output $g(t)$ and the corresponding time t_{max}.

43. Show that if $\psi(t)$ is a positive function, a a real constant, and

$$\lim_{T \to \infty} \frac{1}{2T} \int_{-T}^{T} \psi(t)\, dt = M$$

265

then

$$\lim_{T \to \infty} \frac{1}{2T} \int_{-T+a}^{T+a} \psi(t)\, dt = M$$

44. Show that if the autocorrelation $R(t)$ of a given function $f(t)$ is continuous for $t = 0$

$$\lim_{\Delta t \to 0} R(\Delta t) = R(0)$$

then it is continuous for every t:

$$\lim_{\Delta t \to 0} R(t + \Delta t) = R(t) \tag{i}$$

45. Show that the autocorrelation of $f(t) = e^{jt^2}$ is given by $R(0) = 1$, $R(t) = 0$ for $t \neq 0$. Thus, although $f(t)$ is of Class II, its power spectrum is identically zero.

46. Show that the autocorrelation and power spectrum of $f(t) = e^{j\sqrt{|t|}}$ are given by $R(\omega) = 1$, $S(\omega) = 2\pi\delta(\omega)$.

47. Show that if the autocorrelation of a function $f(t)$ is $R(t)$ and its power spectrum $S(\omega)$, then the autocorrelation and power spectrum of

$$\phi(t) = f(t + t_1) + f(t - t_1)$$

are given by

$$R_\phi(t) = 2R(t) + R(t + 2t_1) + R(t - 2t_1) \tag{i}$$

$$S_\phi(\omega) = 4S(\omega) \cos^2 \omega t_1 \tag{ii}$$

48. (*Thermal noise.*) The network of Fig. P-48a is passive lossless. The power spectrum $S_e(\omega)$ of the source $e(t)$ is given by $S_e(\omega) = KR$ where K is a

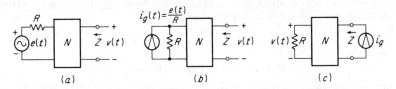

$$(a) \qquad\qquad (b) \qquad\qquad (c)$$

FIGURE P-48

constant (in the case of thermal noise, $K = 2kT$ where k is the Boltzmann constant and T the absolute temperature of the resistance R).

(a) Show that the power spectrum $S_v(\omega)$ of the output $v(t)$ is given by

$$S_v(\omega) = K \operatorname{Re} Z(j\omega) \tag{i}$$

(b) Prove that if the input capacity of Z equals C, then the average power of $v(t)$ equals

$$\overline{v^2}(t) = \frac{1}{2\pi} \int_{-\infty}^{\infty} S_v(\omega)\, d\omega = \frac{K}{2C} \tag{ii}$$

Solutions

40. (i) From

$$1 - \cos \omega t = 2 \sin^2 \frac{\omega t}{2} \geq 2 \sin^2 \frac{\omega t}{2} \cos^2 \frac{\omega t}{2} = \frac{1}{2} \sin^2 \omega t = \frac{1}{4}(1 - \cos 2\omega t)$$

we have

$$R(0) - R(\omega) = \int_{-\infty}^{\infty} f(t)(1 - \cos \omega t) \, dt$$

$$\geq \frac{1}{4} \int_{-\infty}^{\infty} f(t)(1 - \cos 2\omega t) dt = \frac{1}{4} [R(0) - R(2\omega)]$$

(ii) follows by repeated application of (i).

41. With $\phi(t) = f(t) * f(-t)$ we have $\phi(t) \geq 0$ and

$$\phi(t) \longleftrightarrow \overset{*}{F}(\omega)F(\omega) = A^2(\omega)$$

Hence (i) follows from Prob. 40.

42. The mean and dispersion of $f(t)/ET$ are given by $T/2$ and $T^2/12$ and those of $\alpha h(t)/A$ by $1/\alpha$ and $1/\alpha^2$; hence [see (11-17)]

$$t_{\max} \simeq \frac{T}{2} + \frac{n}{\alpha} \qquad g_{\max} \simeq \frac{ETA^n}{\sqrt{2\pi\sigma\alpha^n}}$$

where

$$\sigma^2 = \frac{T^2}{12} + \frac{n}{\alpha^2}$$

The above is approximate. The quantity $T/2 + n/\alpha$ is the exact value of the center of gravity of $g(t)$.

43. For $T > |a|$ we have, since $\psi(t) \geq 0$,

$$\left(1 - \frac{|a|}{T}\right) \frac{1}{2(T - |a|)} \int_{-T+|a|}^{T-|a|} \psi(t) \, dt \leq \frac{1}{2T} \int_{-T+a}^{T+a} \psi(t) \, dt$$

$$\leq \left(1 + \frac{|a|}{T}\right) \frac{1}{2(T + |a|)} \int_{-T-|a|}^{T+|a|} \psi(t) \, dt$$

Since

$$\lim_{T\to\infty} \left(1 \pm \frac{|a|}{T}\right) = 1$$

the first and the last of the above three expressions tend to M with $T \to \infty$; hence the second expression tends also to M.

44. We shall prove (i) for $f(t)$ real; the proof is similar if $f(t)$ is complex. We have

$$|R(t + \Delta t) - R(t)| = \lim_{T\to\infty} \left| \frac{1}{2T} \int_{-T}^{T} [f(t + \Delta t + \tau) - f(t + \tau)] f(\tau) \, d\tau \right|$$

From Schwarz's inequality [see (4-47)] it follows that

$$|R(t + \Delta t) - R(t)|$$

$$\leq \lim_{T-\infty} \frac{1}{2T} \left[\int_{-T}^{T} |f(t) + \Delta t + \tau) - f(t + \tau)|^2 \, d\tau \int_{-T}^{T} |f(\tau)|^2 \, d\tau \right]^{1/2} \quad \text{(ii)}$$

and from Prob. 43

$$\lim_{T\to\infty} \frac{1}{2T} \int_{-T}^{T} |f(t + \Delta t + \tau)]^2 \, d\tau = \lim_{T\to\infty} \frac{1}{2T} \int_{-T}^{T} f(t + \tau)^2 \, d\tau = R(0)$$

$$\lim_{T\to\infty} \frac{1}{2T} \int_{-T}^{T} f(t + \Delta t + \tau) f(t + \tau) \, d\tau = R(\Delta t)$$

Hence, expanding the square in (ii), we have

$$|R(t + \Delta t) - R(t)| \leq \sqrt{R(0)\,[2R(\Delta t) - 2R(0)]} \to 0$$
$$\Delta t \to 0$$

45. It is easy to see that

$$\frac{1}{2T} \int_{-T}^{T} f(t + \tau)\dot{f}(\tau)\,d\tau = \frac{e^{jt^2}}{2T} \int_{-T}^{T} e^{2jt\tau}\,d\tau$$

The right-hand side equals one for $t = 0$ and for $t \neq 0$ tends to zero as $T \to \infty$.

46. Since $f(t)$ is even, it suffices to find the limit of

$$\frac{1}{T} \int_{0}^{T} f(t + \tau)\dot{f}(\tau)\,d\tau = \frac{1}{T} \int_{0}^{T} e^{j\sqrt{t+\tau}}\,e^{-j\sqrt{\tau}}\,d\tau = \frac{1}{T} \int_{0}^{T} e^{j[t/(\sqrt{t+\tau}+\sqrt{\tau})]}\,d\tau$$

as $T \to 0$ and $t > 0$. For a fixed t the last integral tends to one as $\tau \to \infty$. And since the limit of the above integral is determined only from large values of τ, it follows easily that $R(t) = 1$.

47. The function $f(t + t_1) + f(t - t_1)$ can be considered as the output of a system with input $f(t)$ and system function

$$H(\omega) = e^{j\omega t_1} + e^{-j\omega t_1} = 2\cos \omega t_1$$

Since

$$|H(\omega)|^2 = 4\cos^2 \omega t_1$$

we conclude from (12-44) that

$$S_\phi(\omega) = 4S(\omega)\cos^2 \omega t_1 = (2 + e^{j2\omega t_1} + e^{-j2\omega t_1})\,S(\omega)$$

and (ii) follows; (i) is a consequence of (ii).

48. (a) Since [see (12-44)]

$$S_v(\omega) = S_e(\omega) \left|\frac{V(j\omega)}{E(j\omega)}\right|^2$$

it suffices to show that

$$\operatorname{Re} Z(j\omega) = R \left|\frac{V(j\omega)}{E(j\omega)}\right|^2 \tag{iii}$$

We change the voltage into a current source (Fig. P-48b) and apply the reciprocity theorem (Fig. P-48c). Equation (iii) follows by equating the power input to the power dissipated in R (see also Prob. 34).

(b) From the resistance-integral theorem (10-2) we have

$$\frac{1}{\pi} \int_{-\infty}^{\infty} \operatorname{Re} Z(j\omega)\,d\omega = \frac{1}{C}$$

and since the average power of $v(t)$ equals $R_v(0)$, where $R_v(t)$ is the autocorrelation of $v(t)$, (ii) follows from (i) and (12-32).

Appendices

Appendix I. The Impulse Function as Distribution

The impulse function $\delta(t)$ is an important tool in applied mathematics. It simplifies the derivation of many results that would otherwise involve complicated arguments. Its treatment in the technical literature, however, is basically unsound. In fact, the usual definition (I-1) is meaningless; nevertheless, the concept is correctly applied in the solution of many problems. The reason is that only certain properties of $\delta(t)$ are significant, and these properties are assumed independently of the definition of $\delta(t)$, although an attempt is often made to "derive" them from (I-1). The difficulties involving the delta function can be eliminated if one formally recognizes the fact that $\delta(t)$ is not an ordinary function, having definite values for every t, but a new concept, specified by its properties. This is done in mathematics in a relatively new discipline known as theory of distributions.†

The concept of distribution is seldom introduced in applied books, partly because of the difficulty of the general theory, partly because of the reluctance of the applied scientist to accept the description of a physical quantity by a concept that is not an ordinary function, but is specified by certain properties of integral nature. In the following short discussion we shall attempt to show that the above objections are not justified. A simple but, for the ordinary applications, adequate and logically self-consistent theory can be developed. Furthermore, one can, indeed, characterize a physical quantity F by

† M. J. Lighthill, "An Introduction to Fourier Analysis and Generalized Function," Cambridge University Press, New York, 1959; B. Friedman, "Principles and Techniques of Applied Mathematics," John Wiley & Sons, Inc., New York, 1956; I. Halperin, "Introduction to the Theory of Distributions," based on lectures given by Laurent Schwartz, University of Toronto Press, Toronto, 1952.

269

a distribution; in fact, it is the direct way to do so. The universal assumption that F has a meaning only if it is viewed as an ordinary function $f(t)$ is, without a doubt, a very convenient idealization; however, the fact is that no physical quantity has a priori meaning revealed by $f(t)$. It exists only because of the totality of the effects that it causes. It is, therefore, natural to define it through them, and this leads to the concept of distribution, as we shall show in Sec. I-3.

I-1. Definitions

The delta function is usually defined in one of the following three ways:

A. By the equation

$$\int_{-\infty}^{\infty} \delta(t)\, dt = 1 \qquad \delta(t) = 0 \qquad \text{for } t \neq 0 \tag{I-1}$$

B. As a limit

$$\delta(t) = \lim_{n \to \infty} f_n(t) \tag{I-2}$$

of a sequence of functions satisfying

$$\int_{-\infty}^{\infty} f_n(t)\, dt = 1 \qquad \lim_{n \to \infty} f_n(t) = 0 \qquad \text{for } t \neq 0 \tag{I-3}$$

C. By the property

$$\int_{-\infty}^{\infty} \delta(t) f(t)\, dt = f(0) \tag{I-4}$$

where $f(t)$ is an arbitrary function, continuous at the origin.

The above, or any other definition of $\delta(t)$, is meaningless if one insists on viewing $\delta(t)$ as an ordinary function. The difficulties are eliminated only if the delta function is introduced as a new concept, a generalized function, as it is sometimes called, characterized by certain properties. Then Definition C can be given a precise meaning; A or B, however, does not uniquely specify $\delta(t)$ because, as we shall see, there are other generalized functions satisfying (I-1) or (I-2), e.g., the function $\delta(t) + \delta'(t)$.

The concept of distribution. A distribution, or generalized function, (or functional) $g(t)$ is a process of assigning to an arbitrary function $\phi(t)$ of a given Class C a number

$$N_g[\phi(t)]$$

This number could be the value of $\phi(t)$ or its derivatives for some $t = t_0$, the area under $\phi(t)$ in some interval, or any other quantity depending on $\phi(t)$. For reasons that will be presently explained, the number $N_g[\phi(t)]$ is written as a definite integral

$$\int_{-\infty}^{\infty} g(t)\phi(t)\, dt = N_g[\phi(t)] \tag{I-5}$$

The delta function is a distribution assigning to the function $\phi(t)$ the number $\phi(0)$; thus

$$\int_{-\infty}^{\infty} \delta(t)\phi(t)\, dt = \phi(0) \tag{I-6}$$

This definition of $\delta(t)$ seems to be the same as (I-4); there is, however, a fundamental difference. For ordinary functions, an integral has an independent meaning, and (I-4) says that the integral of the function $\delta(t)\phi(t)$ equals $\phi(0)$; in (I-6) the integral and the function $\delta(t)$ are merely *defined* by the number $\phi(0)$ assigned to the function $\phi(t)$, and they have no independent meaning.

The Class C will generally contain all functions that have derivatives of any order and tend to zero more rapidly than any power of t, as t tends to infinity. However, for certain distributions, this class can be extended; e.g., for the delta function it suffices to assume that $\phi(t)$ is continuous at the origin. A function of Class C will be called *test function*.

We now come to the explanation of (I-5); we repeat, the left-hand side has no meaning as an integral but is merely defined by the number $N_g[\phi(t)]$ assigned to the function $\phi(t)$ by the distribution $g(t)$. There are two main reasons for expressing formally this number by an integral.

First Reason. Suppose that $f(t)$ is an ordinary function such that the integral

$$\int_{-\infty}^{\infty} f(t)\phi(t)\, dt \tag{I-7}$$

exists for every test function $\phi(t)$. Clearly this integral is a number $I_f[\phi(t)]$ depending on $\phi(t)$. We define the distribution $f(t)$ as the process of assigning to the arbitrary function $\phi(t)$ the number $I_f[\phi(t)]$. Thus an ordinary function will be interpreted as distribution specified by the ordinary integral (I-7). Consider, for example, the function $f(t) = U(t)$. Since

$$\int_{-\infty}^{\infty} U(t)\phi(t)\, dt = \int_{0}^{\infty} \phi(t)\, dt \tag{I-8}$$

the distribution $U(t)$ is the process of assigning to a function $\phi(t)$ a number equal to the area of $\phi(t)$ from zero to infinity.

The notation (I-5) is thus a formal extension of the above process to distributions that have no meaning as ordinary functions.

Second Reason. We shall require that the number $N_g[\phi(t)]$, which from now on will be written as an integral of the form (I-5), has certain properties. These properties are compressed into the integral notation (I-5), if we assume that the formal operations involving integrals are valid.

In the following we give a number of consequences of the above assumption.

The distribution $g(t)$ is linear, i.e.,

$$\int_{-\infty}^{\infty} g(t)[a_1\phi_1(t) + a_2\phi_2(t)]\,dt = a_1\int_{-\infty}^{\infty} g(t)\phi_1(t)\,dt + a_2\int_{-\infty}^{\infty} g(t)\phi_2(t)\,dt \quad \text{(I-9)}$$

The sum

$$g(t) = g_1(t) + g_2(t)$$

of two distributions is defined by

$$\int_{-\infty}^{\infty} g(t)\phi(t)\,dt = \int_{-\infty}^{\infty} g_1(t)\phi(t)\,dt + \int_{-\infty}^{\infty} g_2(t)\phi(t)\,dt \quad \text{(I-10)}$$

The distributions $g(t - t_0)$ and $g(at)$ are given by

$$\int_{-\infty}^{\infty} g(t - t_0)\phi(t)\,dt = \int_{-\infty}^{\infty} g(t)\phi(t + t_0)\,dt \quad \text{(I-11)}$$

and

$$\int_{-\infty}^{\infty} g(at)\phi(t)\,dt = \frac{1}{|a|}\int_{-\infty}^{\infty} g(t)\phi(t/a)\,dt \quad \text{(I-12)}$$

as one can see, with a formal change in the independent variable.

A distribution $g(t)$ is called even (odd) if, for every odd (even) test function $\phi(t)$, we have

$$\int_{-\infty}^{\infty} g(t)\phi(t)\,dt = 0 \quad \text{(I-13)}$$

A product of a distribution $g(t)$ by an ordinary function $f(t)$ is a distribution $g(t)f(t)$ defined by

$$\int_{-\infty}^{\infty} [g(t)f(t)]\phi(t)\,dt = \int_{-\infty}^{\infty} g(t)[f(t)\phi(t)]\,dt \quad \text{(I-14)}$$

consistent with the associative law; this is possible, of course, if $f(t)\phi(t)$ is a test function. The product of two distributions $g_1(t)$ and $g_2(t)$ is, in general, undefined; however, their convolution is given by

$$\int_{-\infty}^{\infty} \left[\int_{-\infty}^{\infty} g_1(\tau)g_2(t - \tau)\,d\tau\right]\phi(t)\,dt$$
$$= \int_{-\infty}^{\infty} g_1(\tau)\left[\int_{-\infty}^{\infty} g_2(t - \tau)\phi(t)\,dt\right]d\tau \quad \text{(I-15)}$$

as we see, by a formal change of the order of integration.

A distribution $g(t)$ cannot be assigned specific values for a given t; however, it can be equated to an ordinary function $f(t)$ in a certain interval (a,b)

$$g(t) = f(t) \qquad a < t < b$$

if for all test functions that are equal to zero outside the interval (a,b)

$$\phi(t) = 0 \qquad \text{for } t > b, t < a$$

we have
$$\int_a^b f(t)\phi(t)\, dt = \int_{-\infty}^{\infty} g(t)\phi(t)\, dt \tag{I-16}$$

One can similarly define equality of two distributions in a given interval.

The derivative $dg(t)/dt$ of a distribution $g(t)$ is defined by

$$\int_{-\infty}^{\infty} \frac{dg(t)}{dt}\, \phi(t)\, dt = -\int_{-\infty}^{\infty} g(t)\, \frac{d\phi(t)}{dt}\, dt \tag{I-17}$$

This definition is consistent with the formal integration by parts

$$\int_{-\infty}^{\infty} \frac{dg(t)}{dt}\, \phi(t)\, dt = g(t)\phi(t) \Big|_{-\infty}^{\infty} - \int_{-\infty}^{\infty} g(t)\, \frac{d\phi}{dt}\, dt = -\int_{-\infty}^{\infty} g(t)\, \frac{d\phi}{dt}\, dt \tag{I-18}$$

and the assumed behavior of $\phi(t)$ at infinity. The same result can be obtained if the ordinary definition

$$\frac{dg(t)}{dt} = \lim_{\epsilon \to 0} \frac{g(t + \epsilon) - g(t)}{\epsilon} \tag{I-19}$$

is formally applied. Indeed, from (I-11) we obtain

$$\int_{-\infty}^{\infty} \frac{g(t + \epsilon) - g(t)}{\epsilon}\, \phi(t)\, dt = \int_{-\infty}^{\infty} g(t)\, \frac{\phi(t - \epsilon) - \phi(t)}{\epsilon}\, dt$$

and, with $\epsilon \to 0$, (I-18) follows. The nth derivative of a distribution can be similarly defined by a repeated application of (I-18)

$$\int_{-\infty}^{\infty} \frac{d^n g(t)}{dt^n}\, \phi(t)\, dt = (-1)^n \int_{-\infty}^{\infty} g(t)\, \frac{d^n \phi}{dt^n}\, dt \tag{I-20}$$

From (I-17) it follows that, if $g(t)$ is even (odd), then $dg(t)/dt$ is odd (even), because, if $\phi(t)$ is odd (even), then $d\phi(t)/dt$ is even (odd); therefore, according to (I-13), the right-hand side of (I-17) equals zero.

The Delta Function. We shall now apply the above results to the delta function. From (I-11) and (I-12) we see that $\delta(t - t_0)$ is defined by

$$\int_{-\infty}^{\infty} \delta(t - t_0)\phi(t)\, dt = \phi(t_0) \tag{I-21}$$

and $\delta(at)$ is given by

$$\delta(at) = \frac{1}{|a|} \delta(t) \tag{I-22}$$

as we see from [see (I-12)]

$$\int_{-\infty}^{\infty} \delta(at)\phi(t)\,dt = \frac{1}{|a|} \int_{-\infty}^{\infty} \delta(t)\phi(t/a)\,dt = \frac{1}{|a|}\,\phi(0)$$

Clearly, if $\phi(t)$ is odd, then, since it is continuous, we must have $\phi(0) = 0$; therefore [see (I-13)] $\delta(t)$ is an even function.

From (I-14) it follows that if $f(t)$ is continuous at $t = 0$, then

$$f(t)\,\delta(t) = f(0)\,\delta(t) \tag{I-23}$$

To prove (I-23), we observe that [see (I-14) and (I-6)]

$$\int_{-\infty}^{\infty} \delta(t)f(t)\phi(t)\,dt = f(0)\phi(0) = \int_{-\infty}^{\infty} \delta(t)f(0)\phi(t)\,dt$$

From (I-23) it follows that

$$t\,\delta(t) = 0$$

Therefore if

$$tg_1(t) = tg_2(t)$$

then

$$g_1(t) = g_2(t) + C\,\delta(t)$$

From (I-17) and (I-20) we see that $\delta'(t)$ is given by

$$\int_{-\infty}^{\infty} \delta'(t)\phi(t)\,dt = -\phi'(0) \tag{I-24}$$

and the nth derivative $\delta^{(n)}(t)$ of $\delta(t)$ by

$$\int_{-\infty}^{\infty} \delta^{(n)}(t)\phi(t)\,dt = (-1)^n \phi^{(n)}(0) \tag{I-25}$$

Reasoning as in (I-23), we obtain

$$f(t)\,\delta'(t) = f(0)\,\delta'(t) - f'(0)\,\delta(t) \tag{I-26}$$

because

$$\int_{-\infty}^{\infty} \delta'(t)f(t)\phi(t)\,dt = -\int_{-\infty}^{\infty} \delta(t)[f(t)\phi(t)]'\,dt$$

$$= -\int_{-\infty}^{\infty} \delta(t)f(t)\phi'(t)\,dt - \int_{-\infty}^{\infty} \delta(t)f'(t)\phi(t)\,dt = -f(0)\phi'(0) - f'(0)\phi(0)$$

and (I-26) follows.   In particular

$$t\,\delta'(t) = -\delta(t)$$

One can similarly show that

$$f(t)\,\delta^{(n)}(t) = \sum_{k=0}^{n} (-1)^k \frac{n!}{k!\,(n-k)!}\,f^{(k)}(0)\,\delta^{(n-k)}(t) \tag{I-27}$$

The proof is left as an exercise.

Although it makes no sense to talk about the value of $\delta(t)$ for a fixed t, the statement

$$\delta(t) = 0 \qquad \text{for } t < 0 \text{ (or } t > 0) \tag{I-28}$$

is correct in the sense of (I-16). Indeed, suppose that $\phi(t)$ equals zero outside the interval $(-\infty, a)$, where $a < 0$, but arbitrary. Then, since $\phi(t)$ is continuous, we must have $\phi(0) = 0$; hence

$$\int_{-\infty}^{\infty} \delta(t)\phi(t)\, dt = \phi(0) = 0$$

Therefore [see (I-16)] $\delta(t)$ equals the function $f(t) = 0$ in the $(-\infty, a)$ interval and (I-28) follows.

The quantity $\delta^2(t)$ is undefined, however,

$$\delta(t - t_1) * \delta(t - t_2) = \int_{-\infty}^{\infty} \delta(\tau - t_1)\,\delta(t - \tau - t_2)\, d\tau = \delta[t - (t_1 + t_2)] \tag{I-29}$$

as we can easily see from (I-15).

We shall finally define the integral of a distribution over a finite or semi-infinite interval. Consider

$$\int_{a}^{b} g(t)\phi(t)\, dt$$

where $g(t)$ is a distribution and $\phi(t)$ an ordinary function. We form the function

$$\phi_1(t) = \begin{cases} \phi(t) & a < t < b \\ 0 & t > b, t < a \end{cases}$$

The above integral is defined by

$$\int_{a}^{b} g(t)\phi(t)\, dt = \int_{-\infty}^{\infty} g(t)\phi_1(t)\, dt$$

The last integral is, of course, the number that the distribution $g(t)$ assigns to the function $\phi_1(t)$. Applying the above to the delta function, we have

$$\int_{a}^{b} \delta(t)\,\phi(t)\, dt = \begin{cases} \phi_1(0) = \phi(0) & a < 0 < b \\ \phi_1(0) = 0 & b > a > 0 \text{ or } b < a < 0 \end{cases}$$

If one of the limits of integration is zero, then the integral is not defined, because $\phi_1(t)$ is in general discontinuous at the origin; however, the integral in the $(0^-, b)$ interval is given by

$$\int_{0^-}^{b} \delta(t)\phi(t)\, dt = \lim_{\substack{\epsilon \to 0 \\ \epsilon > 0}} \int_{-\epsilon}^{b} \delta(t)\phi(t)\, dt = \phi(0)$$

Generalized derivative of an ordinary function. Suppose that the ordinary function $f(t)$ is differentiable everywhere. As we have seen, $f(t)$ can be considered as a distribution defined by the integral (I-7), and its generalized derivative (I-17) equals the ordinary derivative $df(t)/dt$, interpreted as a distribution; the proof is simple. If $f(t)$ is continuous but has no derivatives at a number of points $t_1, t_2, \ldots, t_n, \ldots$, then its generalized derivative (I-17) exists and equals df/dt everywhere except at the points t_n. Suppose, finally, that $f(t)$ is discontinuous; since it can be written as a sum of a continuous function and a number of steps, it suffices to consider the derivative of $U(t)$. We shall show that the derivative of the distribution $U(t)$ equals the delta function

$$\frac{dU(t)}{dt} = \delta(t) \tag{I-30}$$

Indeed, from (I-17) we have

$$\int_{-\infty}^{\infty} U'(t)\phi(t)\,dt = -\int_{-\infty}^{\infty} U(t)\phi'(t)\,dt = -\int_{0}^{\infty} \phi'(t)\,dt = \phi(0) - \phi(\infty)$$

and since $\phi(\infty) = 0$, (I-30) follows. Thus the generalized derivative of a discontinuous function contains impulses $S_n\,\delta(t - t_n)$, where S_n is the jump at the discontinuity point $t = t_n$.

The Function t^{-m} as Distribution. Using the concept of a generalized derivative, we can define t^{-m} as distribution and give meaning to the integral

$$\int_{-\infty}^{\infty} \frac{\phi(t)}{t^m}\,dt \qquad m \text{ integer}$$

that ordinarily does not exist. The quantity $\ln |t|$, although infinite at the origin, is integrable; hence the ordinary integral

$$\int_{-\infty}^{\infty} \ln |t|\,\phi(t)\,dt$$

exists and defines the distribution $\ln |t|$. The function $1/t$ is defined as the generalized derivative of $\ln |t|$

$$\frac{1}{t} = \frac{d}{dt} \ln |t|$$

satisfying $$\int_{-\infty}^{\infty} \frac{\phi(t)}{t}\,dt = -\int_{-\infty}^{\infty} \ln |t|\,\phi'(t)\,dt$$

[see (I-17)]. We can similarly define the successive derivatives of $1/t$. Thus

$$t^{-m} = \frac{(-1)^{m-1}}{(m-1)!} \frac{d^{(m-1)}(t^{-1})}{dt^{m-1}}$$

is given by

$$\int_{-\infty}^{\infty} \frac{\phi(t)}{t^m} \, dt = \frac{1}{(m-1)!} \int_{-\infty}^{\infty} \frac{\phi^{(m-1)}(t)}{t} \, dt = \frac{-1}{(m-1)!} \int_{-\infty}^{\infty} \ln|t| \, \phi^{(m)}(t) \, dt$$

(I-31)

As an application of (I-31) we shall show that

$$\int_{-\infty}^{\infty} \frac{\cos \omega t}{t^2} \, dt = -\pi \, |\omega| \tag{I-32}$$

Indeed [see (II-57)]

$$\int_{-\infty}^{\infty} \frac{\cos \omega t}{t^2} \, dt = \int_{-\infty}^{\infty} \frac{1}{t} \frac{d(\cos \omega t)}{dt} \, dt = -\omega \int_{-\infty}^{\infty} \frac{\sin \omega t}{t} \, dt = -\pi \, |\omega|$$

I-2. Generalized Limits

Consider the sequence $g_n(t)$ of distributions. If there exists a distribution $g(t)$ such that for every test function $\phi(t)$ we have

$$\lim_{n \to \infty} \int_{-\infty}^{\infty} g_n(t)\phi(t) \, dt = \int_{-\infty}^{\infty} g(t)\phi(t) \, dt \tag{I-33}$$

then we say that $g(t)$ is the limit of $g_n(t)$

$$g(t) = \lim_{n \to \infty} g_n(t) \tag{I-34}$$

We can similarly define the limit of a distribution $g_x(t)$ with respect to a parameter x. If for every $\phi(t)$

$$\lim_{x \to x_0} \int_{-\infty}^{\infty} g_x(t)\phi(t) \, dt = \int_{-\infty}^{\infty} g_{x_0}(t)\phi(t) \, dt \tag{I-35}$$

then $g_{x_0}(t)$ is defined as the limit

$$g_{x_0}(t) = \lim_{x \to x_0} g_x(t) \tag{I-36}$$

An example of a distribution depending on a parameter is the function

$$e^{-j\omega t}$$

The number that it assigns to a test function $\phi(t)$ is, of course, the Fourier transform

$$\int_{-\infty}^{\infty} e^{-j\omega t}\phi(t) \, dt = \Phi(\omega)$$

of $\phi(t)$. We maintain that

$$\lim_{\omega \to \infty} e^{-j\omega t} = 0 \tag{I-37}$$

We note that (I-37) does not exist as an ordinary limit but only as a generalized limit in the sense [see (I-35)]

$$\lim_{\omega \to \infty} \int_{-\infty}^{\infty} e^{-j\omega t}\phi(t) \, dt = 0 \tag{I-38}$$

The proof of (I-38) is based on the following important theorem.

Riemann-Lebesgue Lemma. If a function $\phi(t)$ is absolutely integrable in an interval (a,b), then

$$\lim_{\omega \to \infty} \int_a^b e^{-j\omega t}\phi(t)\,dt = 0 \tag{I-39}$$

where a and b are finite or infinite constants.

To simplify the proof of (I-39), we shall assume that $\phi(t)$ has a bounded derivative and that the interval (a,b) is finite. The general case can be proved with a limiting argument, but the details will be omitted. Integrating (I-39) by parts, we have

$$\int_a^b e^{-j\omega t}\phi(t)\,dt = \frac{1}{j\omega}[\phi(a)e^{-j\omega a} - \phi(b)e^{-j\omega b}] + \frac{1}{j\omega}\int_a^b \phi'(t)e^{-j\omega t}\,dt$$

and, with $\omega \to \infty$, the desired result (I-39) follows. (I-38) is a special case. From the above we also see that

$$\lim_{\cos}^{\sin} \omega t = 0 \qquad \text{for } \omega \to \infty \tag{I-40}$$

The concept of generalized limit (I-36) guarantees the convergence of certain limits that ordinarily do not exist; (I-37) is an example. The same is true of a number of integrals, as the following identity

$$\int_0^\infty \sin \omega t\,dt = \frac{1}{\omega} \qquad \omega \neq 0 \tag{I-41}$$

shows. The above integral obviously does not converge; however, if it is considered as a distribution, then

$$\int_0^\infty \sin \omega t\,dt = \lim_{T \to \infty} \int_0^T \sin \omega t\,dt = \lim_{T \to \infty} \frac{1 - \cos \omega T}{\omega} = \frac{1}{\omega}$$

(see Riemann-Lebesgue lemma), and (I-41) follows.

Distributions as generalized limits. We can now define a distribution as a generalized limit of a sequence $f_n(t)$ of ordinary functions and thus give meaning to (I-2). Suppose that $f_n(t)$ is such that the limit

$$\lim_{n \to \infty} \int_{-\infty}^\infty f_n(t)\phi(t)\,dt$$

exists for every test function. This limit is, obviously, a number depending on $\phi(t)$; it defines, therefore, a distribution $g(t)$ and

$$g(t) = \lim f_n(t) \tag{I-42}$$

in the sense of (I-33). If (I-42) exists as an ordinary limit, then it defines the same function [assuming the validity of interchanging the order of limit and integration in (I-33)].

As an example of (I-42), we shall define $\delta(t)$ as a generalized limit of a sequence of ordinary functions satisfying

$$\lim_{n \to \infty} \int_{-\infty}^{\infty} f_n(t)\phi(t)\,dt = \phi(0) \qquad (\text{I-43})$$

Consider the rectangular pulse

$$r_\epsilon(t) = \frac{1}{\epsilon}\left[U(t) - U(t - \epsilon) \right]$$

of Fig. I-1. From the continuity of $\phi(t)$ it follows that

$$\lim_{\epsilon \to 0} \int_{-\infty}^{\infty} r_\epsilon(t)\phi(t)\,dt = \lim_{\epsilon \to 0} \frac{1}{\epsilon}\int_0^\epsilon \phi(t)\,dt = \phi(0)$$

Therefore
$$\delta(t) = \lim_{\epsilon \to 0} r_\epsilon(t) \qquad (\text{I-44})$$

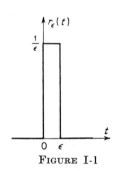

FIGURE I-1

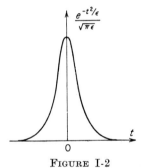

FIGURE I-2

One can use many other sequences to define $\delta(t)$. An interesting example is the Gaussian function of Fig. I-2. From

$$\lim_{\epsilon \to 0} \frac{1}{\sqrt{\epsilon\pi}} \int_{-\infty}^{\infty} e^{-t^2/\epsilon}\phi(t)\,dt \simeq \frac{\phi(0)}{\sqrt{\epsilon\pi}} \int_{-\infty}^{\infty} e^{-t^2/\epsilon}\,dt = \phi(0)$$

it follows that

$$\delta(t) = \lim_{\epsilon \to 0} \frac{e^{-t^2/\epsilon}}{\sqrt{\epsilon\pi}} \qquad (\text{I-45})$$

The function $\delta'(t)$ can be similarly defined; with

$$s_\epsilon(t) = \frac{1}{\epsilon^2}\left[U(t) - 2U(t - \epsilon) + U(t - 2\epsilon) \right]$$

as in Fig. I-3, we have

$$\int_{-\infty}^{\infty} s_\epsilon(t)\phi(t)\,dt = \frac{1}{\epsilon^2}\int_0^\epsilon \phi(t)\,dt - \frac{1}{\epsilon^2}\int_\epsilon^{2\epsilon} \phi(t)\,dt = \frac{1}{\epsilon}\int_0^\epsilon \frac{\phi(t) - \phi(t + \epsilon)}{\epsilon}\,dt$$

The last result tends to $-\phi'(0)$ with $\epsilon \to 0$, as it is easy to see; hence

$$\delta'(t) = \lim_{\epsilon \to 0} s_\epsilon(t) \qquad (\text{I-46})$$

We observe that the functions in (I-44) and (I-45) that were used to define $\delta(t)$ satisfied (I-3). One wonders then whether condition (I-3) can replace (I-43). We maintain that (I-3) is neither sufficient nor necessary. Indeed, with $r_\epsilon(t)$ and $s_\epsilon(t)$ as in Figs. I-1 and I-3, we have

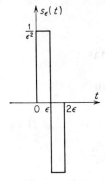

FIGURE I-3

$$\int_{-\infty}^{\infty} [r_\epsilon(t) + s_\epsilon(t)]\, dt = 1$$

$$\lim_{\epsilon \to 0} [r_\epsilon(t) + s_\epsilon(t)] = 0 \text{ for } t \neq 0$$

However, $r_\epsilon(t) + s_\epsilon(t)$ tends not to $\delta(t)$, but to $\delta(t) + \delta'(t)$. One can show that if the functions $f_n(t)$ satisfy (I-3) and are positive, then their generalized limit equals $\delta(t)$; i.e., they satisfy (I-43).

We shall now show that (I-3) is not necessary, i.e., that we can find a sequence of functions tending to $\delta(t)$, but whose limit is not zero for $t \neq 0$. For this purpose we shall prove that

$$\delta(t) = \lim_{\omega \to \infty} \frac{\sin \omega t}{\pi t} \qquad (\text{I-47})$$

This result is basic in the theory of the Fourier integral. To prove (I-47), it suffices to show that

$$\lim_{\omega \to \infty} \frac{\sin \omega t}{\pi t}\, \phi(t)\, dt = \phi(0) \qquad (\text{I-48})$$

We write the integral above as a sum of three terms

$$\int_{-\infty}^{\infty} \frac{\sin \omega t}{\pi t}\, \phi(t)\, dt = \int_{-\infty}^{-\epsilon} + \int_{-\epsilon}^{+\epsilon} + \int_{\epsilon}^{\infty} \frac{\sin \omega t}{\pi t}\, \phi(t)\, dt$$

Since $\phi(t)/t$ is integrable in the $(-\infty, -\epsilon)$ and (ϵ, ∞) intervals, it follows from the Riemann-Lebesgue lemma that the first and last integrals in the right-hand side of the above equation tend to zero with $\omega \to \infty$. To justify the fact that the second integral tends to $\phi(0)$, we observe that, since $\phi(t)$ is continuous, it can be approximated by $\phi(0)$ in the $(-\epsilon, \epsilon)$ interval, provided ϵ is small enough. Thus

$$\int_{-\epsilon}^{\epsilon} \frac{\sin \omega t}{\pi t}\, \phi(t)\, dt \simeq \phi(0) \int_{-\epsilon}^{\epsilon} \frac{\sin \omega t}{\pi t}\, dt = \phi(0) \int_{-\epsilon\omega}^{\epsilon\omega} \frac{\sin x}{\pi x}\, dx$$

But the limits of the last integral tend to $\pm\infty$ with $\omega \to \infty$, and since [see (II-57)]

$$\int_{-\infty}^{\infty} \frac{\sin x}{\pi x}\, dx = 1$$

Eq. (I-48) follows. This equation is extensively discussed in the mathematical literature, where one finds rigorous proofs under general conditions. We remark here only that, although in (I-6) we merely assumed that $\phi(t)$ was continuous, this is not sufficient for (I-48). A sufficient condition is the assumption that $\phi(t)$ is also of bounded variation (see Sec. 2-1); under this condition, the validity of (2-89) in the proof of the inversion formula (2-3) is ensured.

Using (I-47), we can prove that

$$\int_{-\infty}^{\infty} \cos \omega t\, d\omega = 2\pi\, \delta(t) \qquad (\text{I-49})$$

Indeed,

$$\int_{-\infty}^{\infty} \cos \omega t\, d\omega = \lim_{\Omega \to \infty} \int_{-\Omega}^{\Omega} \cos \omega t\, d\omega = \lim_{\Omega \to \infty} \frac{2 \sin \Omega t}{t} = 2\pi\, \delta(t)$$

I-3. Physical Concepts as Distributions

A physical quantity F is represented by a function $f(t)$ [for simplicity, we ignore the possible dependence of $f(t)$ on other variables]. Once F is specified, the existence of $f(t)$ is assumed independently of any experiments. If, for example, F is a voltage source, then it is taken for granted that its voltage $f(t)$ is a well-defined function of time and that a measurement merely reveals its values. The fact that no instrument can measure exactly $f(t)$ in no way shakes one's conviction that a true value of $f(t)$ exists, but cannot be found only because of the imperfection of our instruments. This point of view necessitates the representation of F by an ordinary function having values for every t, and not by a distribution that merely assigns values to other functions.

The above interpretation of a physical quantity is only a useful idealization, and it is naïve to accept it as representing the true meaning of F. The quantity F is nothing more than the totality of the effects R that it produces, and since these effects are certain numbers that F assigns to the function $\phi(t)$ representing the physical system between F and R, it is natural to describe F by these numbers, i.e., by a distribution.

Consider, for example, the case of the voltage source. Its effects are the responses of various transducers T, with input F, as in Fig. I-4. The system T could, for example, be a voltmeter, and R the deviation of a needle.   If T is linear, then, accepting the representation of F by an ordinary function $f(t)$, we can express the response R, at a given time $t = t_0$, as an integral [see (5-1)]

$$\int_{-\infty}^{\infty} f(t)h(t_0 - t)\, dt$$

where $h(t)$ is the inverse of the system function $H(\omega)$.   With $\phi_{t0}(t) = h(t_0 - t)$, the above takes the form

$$\int_{-\infty}^{\infty} f(t)\phi_{t0}(t)\, dt = R[\phi_{t0}(t)] \qquad (\text{I-50})$$

Thus, for a fixed input F, the response R is a number depending on the system function $\phi_{t0}(t)$.   The experimentally measured quantity is

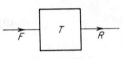

FIGURE I-4

never $f(t)$, but only R; therefore the source can be specified only by the totality of the responses that it causes, and since these responses are numbers as in (I-50), assigned to the test functions $\phi_{t0}(t)$, we conclude that F is indeed described by the distribution $R[\phi_{t0}(t)]$.

If (I-50), interpreted as an ordinary integral equation, has as solution the ordinary function $f(t)$, and that is usually the case, then we say that the voltage of the source F is $f(t)$.   However, it is conceivable that (I-50) is not satisfied by an ordinary function.   Since the results R still exist, we must assume that there is a source F causing them, and the only way to characterize this source is by the distribution (I-50).   We should point out that even if (I-50) has as solution an ordinary function $f(t)$, in order to determine its values, we must measure the readings of instruments with "short" time characteristics and use a limiting argument.   And, since these readings always involve integrals of $f(t)$ and not its specific values, $f(t)$ is never measured directly; it is merely "sensed" as a distribution.

The same reasoning can be applied to most physical quantities because their linear effects are always expressible in the form (I-50), and it is only through linear effects that we determine them quantitatively.

Appendix II. Analytic Functions[1]

For the understanding of the material in Chaps. 9 and 10, some knowledge of the theory of analytic functions is required. In the next three sections we give, for easy reference, a frugally self-contained discussion of the relevant parts of this theory.[†] In Sec. II-4 we develop a special technique of integration, not usually covered in books on function theory, known as saddle-point method. This technique facilitates the evaluation of the asymptotic form of the inversion integral and is related to the stationary-phase method of Sec. 7-7. Finally, in Sec. II-5 we discuss the properties of positive real functions, used extensively in network theory; these functions are, in a sense, the frequency-domain equivalent of the densities introduced in Chap. 11.

II-1. Definitions

A complex variable

$$z = x + jy$$

is a quantity taking values in a region R of the complex plane. If, to every z in R, we assign, according to some rule, a value to a new quantity w, then we say that w is a function of z

$$w = f(z) \tag{II-1}$$

and its domain of definition is the region R. The quantities

$$w = z^2 \qquad w = x \qquad w = |z|$$

are examples of such functions, and their domain of definition is the entire plane.

As z takes values in R, the corresponding values of w form a region R' of the w plane; R' is often called the *image* of R. Thus $f(z)$ establishes the rule of correspondence between the points of R and R'.

[†] K. Knopp, "Theory of Functions," Dover Publications, New York, 1945.

Clearly the real and imaginary parts u and v of w are uniquely determined from z, and since z is specified if x and y are known, we have

$$w = u(z) + jv(z) = u(x,y) + jv(x,y) \qquad \text{(II-2)}$$

Continuity. A function $f(z)$ is continuous at a point z_0 of R if, with z_1 close to z_0, $f(z_1)$ is also close to $f(z_0)$. Stating it more precisely, given $\epsilon > 0$, we can find a $\delta > 0$ such that the image of the circle $|z - z_0| < \delta$ lies inside the circle with center $f(z_0)$ and radius ϵ (Fig. II-1):

$$\text{if } |z - z_0| < \delta \qquad \text{then } |f(z) - f(z_0)| < \epsilon \qquad \text{(II-3)}$$

In the above definition we assumed that if z_0 is in the interior of R, then the entire circle $|z - z_0| < \delta$ of sufficiently small radius δ is also in R. The definition of continuity can be extended to boundary

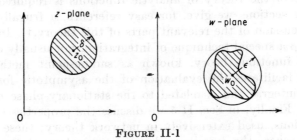

FIGURE II-1

points of R by considering in (II-3) only the points of the circle $|z - z_0| < \delta$ that lie in R.

Analyticity. A function $f(z)$ is called analytic at a point z_0 if it has a derivative $df(z_0)/dz$ at that point:

$$\lim_{z \to z_0} \frac{f(z) - f(z_0)}{z - z_0} = \frac{df(z_0)}{dz} \qquad \text{(II-4)}$$

The meaning of the above limit is similar to (II-3): given $\epsilon > 0$, we can find a $\delta > 0$ such that

$$\left| \frac{f(z) - f(z_0)}{z - z_0} - \frac{df(z_0)}{dz} \right| < \epsilon \qquad \text{for } |z - z_0| < \delta \qquad \text{(II-5)}$$

A function is analytic in a region if it is analytic in every point of that region. It can easily be seen from (II-4) that the function $f(z) = z^n$ is analytic everywhere (except at $z = 0$ if n is negative) and

$$\frac{dz^n}{dz} = nz^{n-1}$$

The limit in (II-4) must exist, regardless of the direction in which z approaches z_0. This is a strong requirement imposing severe restrictions on u and v. Indeed, suppose that z tends to z_0 in a direction parallel to the real axis, i.e., $z - z_0 = \Delta x$; then

$$\frac{df(z_0)}{dz} = \frac{\partial u}{\partial x} + j \frac{\partial v}{\partial x}$$

Similarly if $z - z_0 = j \, \Delta y$, then

$$\frac{df(z_0)}{dz} = \frac{\partial v}{\partial y} - j \frac{\partial u}{\partial y}$$

Therefore
$$\frac{\partial u}{\partial x} = \frac{\partial v}{\partial y} \qquad \frac{\partial u}{\partial y} = -\frac{\partial v}{\partial x} \tag{II-6}$$

Thus u and v must have partial derivatives satisfying the above equations (II-6), known as *Cauchy-Riemann* conditions. It can be shown that the converse is also true: if $f(z)$ is defined by (II-2) and the partial derivatives of u and v are continuous and satisfy (II-6) at every point of a region R, then $f(z)$ is analytic in R.

From (II-4) we see that in the vicinity of a point z_0, the function $f(z)$ is approximately given by

$$f(z) \simeq f(z_0) + f'(z_0)(z - z_0) \tag{II-7}$$

or exactly

$$f(z) = f(z_0) + f'(z_0)(z - z_0) + \eta(z - z_0) \tag{II-8}$$

where η is a function of z and z_0 tending to zero with $z - z_0$; i.e., given $\epsilon > 0$, we can find a δ such that

$$\text{if } |z - z_0| < \delta \qquad \text{then } |\eta| < \epsilon \tag{II-9}$$

Clearly δ depends not only on ϵ, but also on z_0; however, if $f(z)$ is analytic in a closed region, then (Heine-Borel theorem) a positive δ can be found so that (II-9) holds for all points of this region. For the boundary points one must, of course, modify the definition of the derivative by assuming in (II-4) that z tends to z_0 from the interior of R.

Conformality. Consider the points z_1, z_2, z_3 in the z plane. If z_2 and z_3 are sufficiently close to z_1, then [see (II-7)]

$$w_2 - w_1 \simeq f'(z_1)(z_2 - z_1), \qquad w_3 - w_1 \simeq f'(z_1)(z_3 - z_1)$$

Hence if $f'(z_1) \neq 0$, then

$$\frac{w_2 - w_1}{w_3 - w_1} = \frac{z_2 - z_1}{z_3 - z_1}$$

i.e., the triangle $z_1 z_2 z_3$ is similar to the triangle $w_1 w_2 w_3$ (Fig. II-2). From the above we conclude that if two curves Γ_1 and Γ_2 of the z

plane form an angle θ between them, then their images Γ_1' and Γ_2' form the same angle θ, provided that at the intersection point z_1 we have $f'(z_1) \neq 0$.

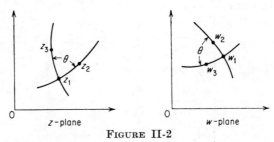

FIGURE II-2

Extremum theorem.   Suppose that $w = f(z)$ is analytic in a region R.   We maintain that the maximum or minimum of $\operatorname{Re} w$, $\operatorname{Im} w$ and the maximum of $|w|$ cannot be attained at an interior point of R.

Indeed, suppose that one of the above extremes is attained at a point z_0.   If z_0 is an interior point of R, then we can find a sufficiently small circle C, lying entirely in R and such that (II-7) holds for every z in C (Fig. II-3).   Since, according to (II-7), the image of

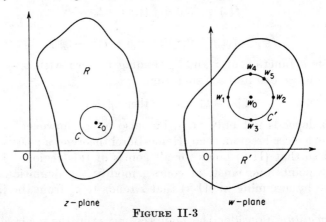

FIGURE II-3

C is a circle C', we conclude that the image R' of R contains C' and the center $w_0 = f(z_0)$ in its interior.   From Fig. II-3 we see that

$$\left. \begin{array}{ll} \operatorname{Re} w_1 < \operatorname{Re} w_0 & \operatorname{Re} w_2 > \operatorname{Re} w_0 \\ \operatorname{Im} w_3 < \operatorname{Im} w_0 & \operatorname{Im} w_4 > \operatorname{Im} w_0 \end{array} \right\} |w_5| > |w_0| \qquad \text{(II-10)}$$

Hence z_0 can only be on the boundary of R.   In the above we assumed $f'(z_0) \neq 0$.   One can similarly show that the theorem is true even if $f'(z_0) = 0$ provided $f(z)$ is not a constant.

A very useful application of the extremum theorem is the following.

Schwarz's Lemma. Suppose that $f(z)$ is analytic in the circle $|z| \le r$ and that $f(0) = 0$. If $|f(z)| \le M$ for $|z| = r$, then

$$|f(z)| \le \frac{M}{r} |z| \qquad \text{for } |z| \le r$$

And for $|z| < r$ the equality sign is possible only if $f(z) = Cz$, where $|C| \le M/r$.

To prove this lemma we observe that since $f(0) = 0$, the function $f(z)/z$ is analytic in the region $|z| \le r$; hence its amplitude is maximum at the boundary of this region. But for $|z| = r$ we have $|f(z)/z| \le M/r$; therefore,

$$\left| \frac{f(z)}{z} \right| \le M/r \qquad \text{for every } |z| \le r$$

The equality is possible for $|z| < r$ only if $f(z)/z$ is a constant, and since $|f(z)| \le M$ for $|z| = r$, the amplitude of this constant must not exceed M/r.

Multivalued functions. Consider the function

$$w = \sqrt{z} \qquad\qquad\qquad (\text{II-11})$$

where by $\sqrt{z}$ we mean a number w such that $w^2 = z$. Since there are two possible values of w whose square equals a given z, we conclude that (II-11) does not define a single-valued function in the sense given to (II-1). We can, of course, agree to accept only one of the two possible values of $\sqrt{z}$, but the resulting function will not be continuous at every point of the z plane. Suppose, for example, that $\sqrt{z}$ is defined by

$$z = re^{j\theta} \qquad -\pi < \theta \le \pi \qquad w = \sqrt{z} = \sqrt{r}e^{j\theta/2} \qquad (\text{II-12})$$

The region of definition of w is the entire z plane, and its image is the right-hand part of the w plane, shown in Fig. II-4. This function is continuous everywhere except for z on the negative real axis. Indeed, the points

$$z_3 = -a^2 \qquad z_4 = -a^2 - j\epsilon \qquad a > 0$$

are neighboring and $z_4 \to z_3$ as $\epsilon \to 0$; however, their images w_3 and w_4 are far apart and

$$w_4 \to -w_3 = -ja \qquad \text{for } \epsilon \to 0$$

One might wonder whether it is possible to specify $\sqrt{z}$ in some other way, so that the resulting function $w = \sqrt{z}$ will be continuous for every z; we maintain that this cannot be done. Indeed, as z takes values along a contour C enclosing the origin (Fig. II-5), the quantity $w = \sqrt{z}$, defined as a continuous function along C starting with $w_1 = \sqrt{z_1}$, would take values along the open curve C'; and with $z_3 \to z_1$ we would have $w_3 \to -w_1$.

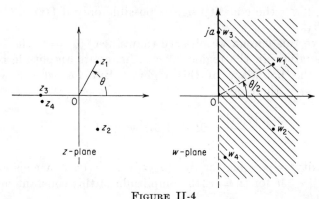

FIGURE II-4

If, however, we limit the definition of w to a region R of the z plane with the property that no closed curve in R contains the origin in its interior, then the above situation will not arise; $\sqrt{z}$ can then be defined as a continuous function in R. The maximum region with the above property is the entire plane from which a line from 0

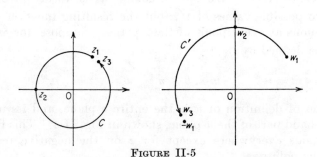

FIGURE II-5

to ∞ is excluded. Such a line is called *branch-cut*, and the origin $z = 0$ is called a *branch point* of w. The cut that guarantees the continuity of $\sqrt{z}$ defined by (II-12) is obviously the negative real axis.

Consider next the function

$$w = \sqrt{(z-a)(z-b)} \tag{II-13}$$

We shall determine the region of the z plane in which w can be defined as a (single-valued) continuous function. If a region contains a closed curve C in whose interior only one of the points a or b lies (Fig. II-6), then as z moves around C starting from z_1, the total increase in the angle of $z - b$ will equal 2π, whereas the angle of $z - a$ will return to its original value; hence the angle of w will increase by π. Thus, as z approaches z_1, w will approach $-w_1$; therefore it cannot be continuous in the above region. If, however, a curve C' contains both points a and b in its interior, then, as z

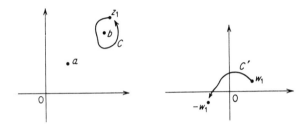

FIGURE II-6

moves around C', the angle of both factors $z - a$ and $z - b$ will increase by 2π; hence the same will be true for the angle of w, and as z returns to z_1, w will also return to w_1. From the above one concludes that, if w is continuous in a region R, then any closed curve in R must contain either both points a and b, or neither of them, in its interior. Such a region is obtained by excluding from the z plane either a line connecting the branch points a and b, or two lines starting from a and b and extending to infinity. These two possibilities are shown in Fig. II-7a and b for the function

$$w = \sqrt{1 + z^2} \tag{II-14}$$

Once the region R is selected, w is completely specified as a continuous function if its value at a single point is given. In Fig. II-7c we have shown the image of the line $AB \cdots HA$ of Fig. II-7b; we have chosen $w(1) = +\sqrt{2}$. We observe that, for $z = jy$, w is given by

On AC: $w = +\sqrt{1-y^2}$ On EG: $w = -\sqrt{1-y^2}$

On CD: $w = +j\sqrt{y^2-1}$ On GH: $w = -j\sqrt{y^2-1}$ (II-15)

One can easily show that the above radicals, (II-11), (II-13), and (II-14), once defined as continuous functions in a region R, are also analytic in R.

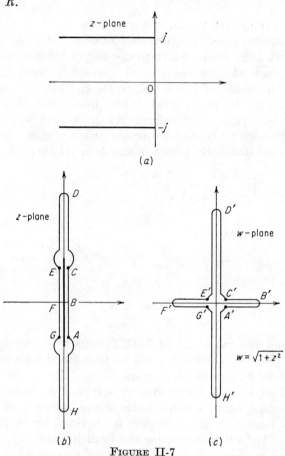

FIGURE II-7

II-2. Integration

Suppose that $f(z)$ is a continuous function in a region R of the z plane, and Γ is a curve in R whose end points are z_0 and z_n (Fig. II-8). We select an arbitrary set of points $z_1, \ldots, z_{n-1}$ on Γ and form the sum

$$\sum_{k=0}^{n} f(z_k)(z_k - z_{k-1})$$

It can be shown that, as $n \to \infty$, this sum tends to a limit depending only on $f(z)$ and Γ, provided all lengths $|z_k - z_{k-1}|$ tend to zero. This limit is called the line integral of $f(z)$ along Γ

$$\int_{\Gamma} f(z)\, dz = \lim_{|z_k - z_{k-1}| \to 0} \sum_{k=1}^{n} f(z_k)(z_k - z_{k-1}) \qquad \text{(II-16)}$$

From the above definition we easily obtain the following simple, but very useful, estimate of the magnitude of a line integral. If

$$|f(z)| < M \qquad \text{for } z \text{ on } \Gamma$$

and the length of Γ equals L, then

$$\left| \int_\Gamma f(z)\, dz \right| \le ML \tag{II-17}$$

Indeed, from the triangle inequality $|a + b| \le |a| + |b|$, we have

$$\left| \sum_{k=1}^{n} f(z_k)(z_k - z_{k-1}) \right| \le \sum_{k=1}^{n} |f(z_k)|\, |z_k - z_{k-1}| \le M \sum_{k=1}^{n} |z_k - z_{k-1}|$$

but the last sum is smaller than L because it equals the length of the inscribed polygon in Γ; hence, with $n \to \infty$, (II-17) follows.

As a trivial example of (II-16), we shall evaluate the line integral of $f(z) = A$; since

$$\sum_{k=1}^{n} f(z_k)(z_k - z_{k-1}) = A \sum_{k=1}^{n} (z_k - z_{k-1}) = A(z_n - z_0)$$

we obtain from (II-16)

$$\int_\Gamma A\, dz = A(z_n - z_0) \tag{II-18}$$

In the definition (II-16) one could replace $f(z_k)$ by the value of $f(z)$ anywhere between z_k and z_{k-1}, either on the part of Γ from z_k to z_{k-1} or on the straight line between these two points; the value of the limit would remain the same. Thus, to evaluate the integral of $f(z) = z$, we can form the sum

$$\sum_{k=1}^{n} \frac{z_k + z_{k-1}}{2} (z_k - z_{k-1}) = \tfrac{1}{2} \sum_{k=1}^{n} (z_k^2 - z_{k-1}^2) = \tfrac{1}{2}(z_n^2 - z_0^2)$$

Hence
$$\int_\Gamma z\, dz = \frac{z_n^2 - z_0^2}{2} \tag{II-19}$$

The integrals in (II-18) and (II-19) depend only on the end points of Γ; however, this is not always the case. In general the value of an integral depends also on the connecting path Γ.

A line integral of a complex variable can be written as an ordinary integral of a real variable. Indeed, suppose that the curve Γ of the $z = x + jy$ plane is described in a parametric form in terms of the real variable t

$$x = x(t) \qquad y = y(t) \qquad t_0 \le t \le t_n$$

We can easily see from (II-16) that

$$\int_{\Gamma} f(z)\,dz = \int_{t_0}^{t_n} f[z(t)]z'(t)\,dt = \int_{t_0}^{t_n} (ux' - vy')\,dt + j\int_{t_0}^{t_n} (uy' + vx')\,dt \quad \text{(II-20)}$$

where
$$f[z(t)] = u[x(t),y(t)] + jv[x(t),y(t)]$$

The above expression often facilitates the evaluation of a complex integral. Suppose, for example, that Γ is an arc of a circle, as in

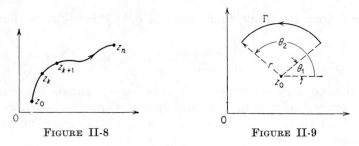

FIGURE II-8   FIGURE II-9

Fig. II-9, with center z_0 and radius r, and

$$f(z) = (z - z_0)^n$$

Then with
$$z - z_0 = re^{j\theta}$$

we have

$$\int_{\Gamma} (z - z_0)^n\,dz = r^{n+1}\int_{\theta_1}^{\theta_2} e^{jn\theta} je^{j\theta}\,d\theta = \begin{cases} r^{n+1}\dfrac{e^{j(n+1)\theta_2} - e^{j(n+1)\theta_1}}{n+1} & n \neq -1 \\[2mm] j(\theta_2 - \theta_1) & n = -1 \end{cases}$$

In particular, if Γ is a complete circle C, then $\theta_2 = \theta_1 + 2\pi$; hence

$$\int_C (z - z_0)^n\,dz = \begin{cases} 0 & n \neq -1 \\ 2\pi j & n = -1 \end{cases} \quad \text{(II-21)}$$

Notice that (II-21) holds if the direction of integration is counterclockwise; only then $\theta_2 = \theta_1 + 2\pi$.

Cauchy's theorem. The most important result in the entire theory of analytic functions is the following fundamental theorem. If a region R is bounded by one or more closed curves

$$C = C_1 + C_2 + \cdots + C_n$$

and the function $f(z)$ is analytic in every point of the region R and its boundary, then

$$\int_C f(z)\,dz = 0 \quad \text{(II-22)}$$

We note that, in general, (II-22) is a sum of contour integrals. If R is simply connected, as in Fig. II-10a, then C consists of only one curve; if R is not simply connected, as in Fig. II-10b, then

$$C = C_1 + C_2 + C_3$$

and (II-22) should read

$$\int_{C_1} f(z)\,dz + \int_{C_2} f(z)\,dz + \int_{C_3} f(z)\,dz = 0 \qquad (\text{II-23})$$

where the direction of integration is such that R is at the same side of all boundary lines (see Fig. II-10b).

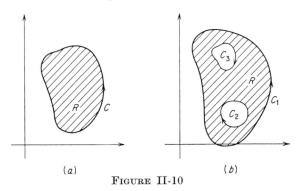

(a)   (b)

FIGURE II-10

Proof.† As we can see from (II-18) and (II-19), the theorem is trivially true for $f(z) = A_1$ and $f(z) = A_0(z - z_0)$

$$\int_C A_1\,dz = 0 \qquad \int_C A_0(z - z_0)\,dz = 0 \qquad (\text{II-24})$$

From the analyticity of $f(z)$ it follows that [see (II-8)], given $\epsilon > 0$, we can find a $\delta > 0$ such that for any two points z and z_0 in R, whose distance is less than δ, we have

$$f(z) = f(z_0) + f'(z_0)(z - z_0) + \eta(z - z_0) \qquad |\eta| < \epsilon \qquad (\text{II-25})$$

We now cover the z plane with squares of size $h = \delta/\sqrt{2}$, as in Fig. II-11. Of these squares, n lie entirely in R and m are partly in R and partly outside R. We denote by S_k the contour of the kth interior square and by B_k the contour of the region that is formed by a portion of a square that lies inside R and a piece of the boundary C of R. We can easily see that, with the direction of integration as in Fig. II-11,

$$\int_{C_1 + C_2} f(z)\,dz = \sum_{k=1}^{n} \int_{S_k} f(z)\,dz + \sum_{k=1}^{m} \int_{B_k} f(z)\,dz \qquad (\text{II-26})$$

† E. T. Whittaker and G. N. Watson, "A Course of Modern Analysis," Cambridge University Press, New York, 1948.

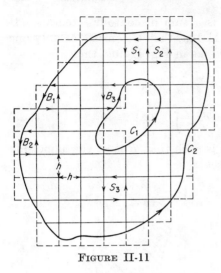

FIGURE II-11

From (II-24) and (II-25) we have, with z_k an arbitrary point inside S_k,

$$\int_{S_k} f(z)\, dz = \int_{S_k} \eta(z - z_k)\, dz \quad \text{(II-27)}$$

since the contour integral of the first two terms of (II-25) equals zero [see (II-24)]. Clearly, the distance of any two points inside the square S_k is less than the diagonal $h\sqrt{2} = \delta$ of this square; hence [see (II-25)] for z on S_k

$$|z - z_k| \le h\sqrt{2} \qquad |\eta| < \epsilon$$

We now apply (II-17) to the integral (II-27); since the length of S_k equals $4h$, we have

$$\left| \int_{S_k} f(z)\, dz \right| = \left| \int_{S_k} \eta(z - z_k)\, dz \right| \le 4hh\sqrt{2}\epsilon \quad \text{(II-28)}$$

A similar estimate can be given for the integral along B_k; however, the length of B_k is no longer $4h$, but it certainly is less than $4h + l_k$ where l_k is the length of the portion of C contained in B_k. Thus

$$\left| \int_{B_k} f(z)\, dz \right| = \left| \int_{B_k} \eta(z - z_k)\, dz \right| < \epsilon h\sqrt{2}(4h + l_k) \quad \text{(II-29)}$$

From (II-26), (II-28), and (II-29) we obtain

$$\left| \int_{C_1 + C_2} f(z)\, dz \right| < 4\epsilon\sqrt{2}(m + n)h^2 + \epsilon\sqrt{2}h \sum_{k=1}^{m} l_k \quad \text{(II-30)}$$

But $(m + n)h^2$ is the total area of all squares that either are inside or partly cover R, and $\sum_{k=1}^{m} l_k$ is the total length of the boundary $C_1 + C_2$ of R. Since these quantities are less than a fixed constant, and ϵ is arbitrary, we conclude that

$$\int_{C_1 + C_2} f(z)\, dz = 0 \quad \text{(II-31)}$$

and the theorem is proved.

Cauchy's formula. Consider a simply connected region R bounded by a curve C as in Fig. II-12. If $f(z)$ is analytic in R, and z_0 is a point inside R, then

$$f(z_0) = \frac{1}{2\pi j} \int_C \frac{f(z)}{z - z_0} \, dz \qquad \text{(II-32)}$$

This amazing formula shows that the value of $f(z)$, anywhere inside R, can be determined if $f(z)$ is known only on the boundary of R. A similar formula holds if R is not simply connected; the integral in (II-31) must then be taken over all contours bounding R.

To prove (II-32), we first observe that, if C_1 is a circle in R enclosing the point z_0 (Fig. II-12), then, since the integrand of (II-32) is analytic in the region between C and C_1, we have from Cauchy's theorem

$$\int_C \frac{f(z)}{z - z_0} \, dz = \int_{C_1} \frac{f(z)}{z - z_0} \, dz$$

We now form the function

$$\phi(z) = \begin{cases} \dfrac{f(z) - f(z_0)}{z - z_0} & z \neq z_0 \\ f'(z_0) & z = z_0 \end{cases}$$

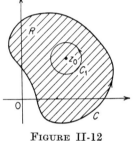

FIGURE II-12

This function is obviously continuous everywhere in R including the point z_0; hence it is bounded for every z in C_1 and its interior; in fact, if the radius r of C_1 is small enough, we have

$$|\phi(z)| < 1 + |f'(z_0)| = M \qquad \text{for } |z - z_0| \le r$$

Therefore from (II-17) we obtain

$$\left| \int_{C_1} \phi(z) \, dz \right| < 2\pi r M$$

and since r can be arbitrarily small and the above integral has a fixed value, we conclude that this value must be zero

$$\int_{C_1} \frac{f(z) - f(z_0)}{z - z_0} \, dz = 0 = \int_{C_1} \frac{f(z)}{z - z_0} \, dz - \int_{C_1} \frac{f(z_0)}{z - z_0} \, dz$$

and (II-32) follows [see (II-21)].

It is not difficult to show that (II-32) can be differentiated under the integral

$$f'(z_0) = \frac{1}{2\pi j} \int_C \frac{f(z)}{(z - z_0)^2} \, dz \qquad \text{(II-33)}$$

The proof will be omitted. Repeating this process k times, we obtain

$$f^{(k)}(z_0) = \frac{k!}{2\pi j} \int_C \frac{f(z)}{(z - z_0)^{k+1}} \, dz \qquad \text{(II-34)}$$

It thus follows that $f(z)$ has derivatives of every order.

II-3. Calculus of Residues

Suppose that a function $f(z)$ is analytic everywhere in a region R and its boundary C, except at certain isolated points $z_1, z_2, \ldots, z_n$. If we enclose each point z_k by a contour C_k as in Fig. II-13, then, since $f(z)$ is analytic in the region between $C_1, C_2, \ldots, C_n$ and C, we have from Cauchy's theorem

$$\int_C f(z) \, dz = \sum_{k=1}^{n} \int_{C_k} f(z) \, dz \qquad \text{(II-35)}$$

with the direction of integration as in the figure. In certain cases it is easy, with the help of (II-34), to evaluate the integrals along each

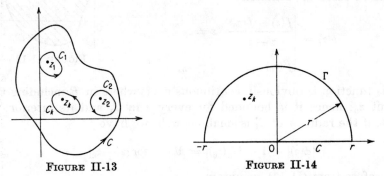

FIGURE II-13   FIGURE II-14

contour C_k; thus a method for evaluating the integral along the original contour C results, known as the calculus of residues. We now assume that z_k is a *pole* of $f(z)$ of order m; this means that there exists an integer m such that the function

$$f_k(z) = f(z)(z - z_k)^m \qquad \text{(II-36)}$$

is analytic at z_k, and m is the smallest such integer. From (II-34) we obtain

$$\int_{C_k} f(z) \, dz = \int_{C_k} \frac{f_k(z)}{(z - z_k)^m} \, dz = 2\pi j \frac{f_k^{(m-1)}(z_k)}{(m - 1)!} \qquad \text{(II-37)}$$

since $f_k(z)$ is analytic everywhere in C_k; the quantity

$$R_k = \frac{f_k^{(m-1)}(z_k)}{(m - 1)!} \qquad \text{(II-38)}$$

is known as the *residue* of $f(z)$ at the pole z_k. From (II-35) and (II-37) it follows that

$$\int_C f(z)\,dz = 2\pi j \sum_{k=1}^{n} R_k \qquad (\text{II-39})$$

and the theorem of residues follows: if $f(z)$ is analytic in a contour C and its interior except at a number of poles, then the contour integral along C is given by (II-39), where the constants R_k are the residues of $f(z)$ at the poles that are inside C.

Applications. We shall now apply the calculus of residues to evaluate certain definite integrals.

A. Consider the real integral

$$I = \int_{-\infty}^{\infty} f(x)\,dx \qquad (\text{II-40})$$

where the function $f(z)$ is analytic in the upper half $\operatorname{Im} z \geq 0$ of the z plane, except at a finite number of poles $z_k,\ k = 1, \ldots, n$. If

$$zf(z) \to 0 \qquad \text{for } |z| \to \infty \qquad \operatorname{Im} z \geq 0 \qquad (\text{II-41})$$

then I can be found from (II-39). Indeed, if the radius r of the semicircle Γ of Fig. II-14 is chosen larger than the maximum $|z_k|$, then the integral along the contour C consisting of Γ and the portion $(-r,r)$ of the real axis is given by [see (II-39)]

$$\int_C f(z)\,dz = 2\pi j \sum_{k=1}^{n} R_k = \int_{-r}^{r} f(x)\,dx + \int_\Gamma f(z)\,dz$$

But from (II-41) and (II-17) it follows that

$$\left| \int_\Gamma f(z)\,dz \right| \leq \max |f(z)|\, \pi r \to 0 \qquad \text{for } r \to \infty \qquad (\text{II-42})$$

Therefore
$$\int_{-\infty}^{\infty} f(x)\,dx = 2\pi j \sum_{k=1}^{n} R_n \qquad (\text{II-43})$$

where R_n are the residues of $f(z)$ in the upper half plane. Clearly a similar result can be obtained if the above assumptions hold in the lower half $\operatorname{Im} z \leq 0$ of the z plane.

Example II-1

$$I = \int_{-\infty}^{\infty} \frac{dx}{1 + x^4}$$

The poles of $f(z) = 1/(1 + z^4)$ in the region $\operatorname{Im} z \geq 0$ are given by

$$\left.\begin{array}{c} z_1 \\ z_2 \end{array}\right\} = \pm \frac{1}{\sqrt{2}} + j\,\frac{1}{\sqrt{2}}$$

and their residues by [see (II-38)]

$$R_1 = \tfrac{1}{4}e^{-j3\pi/4} \qquad R_2 = \tfrac{1}{4}e^{-j\pi/4}$$

Hence
$$\int_{-\infty}^{\infty} \frac{dx}{1 + x^4} = 2\pi j(R_1 + R_2) = \pi/\sqrt{2}$$

B. In (II-20) we transformed a contour integral to a real integral; the reverse process permits the evaluation of the real definite integral

$$\int_0^{2\pi} F(\sin \theta, \cos \theta) \, d\theta \tag{II-44}$$

with the method of residues. Indeed, with

$$z = e^{j\theta} \tag{II-45}$$

we have
$$\cos \theta = \frac{z + z^{-1}}{2} \qquad \sin \theta = \frac{z - z^{-1}}{2j}$$

Hence
$$\int_0^{2\pi} F(\sin \theta, \cos \theta) \, d\theta = \int_C F\left(\frac{z - z^{-1}}{2j}, \frac{z + z^{-1}}{2}\right) \frac{dz}{jz} \tag{II-46}$$

where C is the unit circle $|z| = 1$. If the last integrand is analytic inside C except at a number of poles, then (II-39) can be used to find the above integral.

Example II-2. We shall evaluate the integral

$$I = \int_0^{2\pi} \frac{d\theta}{\tfrac{5}{4} + \cos \theta}$$

From (II-46) we have

$$I = \int_C \frac{2dz}{j(z^2 + \tfrac{5}{2}z + 1)}$$

The only pole inside C is $z_1 = -\tfrac{1}{2}$ and its residue $R_1 = -\tfrac{4}{3}j$; hence

$$\int_0^{2\pi} \frac{d\theta}{\tfrac{5}{4} + \cos \theta} = \frac{8\pi}{3}$$

C. *Branch Points.* Cauchy's theorem and the calculus of residues are valid also for functions with branch points provided, of course, that $f(z)$ is analytic in R, i.e., that the branch cuts of $f(z)$ are either outside or on the boundary of R. Examples were given in Sec. 9-4. An interesting application is the evaluation of certain real integrals of the form

$$\int_0^{\infty} x^a f(x) \, dx \tag{II-47}$$

The method will be presented in an example; the general case can be handled similarly.

Example II-3. We shall evaluate the real integral

$$\int_0^\infty \frac{\sqrt{x}}{1 + x^2}\, dx$$

We define $\sqrt{z}$ by

$$z = re^{j\theta} \qquad 0 \le \theta < 2\pi \qquad \sqrt{z} = \sqrt{r}\, e^{j\theta/2} \tag{II-48}$$

$\sqrt{z}$ is thus continuous everywhere in the z plane excluding the positive real axis; hence

$$\frac{\sqrt{z}}{1 + z^2}$$

is analytic in the contour C of Fig. II-15 except at the points $\pm j$, where it has simple poles, with residues

$$R_1 = \frac{\sqrt{j}}{2j} = \frac{e^{j\pi/4}}{2j} \qquad R_2 = \frac{\sqrt{-j}}{-2j} = \frac{e^{j3\pi/4}}{-2j}$$

Hence [see (II-39)]

$$\int_C \frac{\sqrt{z}}{1 + z^2}\, dz = \pi(e^{j\pi/4} - e^{j3\pi/4}) = \pi\sqrt{2}$$

But on AB, $\sqrt{z} = \sqrt{x}$, and on GD, $\sqrt{z} = -\sqrt{x}$; therefore

$$\int_\Gamma \frac{\sqrt{z}}{1 + z^2}\, dz + \int_{\Gamma_1} \frac{\sqrt{z}}{1 + z^2}\, dz - \int_{r_1}^{r_0} \frac{\sqrt{x}}{1 + x^2}\, dx + \int_{r_0}^{r_1} \frac{\sqrt{x}}{1 + x^2}\, dx = \pi\sqrt{2}$$

and since the first two integrals tend to zero with $r_0 \to 0$, $r_1 \to \infty$, we obtain

$$2\int_0^\infty \frac{\sqrt{x}}{1 + x^2}\, dx = \pi\sqrt{2}$$

The integral along GD should have been taken along $z = x - j\epsilon$; but this can easily be settled with a limiting argument.

D. As a last application we shall discuss the integral

$$\int_{Br} e^{tz} f(z)\, dz \tag{II-49}$$

where Br is a vertical line Re $z = \alpha$ as in Fig. II-16, and α a constant. The above is mainly used to evaluate the inverse Laplace transform; however, it can also be used to compute certain real integrals of the form

$$\int_{-\infty}^\infty f(\sin x, \cos x, x)\, dx \tag{II-50}$$

(see Example II-4).

Consider the arc Γ of Fig. II-16; for $t < 0$ we have

$$|e^{tz}| = e^{tx} \leq e^{\alpha t} \qquad \text{for } z \text{ on } \Gamma$$

Therefore, if

$$zf(z) \to 0 \qquad \text{for } z \to \infty \tag{II-51}$$

then it follows easily from (II-17) that

$$\int_{\Gamma} e^{tz} f(z)\, dz \to 0 \qquad \text{with } r \to \infty \tag{II-52}$$

as in (II-42).   However, for the validity of (II-52) condition (II-51)

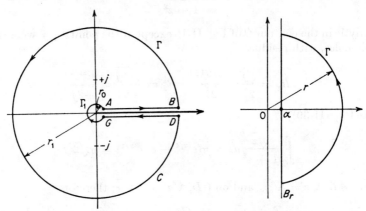

FIGURE II-15   FIGURE II-16

is not necessary; because of the presence of e^{tz}, it is enough that $f(z)$ merely tends to zero as z tends to infinity, as we shall next show.

Jordan's lemma.† If $t < 0$ and

$$f(z) \to 0 \qquad \text{with } z \to \infty \tag{II-53}$$

then

$$\int_{\Gamma} e^{tz} f(z)\, dz \to 0 \qquad \text{with } r \to \infty \tag{II-54}$$

where Γ is the arc of Fig. II-16.

Proof. We can assume that the angle of Γ does not exceed π; this is not true if $\alpha < 0$, however, the portion of Γ in the Re $z < 0$ region will have a length not exceeding $\pi|\alpha|$; hence, because of (II-53), the integration over this portion will tend to zero.   From (II-53) it follows that, given $\epsilon > 0$, we can find a constant r_0 such that

$$|f(z)| < \epsilon \qquad \text{for } |z| > r_0$$

Therefore, with $z = re^{j\theta}$, $r > r_0$,

$$\left| \int_{\Gamma} e^{tz} f(z)\, dz \right| = \left| \int_{-\pi/2}^{\pi/2} e^{tr(\cos\theta + j\sin\theta)} f(re^{j\theta}) jre^{j\theta}\, d\theta \right| < \epsilon r \int_{-\pi/2}^{\pi/2} e^{tr\cos\theta}\, d\theta$$

† E. T. Whittaker and G. N. Watson, *op. cit.*

But for $|\theta| \leq \pi/2$, $\cos\theta \geq 1 - 2|\theta|/\pi$,
as we easily see from Fig. II-17; hence,
since $t < 0$,

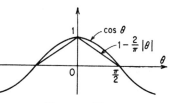

$$\int_{-\pi/2}^{\pi/2} e^{tr\cos\theta}\, d\theta < 2\int_0^{\pi/2} e^{tr(1-2\theta/\pi)}\, d\theta$$

$$= \frac{\pi}{|t|r}(1 - e^{rt}) < \frac{\pi}{|t|r}$$

FIGURE II-17

Finally

$$\left| \int_\Gamma e^{tz} f(z)\, dz \right| < \frac{\pi\epsilon}{|t|}$$

and since ϵ is arbitrarily small, the lemma is proved.

From the above lemma it follows that if $f(z)$ is analytic everywhere in the Re $z \geq \alpha$ region except at a number of poles, then

$$\int_{Br} e^{tz} f(z)\, dz = -2\pi j \sum_{k=1}^n R_k \qquad t < 0 \tag{II-55}$$

where the constants R_k are the corresponding residues; the minus sign occurs because of the direction of integration along the Br line. With a trivial modification of the proof of (II-54), one can conclude that if $t > 0$ and Γ is an arc lying on the Re $z < \alpha$ plane (see Fig. II-18), then

$$\int_\Gamma e^{tz} f(z)\, dz \to 0 \qquad \text{with } r \to \infty \tag{II-56}$$

Again, if $f(z)$ has only poles in the Re $z < \alpha$ plane, with residues R_k', then

$$\int_{Br} e^{tz} f(z)\, dz = 2\pi j \sum_{k=1}^n R_k' \qquad t > 0$$

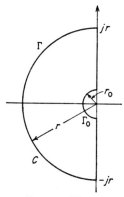

Example II-4. In Sec. 9-4 we gave a number of applications of Jordan's lemma. In this example we shall evaluate, with its help, the integral of the Fourier kernel $(\sin x)/x$

$$\int_0^\infty \frac{\sin x}{x}\, dx$$

Consider the integral

$$\int_C \frac{e^z}{z}\, dz$$

FIGURE II-18

along the contour C of Fig. II-18. From (II-56) we see that the integral along Γ tends to zero with

$r \to \infty$. But for r_0 small enough, we have

$$\int_{\Gamma_0} \frac{e^z}{z}\, dz \simeq \int_{\Gamma_0} \frac{dz}{z} = -j\pi$$

and since e^z/z is analytic in C, we conclude that

$$\int_{-r}^{-r_0} \frac{e^{jy}}{jy} j\, dy + \int_{r_0}^{r} \frac{e^{jy}}{jy} j\, dy - j\pi \to 0 \qquad \text{for } r_0 \to 0,\, r \to \infty$$

Hence
$$\int_0^\infty \frac{e^{jy} - e^{-jy}}{y}\, dy - j\pi = 0$$

and finally
$$\int_0^\infty \frac{\sin y}{y}\, dy = \frac{\pi}{2} \tag{II-57}$$

II-4. Saddle-point Method of Integration†

In this section we shall evaluate integrals of the form

$$\int_C g(z) e^{kh(z)}\, dz \tag{II-58}$$

for large values of the parameter k. This will be done by modifying the path of integration in such a way that only a small portion of the new path will contribute significantly to the value of the integral. As a preparation, we shall first assume that C is the real axis.

Method of Laplace.‡   Consider the real integral

$$I = \int_{-\infty}^\infty g(x) e^{kh(x)}\, dx \tag{II-59}$$

We assume that the second derivative $h''(x)$ of $h(x)$ exists, that $h(x)$ has only a single maximum $x = a$ on the entire real axis

$$h'(a) = 0 \qquad h''(a) < 0 \tag{II-60}$$

and that $g(x)$ is continuous, at least near $x = a$. We shall show that, under these assumptions, I is given by

$$\int_{-\infty}^\infty g(x) e^{kh(x)}\, dx \sim g(a) e^{kh(a)} \sqrt{\frac{-2\pi}{kh''(a)}} \tag{II-61}$$

for large values of k.

† This is only a short introduction of the method; for a detailed discussion see: L. B. Felsen and N. Marcuvitz, "Modal Analysis and Synthesis of Electromagnetic Fields," AFCRC-TN-59-991, Microwave Research Institute, Polytechnic Institute of Brooklyn; A. Erdelyi, "Asymptotic Expansions," Dover Publications, New York, 1956; N. G. De Bruijn, "Asymptotic Methods in Analysis," North-Holland Publishing Company, Amsterdam, 1958.

‡ D. V. Widder, "The Laplace Transform," Princeton University Press, Princeton, N.J., 1941.

Proof. The given integral can be written in the form

$$I = e^{kh(a)} \int_{-\infty}^{\infty} g(x)e^{k[h(x)-h(a)]} \, dx \quad \text{(II-62)}$$

From (II-60) we have,

$$h(x) - h(a) < 0 \qquad \text{for } x \neq a$$

(Fig. II-19a); hence, for sufficiently large k, the quantity

$$e^{k[h(x)-h(a)]}$$

is close to zero everywhere except near the point $x = a$. Therefore only the integration in an interval close to a determines the value of I. But in such an interval we have

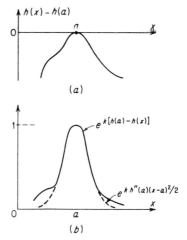

FIGURE II-19

$$g(x) \simeq g(a)$$

$$h(x) - h(a) \simeq \frac{h''(a)}{2} (x - a)^2$$

Hence $$I \sim e^{kh(a)}g(a) \int_{-\infty}^{\infty} e^{kh''(a)(x-a)^2/2} \, dx$$

The last integral is given by [see (2-65)]

$$\int_{-\infty}^{\infty} e^{kh''(a)(x-a)^2/2} \, dx = \sqrt{\frac{-2}{kh''(a)}} \int_{-\infty}^{\infty} e^{-y^2} \, dy = \sqrt{\frac{-2\pi}{kh''(a)}}$$

and (II-61) is proved.

From the above proof we easily conclude that, if $g(x)$ is discontinuous at $x = a$ but the limits $g(a^+)$ and $g(a^-)$ exist, then (II-61) is still valid provided $g(a)$ is replaced by $[g(a^+) + g(a^-)]/2$. We also see that, since only the value of the integrand in the vicinity of $x = a$ matters, (II-61) holds even if the integration is performed over a finite or semi-infinite interval as long as $x = a$ is an interior point of this interval. If $x = a$ coincides with one of the end points, then the right-hand side of (II-61) must be divided by 2.

Comment. The method of Laplace reminds us of the stationary-phase method discussed in Sec. 7-7. In (7-89) the integration away from the stationary point ω_0 was negligible because of the rapid oscillation of the integrand; in (II-61) this is so because the integrand tends to zero for $x \neq a$.

Example II-5. We shall apply (II-61) to determine the value of $n!$ for large n. We have from (9-5)

$$\int_0^\infty e^{-pt} t^n \; dt = \frac{n!}{p^{n+1}}$$

and with $p = 1$, $t = nx$,

$$n! = n^{n+1} \int_0^\infty e^{n(-x+\ln x)} \; dx \qquad \text{(II-63)}$$

The above integral is of the form (II-59) with $k = n$, $h(x) = -x + \ln x$, and $g(x) = 1$. Since

$$h'(x) = -1 + \frac{1}{x}$$

we have $\quad a = 1 \quad h(a) = -1 \quad h''(a) = -1$

Therefore [see (II-61)]

$$n! \sim n^{n+1} e^{-n} \sqrt{\frac{2\pi}{n}} = \sqrt{2\pi} n^{n+\frac{1}{2}} e^{-n} \qquad \text{(II-64)}$$

This result is known as *Stirling's formula*.

Example II-6. Using (II-61), we shall show that

$$\lim_{k \to \infty} \left[\int_c^d |f(x)|^k \; dx \right]^{1/k} = |f|_{\max}$$

where $|f|_{\max} = |f(a)|$ is the maximum of $|f(x)|$ in the interval (c,d), with zero slope at $x = a$. The result holds even if a is an end point of the above interval, provided $f'(a)$ is zero. We can assume $f(x) \geq 0$. From (II-61) we have

$$\int_c^d f^k(x) \; dx = \int_c^d e^{k \ln f(x)} \; dx \underset{k \to \infty}{\sim} e^{k \ln f(a)} \sqrt{\frac{-2\pi}{kh''(a)}} = f_{\max}^k \frac{A}{\sqrt{k}}$$

where the constant $A = \sqrt{-2\pi/h''(a)}$ is independent of k and $h(x) = \ln f(x)$.

Thus

$$\left[\int_c^d f^k(x) \; dx \right]^{1/k} \underset{k \to \infty}{\sim} f_{\max} \left(\frac{A}{\sqrt{k}} \right)^{1/k}$$

It is easy to see that $\lim (A/\sqrt{k})^{1/k} = 1$ for $k \to \infty$, and the desired result follows. If a equals c or d, then the constant A will be reduced by a factor of 2; however, the value of the limit will not be affected.

We now return to the integral (II-58) and we assume, without loss of generality, that k is real. Suppose that the derivative of $h(z)$ has a simple zero at a point $z = z_0$ of the plane

$$h'(z_0) = 0$$

[saddle point of $h(z)$]. We attempt to find a path C_1 with the property that the imaginary part of $h(z)$ is constant everywhere on C_1

(path of steepest descent), its real part is maximum at z_0

$$\operatorname{Im} h(z) = \operatorname{Im} h(z_0) \qquad \operatorname{Re} h(z) \leq \operatorname{Re} h(z_0) \qquad z \text{ on } C_1 \qquad \text{(II-65)}$$

and such that

$$\int_C g(z) e^{kh(z)} \, dz = \int_{C_1} g(z) e^{kh(z)} \, dz \qquad \text{(II-66)}$$

Because of (II-65), the last integral above is essentially of the form (II-59); therefore the estimate (II-61) can be used. We thus have for large k

$$\int_{C_1} g(z) e^{kh(z)} \, dz \sim e^{kh(z_0)} g(z_0) \sqrt{\frac{2\pi}{-kh''(z_0)}} \qquad \text{(II-67)}$$

This is the saddle-point method of integration. The choice of the value of the square root in (II-67) depends on the direction of integration along C_1. Often it is necessary to integrate along more than one path of steepest descent, in order to satisfy (II-66). It is then necessary to add the contributions to the integral from each saddle point on C_1, in the estimate (II-67). This will be shown in the next example, where we shall also explicitly specify C_1 and will clarify the details of the method and the choice of the radical sign.

If the function $g(z) e^{kh(z)}$ is not analytic in the region between C and C_1, then (II-66) is not true; however, if its singularities are known poles, then the difference of the two integrals in (II-66) can be evaluated.

Comment. The saddle-point method of integration is an extension of Laplace's real integral (II-61) to a contour integration. The stationary-phase method of Sec. 7-7 can be similarly extended. If a path C_1' is found such that for z on C_1' the real part of $h(z)$ is constant, then the amplitude of $e^{kh(z)}$ remains constant but its phase varies rapidly as we move away from z_0. We thus obtain an integral similar to (7-89) and the estimate (7-90) can be used.

Example II-7. We shall use the saddle-point method to evaluate the Bessel function

$$J_n(t) = \frac{1}{2\pi} \int_0^{2\pi} e^{j(t \sin \phi - n\phi)} \, d\phi \qquad \text{(II-68)}$$

[see (6-91)] for large values of t. With

$$z = e^{j\phi}$$

as in (II-45), the above becomes a contour integral

$$J_n(t) = \frac{1}{2\pi j} \int_C \frac{e^{(z - z^{-1})t/2}}{z^{n+1}} \, dz \qquad \text{(II-69)}$$

along the unit circle $|z| = 1$ (Fig. II-20a). It is of the form (II-58) with

$$h(z) = \tfrac{1}{2}(z - z^{-1}) \qquad g(z) = \frac{1}{z^{n+1}}$$

The saddle points of $h(z)$ are given by

$$h'(z) = \frac{1}{2}\left(1 + \frac{1}{z^2}\right) = 0 \qquad z = \pm j$$

The path C_1 of steepest descent passing through j is shown in Fig. II-20b. As we shall presently explain, it starts from the origin tangent to the real axis, crosses the imaginary line at a $3\pi/4$ angle, and extends to infinity, with the line $z = j2$ as its asymptote. The path C_2 through $-j$ is the conjugate of

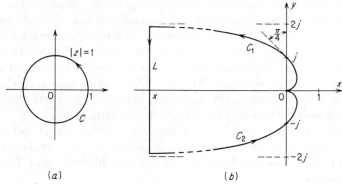

(a)   (b)

FIGURE II-20

C_1. We now form a closed contour as in Fig. II-20b, consisting of C_1, C_2, and the vertical segment L. This contour encloses in its interior the only singularity $z = 0$ of the integrand in (II-69); and since the same is true for the unit circle, we conclude from Cauchy's theorem that

$$\int_C \frac{e^{(z-z^{-1})t/2}}{z^{n+1}}\, dz = \int_{C_1+C_2+L} \frac{e^{(z-z^{-1})t/2}}{z^{n+1}}\, dz \tag{II-70}$$

The length of L is less than 4 and the integrand for z on L tends to zero with $x \to -\infty$; therefore the integration along L tends also to zero. From (II-67) we have for large t

$$\int_{C_1} \frac{e^{(z-z^{-1})t/2}}{z^{n+1}}\, dz \sim e^{jt}\, \frac{1}{j^{n+1}} \sqrt{\frac{2\pi}{-tj}} \tag{II-71}$$

since $h(j) = j \qquad h''(j) = -j \qquad g(j) = \dfrac{1}{j^{n+1}}$

Similarly $\displaystyle\int_{C_2} \frac{e^{(z-z^{-1})t/2}}{z^{n+1}}\, dz \sim e^{-jt}\, \frac{1}{(-j)^{n+1}} \sqrt{\frac{2\pi}{tj}}$ \hfill (II-72)

From (II-69) and the above equations we finally obtain for large t

$$J_n(t) \sim \frac{1}{2\pi j}\left[\frac{e^{jt}}{j^{n+1}}\, e^{j3\pi/4} + \frac{e^{-jt}}{(-j)^{n+1}}\, e^{-j\pi/4}\right]\sqrt{\frac{2\pi}{t}} = \sqrt{\frac{2}{\pi t}}\cos\left(t - \frac{2n+1}{4}\,\pi\right)$$

$$\text{(II-73)}$$

The choice in the selection of the value of the radical will presently become clear, but first the equation of the curves C_1 and C_2: with

$$h(z) = u + jv$$

we have $$h(x + jy) = \frac{1}{2}\left(x - \frac{x}{x^2 + y^2}\right) + \frac{j}{2}\left(y + \frac{y}{x^2 + y^2}\right)$$

Hence for (II-65) to be true, we must have

$$y + \frac{y}{x^2 + y^2} = \pm 2 \qquad\qquad \text{(II-74)}$$

subject to the condition

$$x - \frac{x}{x^2 + y^2} = u \leq 0 \qquad\qquad \text{(II-75)}$$

Thus the two curves C_1 and C_2 of Fig. II-20b result. We now come to the sign of the radical. Near the stationary point $z = j$, we have

$$h(z) \simeq h(j) + \frac{h''(j)}{2}\,(z - j)^2$$

Since $h''(j) = -j$, and $h(z) - h(j) < 0$ [see (II-65)], we obtain with

$$z - j = re^{j\delta} \qquad \delta \text{ constant}$$

where r is positive in the direction of integration,

$$r^2 e^{j2\delta} j < 0$$

Hence $$\delta = 3\pi/4$$

and for large t

$$\int_{C_1} e^{h''(j)(z - z_0)^2 t/2}\, dz \simeq e^{j3\pi/4}\int_{-\infty}^{\infty} e^{-r^2 t/2}\, dr = \sqrt{\frac{2\pi}{-tj}}$$

Thus the first term in (II-73) is justified. For the integration near $z = -j$ we similarly obtain $\delta = -\pi/4$.

II-5. Positive Real Functions†

We shall now discuss the properties of a class of analytic functions that are used in network theory to characterize the input impedance of linear passive systems. They are unilateral Laplace transforms of

† W. Cauer, "Synthesis of Linear Communication Networks" (translated from German), 2d ed., McGraw-Hill Book Company, Inc., New York, 1958.

certain time functions and can be defined either in the p domain or by
the properties of their inverse transform. We shall first consider
rational functions of p, and at the end of the section we shall extend
the concept to more general functions.

Definition. A rational function

$$F(p) = \frac{N(p)}{D(p)}$$

is called positive real (p.r.) if
1. it is real for p real
2. Re $F(p) \geq 0$ for Re $p \geq 0$
With

$$F(\alpha + j\omega) = R(\alpha,\omega) + jX(\alpha,\omega) \qquad p = \alpha + j\omega \qquad \text{(II-76)}$$

Condition 2 is equivalent to

$$R(\alpha,\omega) \geq 0 \qquad \text{for } \alpha > 0$$

Properties.   1. If $F(p)$ is p.r., then $1/F(p)$ is also p.r.

Proof. $\qquad \operatorname{Re} \dfrac{1}{F(p)} = \dfrac{R(\alpha,\omega)}{R^2 + X^2} \geq 0 \qquad \text{for } \alpha \geq 0$

2a. $N(p)$ is a Hurwitz polynomial; i.e., it has no zeros in the
region Re $p > 0$.

Proof. If $F(p_0) = 0$ and Re $p_0 > 0$, then a circle C with center
p_0 can be found, lying in the region Re $p > 0$ and such that all the
poles of $F(p)$ are outside C. Thus $F(p)$ is analytic in C and, according to the extremum theorem [see (II-10)], Re $F(p)$ cannot be minimum at the point p_0 interior to C; thus for some p on C we have

$$\operatorname{Re} F(p) < \operatorname{Re} F(p_0) = 0$$

and this is impossible because $F(p)$ is p.r.

2b. $D(p)$ is a Hurwitz polynomial.

Proof. It follows from 1 and 2a.

3a. If $F(p)$ has imaginary poles, they must be simple with positive
residues.

Proof. Suppose that $j\omega_i$ is a root of $D(p)$; then for p close enough
to $j\omega_i$, we have

$$F(p) \simeq \frac{A_i}{(p - j\omega_i)^m}$$

where m is the multiplicity of $j\omega_i$. If $m > 1$, then the image of the
semicircle Γ of Fig. II-21a would be a large circle with center at the

origin, and $F(p)$ would have values with negative real part; but this is impossible; hence $m = 1$. If A_i is not real and positive, then the image Γ' of Γ would be a semicircle as in Fig. II-21b, having points in the region Re $F(p) < 0$.

3b. The pole at infinity must be simple with positive residue.

Proof. For p on a sufficiently large semicircle, we have

$$F(p) \simeq A_\infty p^m$$

Reasoning as in 3a, we conclude that $m = 1$ and $A_\infty > 0$. From 1 and 3b it follows that the degrees of $N(p)$ and $D(p)$ differ by at most one.

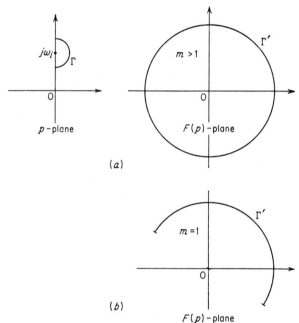

FIGURE II-21

4. If $F(p)$ is p.r., then

$$\left| \frac{F(p) - 1}{F(p) + 1} \right| \leq 1 \qquad \text{for Re } p \geq 0 \qquad \text{(II-77)}$$

Proof. $|F(p) \pm 1|$ is the distance from $F(p)$ to the points -1 and $+1$ (Fig. II-22), and since for Re $p \geq 0$ $F(p)$ is in the right-hand plane, we have

$$|F(p) - 1| \leq |F(p) + 1|$$

5. If $F(p)$ satisfies (II-77), then it is p.r.

Proof. From $|F(p) - 1| \leq |F(p) + 1|$ it follows that (see Fig. II-22) Re $F(p) \geq 0$.

6a. If

    I. $F(p)$ has no poles for Re $p > 0$

    II. Re $F(j\omega) = R(0,\omega) \geq 0$

    III. the imaginary poles of $F(p)$ are simple with positive residues

then $F(p)$ is p.r.

Proof. Suppose first that $F(p)$ has no imaginary poles; then it is the Laplace transform of a causal function $f(t)$ whose Fourier transform equals $F(j\omega)$ [see (9-11)]; therefore from (9-15) we have

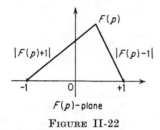

F(p)-plane

FIGURE II-22

$$R(\alpha,\omega) = \frac{1}{\pi} R(0,\omega) * \frac{\alpha}{\alpha^2 + \omega^2} \quad \text{(II-78)}$$

and since the convolution of two positive functions is positive, property 6a is proved. The above proof is valid even if $F(p)$ contains imaginary poles; indeed if $F(p) = A_i/(p - j\omega_i)$, then the real part of the corresponding Fourier transform is given by $A_i[\delta(\omega - \omega_i)]$ and is positive. If all such terms are added to $R(0,\omega)$, then $R(\alpha,\omega)$ is again given by (II-78) and its positiveness is thus established. The above property is usually proved with the help of the extremum theorem.

6b. If $F(p)$ satisfies I and III and

    IV. $\left| \dfrac{F(j\omega) - 1}{F(j\omega) + 1} \right| \leq 1$

then it is p.r.

Proof. From IV we conclude that II is true (see Fig. II-22).

The following two properties are usually proved with the help of Schwarz's lemma; however, a very simple derivation is possible if use is made of (9-16)

$$F(p) = \frac{2p}{\pi} \int_0^\infty \frac{R(y)}{p^2 + y^2}\, dy \quad \text{Re } p > 0 \quad \text{(9-16)}$$

where $R(\omega) = \text{Re } F(j\omega) = R(0,\omega)$.

7. If $F(p)$ is p.r., then

$$|\arg p| \geq |\arg F(p)| \quad \text{for Re } p \geq 0 \quad \text{(II-79)}$$

8. If $F(p)$ is p.r., then function $F(p)/p$ decreases and the function $pF(p)$ increases monotonically as p increases from zero to infinity

along the real axis; i.e., if $0 < p_1 < p_2$, then

$$\frac{F(p_1)}{p_1} > \frac{F(p_2)}{p_2} \qquad p_1 F(p_1) < p_2 F(p_2) \qquad \text{(II-80)}$$

Since $R(y) \geq 0$, it follows from (9-16) that it suffices to prove the above properties for the function $G(p) = p/(p^2 + y^2)$, where y is a constant, because these properties are obviously additive. For this simple function the proof is trivial and will be omitted.

We remark that (II-80) is true if $R(y)$ is not identically zero for $y \neq 0$, i.e., if $F(p)$ is not equal to kp or k/p. In the latter case (II-80) still holds, but with the equality sign in one of these expressions. Similarly if p is complex and Re $p > 0$, then the equality sign in (II-79) is possible only if $F(p) = kp$ or k/p.

Corollary. If $F(p)$ is p.r., $F(0) \neq 0$, $F(\infty) \neq \infty$, and a is a positive constant, then the equation

$$F(p) - ap = 0$$

has no complex roots and only one real root for Re $p > 0$.

Proof. Since $F(p)/p$ is monotonically decreasing from ∞ to 0 as p increases from 0 to ∞ [see (II-80)], it will for one and only one positive p equal a. If p is complex, then $|\text{arg } F(p)| < |\text{arg } p|$; hence $F(p)/p$ cannot equal the real constant a.

The extension of the concept of positive real functions to a more general class is suggested by 6a:

A function $F(p)$ will be called p.r. if it is the Laplace transform of a real causal function $f(t)$ and if the Fourier integral of $f(t)$ has a positive real part $R(\omega)$.

$F(p)$ can be specified in terms of the properties of its inverse transform $f(t)$. Indeed, since the even function $\phi(t) = f(t) + f(-t)$ has a positive Fourier transform

$$\phi(t) \leftrightarrow 2R(\omega)$$

we conclude as in (11-6) that $\phi(t)$ is positive definite. Thus if $h(t)$ is causal and $h(t) + h(-t)$ positive definite in the sense of (11-6), then $F(p)$ is positive real.

Index